ROYAL ENFIELD

The Early History - 1851 to 1930

Peter Miller

FWtoC Publishing

Also by Peter Miller

From Welbike to Corgi (FWtoC Publishing, 1997, Rev. Ed. 2006)
Morgan Three-wheeler: The Complete Story (The Crowood Press, 2004)
Brough Superior: The Complete Story (The Crowood Press, 2010)
The Famous James Military Lightweight (Amberley Publishing, 2016)

First published 2019
FWtoC Publishing,
3 Parklands,
Blandford Forum,
Dorset DT11 7BA

Print Management by Jarrold Publishing
www.jarrold-publishing.co.uk

ISBN: 978-0-85101-644-3

Contents

Preface

The origins of the Enfield Cycle Company Ltd can be traced back to 1851 and the establishment of a factory by George Townsend for the manufacture of needles at Hunt End, three miles south of Redditch, Worestershire, United Kingdom. George Townsend died before the development of the cycle industry and it was his half-brother, Foster, and his son, also George, who were responsible for establishing the company as a cycle manufacturer in the late Eighteen Eighties. George Jr. lacked the business skills of his father and financial difficulties gave the opportunity for Albert Eadie, assisted by Robert Walker, to acquire control of the company. Eadie was a larger than life entrepreneur and salesman whilst Walker was a talented engineer and designer; the perfect mix of skills to ensure the successful growth of the business.

The Royal Enfield name, which dates back to 1893, is nowadays associated with motorcycles, but was first used on the company's bicycles. The first powered machine was a quadricycle developed in 1897 by Robert Walker for his own use, This was followed two years later by the manufacture of a range of quadricycles and tricycles. Bicycle manufacture was the mainstay of the Company's business during the early years and would remain important through to the closure of the firm in the Nineteen Seventies. The first car to carry the Royal Enfield name was a re-badged Ader, which appeared in 1901. The first motorcycles also appeared at about this time, but manufacture was discontinued after a few years to allow effort to be concentrated on motorcars. Despite good quality products, motorcar manufacture did not prove profitable and the motorcar business was sold to Alldays & Onions in 1908. Motorcycle manufacture was recommenced at the end of 1909.

In preparing this book, reliance has been placed upon the various Enfield Cycle Company sales brochures and road tests for many of the illustrations and for the determination of the dates new models were introduced and changes made. There is considerable scope for error in this as often the catalogue illustrations may have been touched up and changes or improvements may have been introduced mid-season. Also the catalogues are seldom dated. A few identify the model year but none give the date of issue. Often the catalogue date must be inferred from the models described or from the dates of the various testimonials published in the brochure.

The book is concerned with the Royal Enfield machines and the men who were responsible for their design and construction. Competition is discussed only when relevant to the ongoing story of the machines' development.

The units used throughout are the Imperial units current during the period of manufacture. The following conversions may prove helpful to those readers more familiar with metric units.

Distance:	1 mile	=	1.61 km
	1 foot	=	30.5 cm
	1 inch	=	25.4 mm
Volume:	1 gal	=	4.546 litre
Weight:	1 lb	=	0.454 kg
	1 cwt	=	50.85 kg
Speed:	1 mph	=	1.61 kph
Currency:	£1	=	20 s
	1s	=	12 d
	1 guinea	=	£1 1s

Inflation: the approximate equivalent purchasing power of £1 compared to 2018 figures is as follows:

1851	£129
1890	£123
1900	£123
1910	£113
1920	£49
1930	£60

CHAPTER ONE

Early History 1851 - 1900

Royal Enfield is the world's oldest surviving motorcycle firm. The Enfield brand name was first introduced in October 1892 and the Royal appendage added the following year. However, the company making the machines bearing this now famous name was already forty-one years old by this time, having been established at Hunt End near Redditch in 1851 for the manufacture of sewing needles.

Up until the 16th Century needles made in England were being hand crafted from sheet steel which was cut into strips before being rolled and filed to shape. They were of variable quality and the necessary skills to produce superior products were not acquired until foreign needle makers immigrated to England in order to flee political unrest in Europe. The first record of needle manufacture in the Redditch area was in 1639 but it was not long before a flourishing cottage industry was established in the town and the surrounding villages. The utilisation of water power using the River Arrow, which also provided a source of emery stones, and the introduction of new machinery in the early 18th Century resulted in the domestic manufacturing system being replaced by industrial methods. In Redditch, corn mills were converted to needle mills with the aim of improving the scouring and pointing processes. The use of water power to polish the needles gave the Redditch manufacturers a technical advantage and the needles produced were both inexpensive and of the highest quality. It was not long before needle manufacture in Britain became concentrated in the Redditch area and by the close of the 18th Century nearly 2,000 people were employed in the trade. By the middle of the 19th Century the number of workers in the industry had grown to 8,000 and the industry was producing 200 million needles a week. Related industries, such as the manufacture of fishing hooks and springs, were also flourishing in the region by this time.

Establishment of the Givry Works

George Townsend, who was born in 1815, was the son of a needle manufacturer in nearby Alcester. By 1851 he had saved sufficient to set up as an independent manufacturer and moved to Hunt End, then a small country village three miles south of Redditch. His mill, which he established with his half-brother, Foster Townsend, was known as the Givry Works. The mill comprised a red-bricked three-storey factory with a small cottage attached, which was occupied by Townsend and his family. The village. which already supported several small mills manufacturing needles and fish hooks, provided a ready supply of skilled labour. The Givry Works was quickly up and running and the 1851 census records a workforce of sixteen men and two boys.

The growth of the sewing manufacturing trade in the 1860s, which resulted in a growth in the demand for needles and precision-made light components, enabled George Townsend & Co to flourish. This was in no short measure due to the ingenuity of George Townsend who was constantly striving to improve the quality of his products and to expand the range of sewing machine components being manufactured. Amongst his products were the machines used to manufacture needles. The quality of his needles was such as to win prize medals in London,1862, Paris,1867, and Havre, 1868. By 1871 the Givry Works was considered the most important engineering works in the area offering employment to 170 workers. The firm's products were widely exported and the company could boast offices in London, Paris and Canada.

With the death of George Townsend in 1879, at the age of 64, the running of George Townsend & Company passed to his half-brother, Foster, and his son George. George Jr. appears to have been the senior partner and took residence with his family in Givry Works House, the works' cottage.

Entering the Bicycle Market

Before Foster retired from an active role in the management of the firm he brought into the factory one of the first boneshaker bicycles; a machine with an iron backbone, wooden wheels with iron tyres and wooden, triangular-shaped, pedals. This machine must have provided considerable food for thought as cycling was gaining in popularity at this time and most companies with the appropriate manufacturing skills were busy designing cycles for manufacture. George Jr. was provided with the perfect opening into this market in 1885 when William Cook, in nearby Headless Cross, invented a new type of bicycle saddle. This was constructed using a single length of wire to form both the framework of the saddle and its two springs. The design was quickly patented and marketed as the 'Townsend Cyclists Saddle and Spring'. Very soon the firm was producing saddles and other cycle components for several manufacturers.

By 1888 Townsend could supply all the components for bicycle manufacture and it is of little surprise that the company began producing its own brand of cycles, initially under the name 'The Ecossais'. Over the next three years bicycle manufacture became the mainstay of the company's output and the Givry Works was rapidly expanded to cope with the extra workload. By 1891 the Givry Works had been extended along half of one side of Enfield Road and a stamp shop erected on the other side of the Wharrage Brook. More than two dozen different designs of cycle were being produced by this time and were exhibited at London's Crystal Palace.

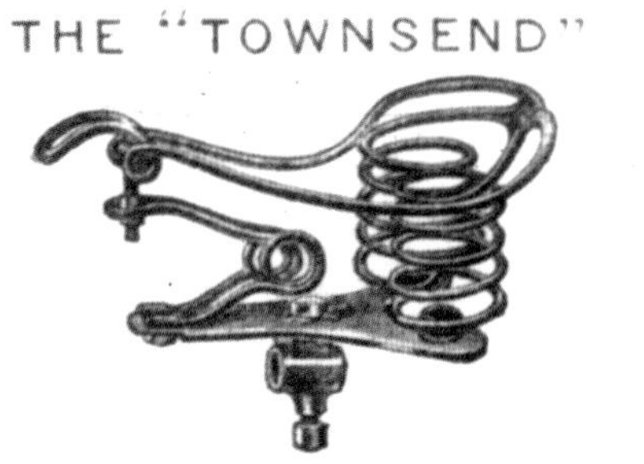

The saddle devised by William Cook and manufactured by George Townsend & Co.

In 1890 the firm was made into a limited liability company under the name George Townsend & Co Ltd.

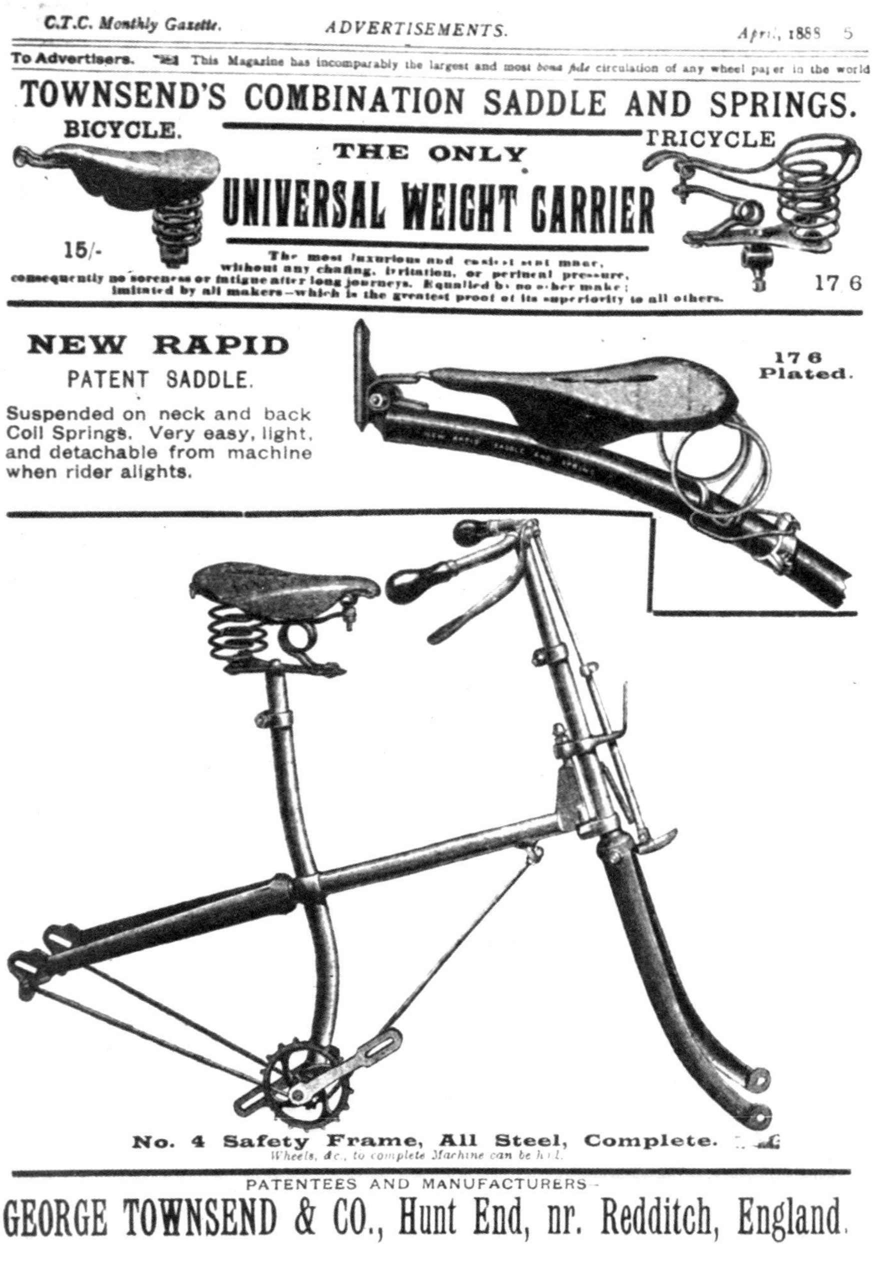

An advert from April 1885 illustrating the various forms of the patent saddle. Already by this date the company was able to offer all the components for the manufacture of bicycles.

Financial Problems and the Arrival of Albert Eadie and Robert Walker Smith

The Townsend manufactured bicycles gained a good reputation through successes in the early cycle races and the company should have been able to look forward to good sales and profitability. However, it would appear that the rapid growth of the company, as it entered into cycle manufacture, had created problems. By the time it was registered as a limited company the Townsend brothers were already in financial difficulties and further capital was urgently required if the company was to survive. News of the company's financial problems came to the attention of Albert Eadie, the thrusting and ambitious sales manager of the Perry's bicycle component business. He recognised a possible business opportunity and, through his contacts arranged for a group of Birmingham financiers to invest in new stock and to provide business advice. The business remedies they proposed did not, however, find favour with the Townsend brothers, who duly decided to leave the business. George Townsend was only 36 years old at this time and decided to set up a new cycle manufacturing firm at Aston Cantlow, near Stratford-upon-Avon, some thirteen miles from Redditch. The departure of the Townsend brothers provided the opening for Albert Eadie to acquire control.

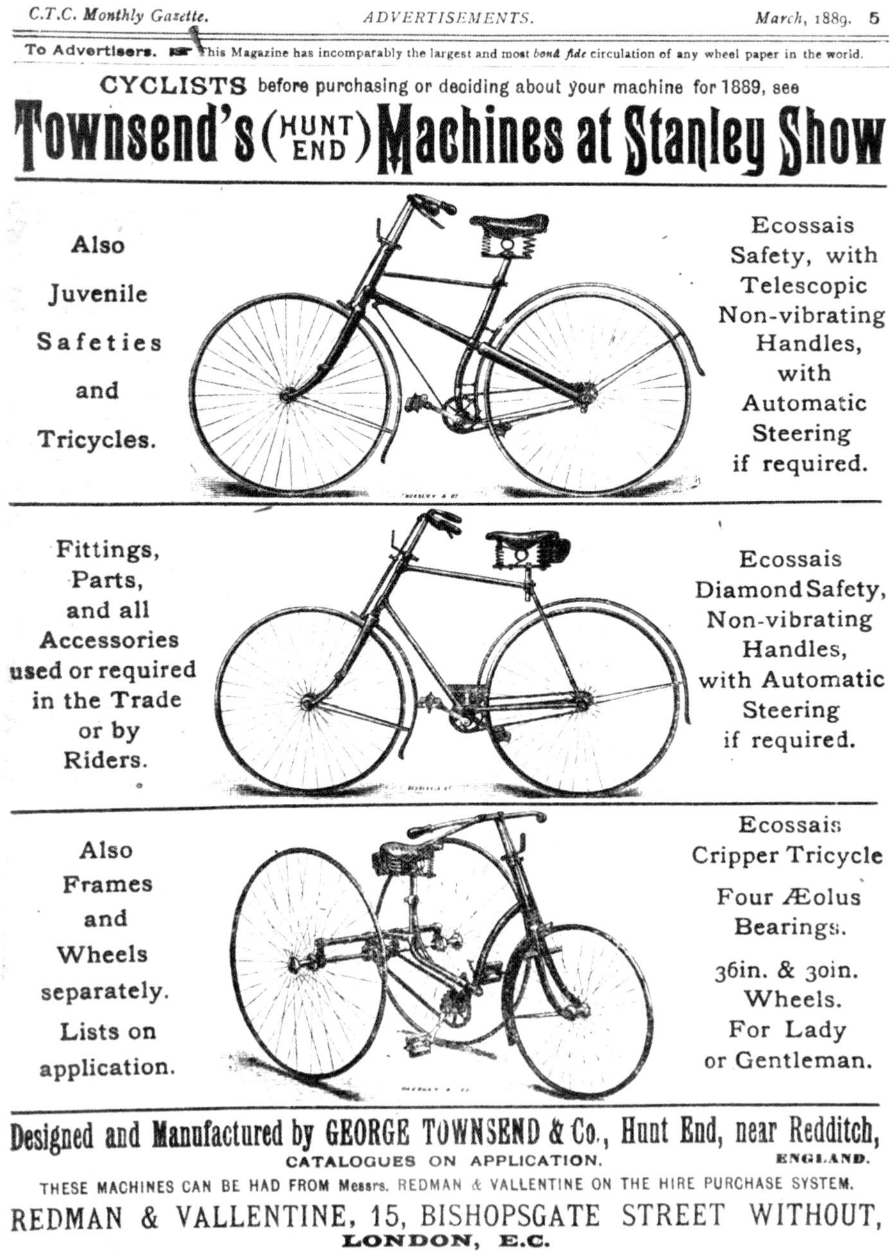
C.T.C. Monthly Gazette. *ADVERTISEMENTS.* *March*, 1889. 5

To Advertisers. This Magazine has incomparably the largest and most *bonâ fide* circulation of any wheel paper in the world.

CYCLISTS before purchasing or deciding about your machine for 1889, see

Townsend's (HUNT END) Machines at Stanley Show

Also Juvenile Safeties and Tricycles.

Ecossais Safety, with Telescopic Non-vibrating Handles, with Automatic Steering if required.

Fittings, Parts, and all Accessories used or required in the Trade or by Riders.

Ecossais Diamond Safety, Non-vibrating Handles, with Automatic Steering if required.

Also Frames and Wheels separately. Lists on application.

Ecossais Cripper Tricycle

Four Æolus Bearings.

36in. & 30in. Wheels. For Lady or Gentleman.

Designed and Manufactured by GEORGE TOWNSEND & Co., Hunt End, near Redditch,

CATALOGUES ON APPLICATION. ENGLAND.

THESE MACHINES CAN BE HAD FROM Messrs. REDMAN & VALLENTINE ON THE HIRE PURCHASE SYSTEM.

REDMAN & VALLENTINE, 15, BISHOPSGATE STREET WITHOUT, LONDON, E.C.

The advertisement shows three of the range of cycles produced by Townsend's and to be displayed at the 1889 Stanley Show. The machines all carry the Ecossais brand name.

Robert Walker Smith, usually referred to as R.W.Smith, the assistant works manager at Rudge & Co., and Albert Eadie had become acquainted though business dealings. Both men were equally keen to enter the cycle manufacturing industry on their own account. They decided to obtain financial backing and in November 1891 were able take over control of George Townsend & Co Ltd and the Givry Works, Hunt End. The acquisition was a major gamble for both men, who were already in good well-paid jobs, but they felt confident of their ability to turn around the ailing company. In April 1892 Eadie severed his connection with Perry & Co in order to assume the position of managing director of Townsend & Co. At the same time R.W.Smith resigned from Rudge & Co. to become works manager at Townsend & Co.

Eadie Manufacturing Company Ltd.

A meeting of the Company was held at 9 Bennetts Hill, Birmingham on 5 August 1892 at which a resolution was passed to change the company name to the Eadie Manufacturing Company Limited. The change was implemented on 13 August 1892. Albert Eadie was appointed managing director, with R.W.Smith as a director and William Steven Tunbridge as company secretary. The firm initially continued to make needles and sewing machine parts as well as bicycles and bicycle components. The bicycle trade in general was experiencing difficult trading conditions at this time and it speaks highly of Eadie's sales skills and enterprise that he was able to expand the bicycle side of the business. Nevertheless the firm continued to have severe financial problems and Eadie struggled to turn the company around. Fortunately George Cartland, the wealthy third son of a Birmingham brass founder, who had helped fund the initial acquisition, retained sufficient faith in the company to provide additional backing. Additional income was obtained through the sale of the needles and sewing machine parts manufacturing business to Alfred Shrimpton & Sons Ltd.

R.W.Smith had been able to persuade a number of his skilled ex-colleagues from Rudge to join him and the new company quickly established a reputation for its precision engineering capabilities. This was recognised towards the end of 1892 when the Eadie Manufacturing Co gained a sizeable contract for the supply of Enfield rifle components to the Royal Small Arms factory at Enfield in Middlesex. The contract was to prove highly lucrative and helped guarantee the future of the company.

The First Royal Enfield Cycles

John Boyd Dunlop's newly developed pneumatic tyre was first demonstrated in May 1889 by Willie Hume, captain of the Belfast Cruisers Cycling Club, who was the first person to win a race using the tyres. By the spring of 1890, despite initial opposition, the supremacy of the Dunlop pneumatic tyres over the old solid rubber type had been recognised. R.W.Smith wasted little time following the creation of the Eadie Manufacturing Company in designing and bringing into production a new range of safety cycles fitted with pneumatic tyres. In recognition of the successful completion of the contract with the Royal Small Arms factory at Enfield, it was decided these should be marketed under the Enfield name. This was not the first use of the Enfield name for a bicycle as it had already been used for a series of bicycles made at the Birmingham Small Arms factory in Enfield, Middlesex. These were now out of production and there appears to have been no objection to Eadie and Walker adopting the name.

The new Enfield machines, which were announced in October 1892, proved an instant success. The announcement was made by The Enfield Manufacturing Company, a subsidiary sales company of the Eadie Manufacturing Company. It was registered as a limited company on 24 February 1893 - Company Reg. No. 170,951 - with headquarters and a showroom established at 166 Edmund Street, Birmingham.

The company's trademark, a shield with a smaller shield inset containing a field gun facing left, first appeared in February 1893 and was registered in June 1893.

In the same year, 1893, Enfield cycles of superior quality were made available and were the first to be given the Royal Enfield name. They were displayed together with Enfield machines on Stand 79 at the Stanley Show in November 1893. Of the total of twenty-four displayed, eight were Royal Enfield racers and two were Royal Enfield road-racers.

On 11 November 1893 the registered office and showrooms of the Enfield Manufacturing Company Limited were transferred to new premises shared with the Eadie Manufacturing Company at 94 Snow Hill, Birmingham. A further change of address took place on 10 January 1895, this time to 2 William Street, Redditch

The company had showrooms both in London at 60 Sloane Street and in Dublin at 73 Grafton Street.

THE

ENFIELD NO. 1 FRONT DRIVER,

FITTED WITH R. W. SMITH'S PATENT GEAR.

Full Roadster, weight 32-lbs. Light Roadster, weight, 27-lbs.

Racer, weight, 21-lbs. PRICE, £24 0 0.

New Redditch Premises

The advent of the smaller-wheeled safety cycles with their lower riding height, the pneumatic tyre and the birth of mass production on assembly lines resulted in a bicycle craze during the last decade of the nineteenth century. The cycle industry was booming and in 1895 Eadie made the decision to separate off the various cycle manufacturing activities of his companies Plans were drawn up to construct a new factory complex on Lodge Road, Redditch. The first building was completed and occupied in September 1895 by the newly formed Albert Eadie Chain Co. Ltd. The Eadie Manufacturing Co Ltd moved into its new premises the following year. The move allowed the plant and machinery to be better laid out for more efficient and larger-scale cycle production. The combination of sound management and a boom market had allowed the Eadie group of companies to become firmly established and financially secure.

The New Enfield Cycle Company Limited

With no sign of the bicycle craze lessening, Eadie decided to raise additional capital in order further to expand the business. A new public company, the New Enfield Cycle Company Ltd, which was registered on 1 July 1896, was established to combine the Enfield Manufacturing Company Ltd and the cycle manufacturing portion of the Eadie Manufacturing Company Ltd. It was set up with a nominal share capital of

£125,000, comprising 100,000 ordinary shares and 25,000 preference shares, each of £1 value. The directors were listed as Lord Ernest Seymour (Chairman), T Evans, G H Cartland, R W Smith and A Eadie, who were each to receive a remuneration of £600 per annum. Eadie, the managing director, and Smith, the works manager, were the only members of the board with cycle making experience. Cartland was a barrister at law and Evans a local publisher.

Both Eadie and Smith were well respected in the cycle trade and there was unlikely to be a shortage of investors keen to acquire shares in the new company. This was ensured by the agreement with the Eadie Manufacturing Company Limited to guarantee dividend whilst the business became established since 'profits of that portion of the business it is proposed to acquire from the Eadie Manufacturing Company Limited cannot with accuracy be stated separately from those of the Cycle Components department'. The guarantee was for a dividend of 7% on preference shares and 10% on ordinary shares for a period of three years.

Eadie purchased 1,500 and Smith 75 shares. Cartland and Evans were major share holders with 500 and 1,500 ordinary shares and 250 and 275 preference shares, respectively. The Eadie Manufacturing Co acquired 40,475 shares. The purchase of 166 shares by Charles Comiot, a cycle dealer in Paris and 1,000 shares by Ivan Alexieff, an Enfield agent in Moscow, clearly indicated that Enfield cycles had already firmly established an international market. Indeed the prospectus had claimed agents in more than thirty countries as well as every important town in the United Kingdom.

The new company took over the bicycle interests of the Eadie Manufacturing Company and the Enfield Manufacturing Company on the 31 August 1896. The purchase price of £90,000 included the land, factory building, machinery and licences for patents and trademarks relating to Enfield cycles. The stock in process of manufacture was valued at £14,000. With the takeover the company inherited a modern factory occupying a 2½ acre site on the Common. It was fully equipped for cycle manufacture with recently purchased machinery of English or American manufacture, which was powered by three Crossley gas engines developing 120 horse power. Electric

RIGHT: The front cover of the prospectus to establish the New Enfield Cycle Company as a new public company presented at a meeting held on the 1st July 1896.

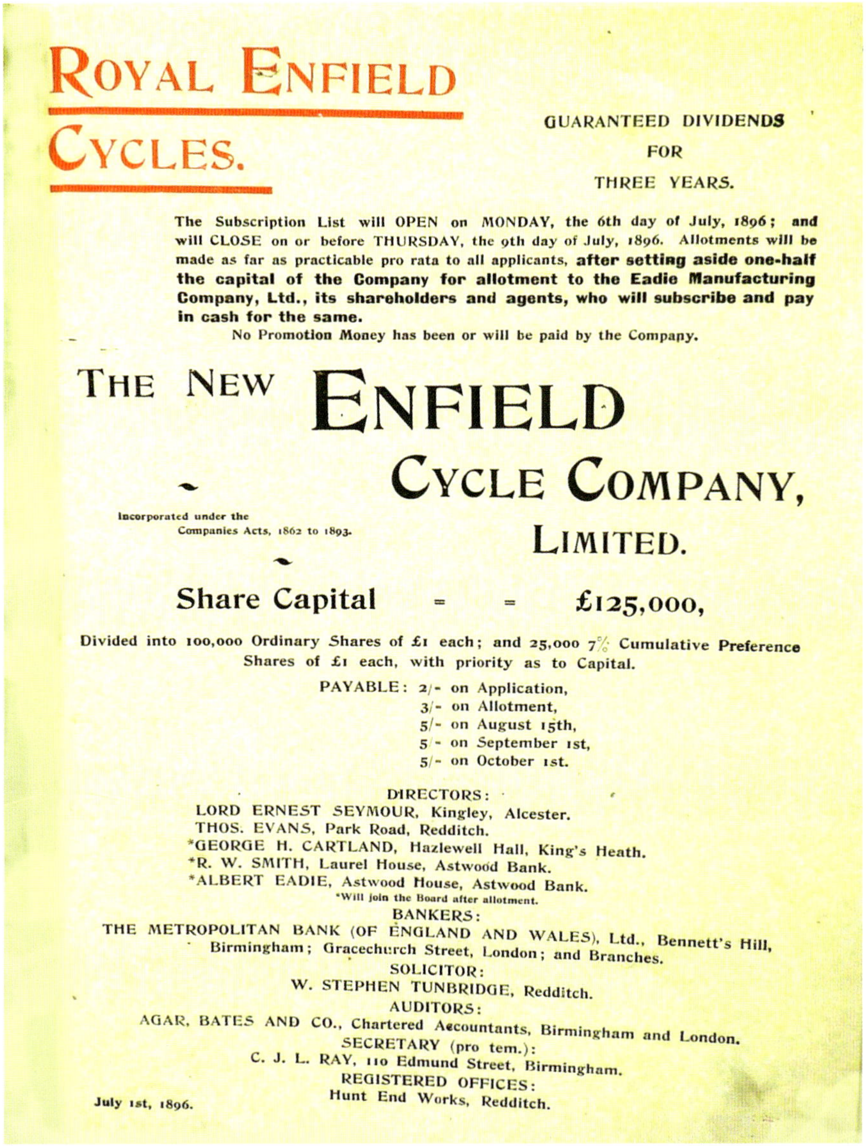

ROYAL ENFIELD CYCLES.

GUARANTEED DIVIDENDS FOR THREE YEARS.

The Subscription List will OPEN on MONDAY, the 6th day of July, 1896; and will CLOSE on or before THURSDAY, the 9th day of July, 1896. Allotments will be made as far as practicable pro rata to all applicants, **after setting aside one-half the capital of the Company for allotment to the Eadie Manufacturing Company, Ltd., its shareholders and agents, who will subscribe and pay in cash for the same.**

No Promotion Money has been or will be paid by the Company.

THE NEW ENFIELD CYCLE COMPANY, LIMITED.

Incorporated under the Companies Acts, 1862 to 1893.

Share Capital = = £125,000,

Divided into 100,000 Ordinary Shares of £1 each; and 25,000 7% Cumulative Preference Shares of £1 each, with priority as to Capital.

PAYABLE: 2/- on Application,
3/- on Allotment,
5/- on August 15th,
5/- on September 1st,
5/- on October 1st.

DIRECTORS:
LORD ERNEST SEYMOUR, Kingley, Alcester.
THOS. EVANS, Park Road, Redditch.
*GEORGE H. CARTLAND, Hazlewell Hall, King's Heath.
*R. W. SMITH, Laurel House, Astwood Bank.
*ALBERT EADIE, Astwood House, Astwood Bank.
*Will join the Board after allotment.

BANKERS:
THE METROPOLITAN BANK (OF ENGLAND AND WALES), Ltd., Bennett's Hill, Birmingham; Gracechurch Street, London; and Branches.

SOLICITOR:
W. STEPHEN TUNBRIDGE, Redditch.

AUDITORS:
AGAR, BATES AND CO., Chartered Accountants, Birmingham and London.

SECRETARY (pro tem.):
C. J. L. RAY, 110 Edmund Street, Birmingham.

REGISTERED OFFICES:
Hunt End Works, Redditch.

July 1st, 1896.

This page is reproduced from the prospectus and identifies the directors of the company and the share capital.

PROSPECTUS
Objectives and Developments

The Enfield Manufacturing Company Limited was formed in the year 1893 for the sale of the well-known Enfield and Royal Enfield Cycles in the British Isles, while on the Continent of Europe and other parts of the Globe the sale of these cycles remained an exclusive property of the Eadie Manufacturing Company Limited. The latter Company manufacture the celebrated 'Eadie Fittings' as well as the Enfield Cycles, but the enormous increase of the demand for both Enfield Cycles and Eadie Fittings has at last compelled a rearrangement of the business.

The New Enfield Cycle Company Limited is formed with the object of separating the manufacture and sale of Enfield Cycles from that of Enfield Fittings and Cycle Components. An agreement has therefore been made by which the New Enfield Cycle Company Limited acquires the right to the manufacture, and to the whole trade throughout the world, of the well-known Enfield and 'Royal Enfield' Cycles together with the very complete Works and highly effective plant which have been used for their manufacture. The business has shown, since it was established by the Eadie Manufacturing Company Limited, a very rapid and continuous development, and has NEVER BEEN IN A POSITION TO SATISFY DEMANDS made upon it by its agents. Owing to the very favourable position created by these excessive demands the Eadie Manufacturing Company Limited has been able to use, and has used every discrimination in the appointment of reliable and responsible agents. The cycles manufactured are of the highest grade, commanding high prices, and more than sufficient orders, BASED ON AN OUTPUT FOR 1897 OF NOT LESS THAN 400 CYCLES WEEKLY, ARE ALREADY AT THIS EARLY PERIOD ASSURED FOR NEXT SEASON, a fact which demonstrates the confidence agents have in the Enfield Cycles, and their eagerness and anxiety to secure future deliveries. It must be specially borne in mind THAT THE BUSINESS IS A GOING CONCERN IN FULL WORK.

The prospectus statement made at the launch the New Enfield Cycle Company Limited

lighting had been installed by the Edison Swan Company in part of the works. The new plant provided employment for 600 workers and produced an output of 300 or more cycles per week and a range of spare parts for other manufacturers.

Ernest Godbold

The appointment of Ernest Godbold as general manager was a shrewd move. He was a keen cyclist and a journalist, who at the time of his appointment was in partnership with Charles Sisley in publishing The Cycle. He remained with Enfields until July 1902 when he resigned to take up a similar appointment with the Premier Cycle Company Ltd.

The New Eadie Manufacturing Company

The New Enfield Cycle Company Ltd was very quickly working to capacity and struggling to meet the ever increasing demands for its products. To alleviate the pressure, Eadie decided in 1897 to set up his own factory in Lodge Road, Redditch for the supply of cycle components and fittings to the overseas market. The company was to be known as the New Eadie Manufacturing Company, the word 'New' appears to have been included to distinguish the company from the original company established in 1892, but was little used and was officially dropped the following year. Amongst the company's products was the famous Eadie Coaster Hub. This was a rear hub brake, developed in France, and for which Eadie had been successful in acquiring the patents and manufacturing rights.

Although independent the Eadie and Enfield companies worked in close cooperation. Eadie manufactured all the components required to produce a finished machine and it is hardly surprising that machines appeared under the Eadie name. This was most apparent in the case of the early motorised vehicles, which were available under both the Eadie and Enfield names.

Albert Eadie sold all assets of the Eadie Manufacturing Company, including its Redditch factory to the Birmingham Small Arms Company in 1907. He resigned from Enfield's to become a manager, and subsequently managing director of BSA.

The Enfield Cycle Company Ltd

The use of the word 'New' in the company's title was a cause of confusion. An extraordinary meeting was called on the 13 October 1897 to agree the change of name to the Enfield Cycle Company Limited from 1 November 1897. The name was to remain in use for seventy years, the entire period of the company's independent operation in the United Kingdom

Survival Through Hard Times

The Enfield Cycle Company had been created during the cycling boom of 1896 and 1897, but was soon experiencing difficult times owing to the subsequent slump in demand for cycles during 1897, which found many firms experiencing financial difficulties and bankruptcy. Fortunately the company was in a sound financial position as the Eadie Manufacturing Company had retained a major shareholding. Furthermore it had given assurance that the Enfield Cycle Company would pay a dividend of 10% per annum for the first three years of its existence. With good products and sound management the Enfield and Eadie companies were able to survive and maintain employment for the workers.

Albert Eadie's and R.W.Smith's achievements in making a financial success of the ailing George Townsend business, renamed the Eadie Manufacturing Company, and founding the Enfield Cycle Company were recognised by the employees of the companies. On the 16 November 1901 the two men were invited to be guests of honour at a dinner at the

General Views of the Works

These images of the Hunt End Works of the New Enfield Cycle Company Limited were reproduced in the prospectus published on July 1st 1896. At this period the works were amongst the best equipped in the country.

General View of Works.

Offices and Packing Room.

Automatic Machine Department.

Self-Acting Drilling Machinery.

Automatic Profiling Machiney.

Toolmaking Department.

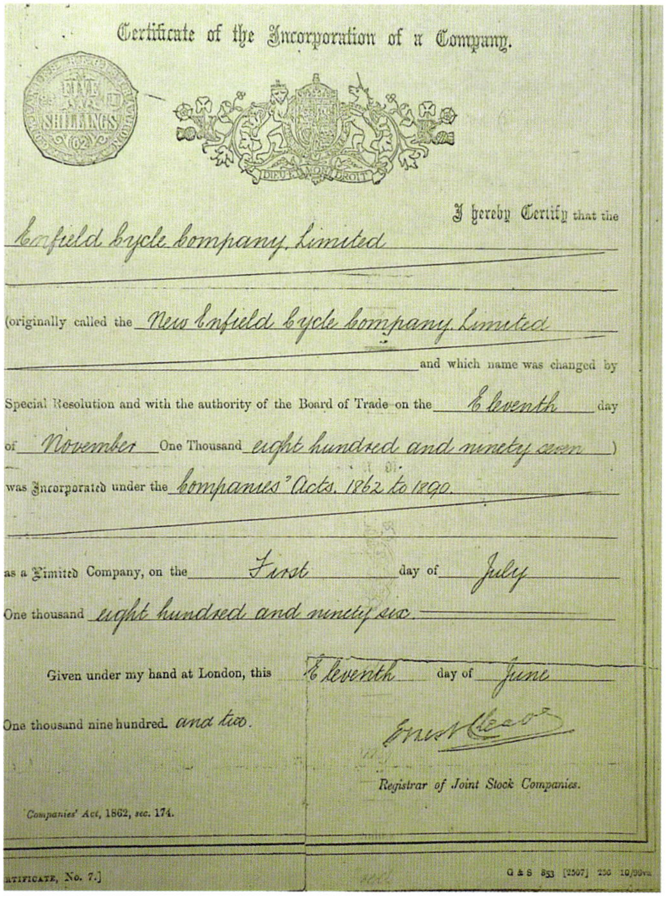

Certificate of the Incorporation of a Company.

I hereby Certify that the Enfield Cycle Company Limited

(originally called the New Enfield Cycle Company Limited

and which name was changed by Special Resolution and with the authority of the Board of Trade on the Eleventh day of November One Thousand eight hundred and ninety seven)

was Incorporated under the Companies' Acts, 1862 to 1890

as a Limited Company, on the First day of July One thousand eight hundred and ninety six.

Given under my hand at London, this Eleventh day of June One thousand nine hundred and two.

Registrar of Joint Stock Companies.

Companies' Act, 1862, *sec.* 174.

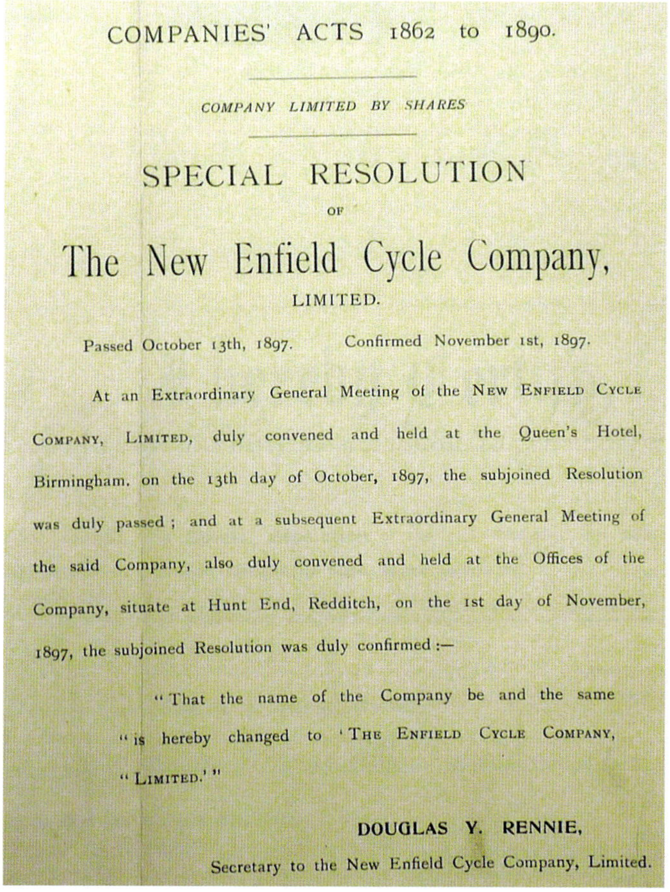

COMPANIES' ACTS 1862 to 1890.

COMPANY LIMITED BY SHARES

SPECIAL RESOLUTION

OF

The New Enfield Cycle Company,

LIMITED.

Passed October 13th, 1897. Confirmed November 1st, 1897.

At an Extraordinary General Meeting of the New Enfield Cycle Company, Limited, duly convened and held at the Queen's Hotel, Birmingham, on the 13th day of October, 1897, the subjoined Resolution was duly passed; and at a subsequent Extraordinary General Meeting of the said Company, also duly convened and held at the Offices of the Company, situate at Hunt End, Redditch, on the 1st day of November, 1897, the subjoined Resolution was duly confirmed:—

"That the name of the Company be and the same "is hereby changed to 'The Enfield Cycle Company, "Limited.'"

DOUGLAS Y. RENNIE,

Secretary to the New Enfield Cycle Company, Limited.

Announcement of the renaming of the company to become 'The Enfield Cycle Company Limited'.

White Hart Assembly Rooms, Headless Cross, for the presentation of congratulatory addresses.

First Powered Machine

Motor car development in Great Britain had been stifled by the Highways and Locomotives (Amendment) Act of 1878 with its requirement for a person to walk ahead of any motor vehicle on the road to give warning of approach. This restrictive legislation was replaced in November 1896 by the Locomotives on the Highway Act, which raised speed limits from 4mph to 12mph. The passing of the act created considerable enthusiasm amongst British companies for the development of powered vehicles and it was not long before several cycle manufacturers began to experiment with motorised versions of their tricycles or quadricycles. Albert Eadie immediately recognised the business opportunities this represented and was keen to diversify into the powered vehicle market

R.W.Smith drew up plans for a motorised quadricycle and in 1897, less than a year from the repeal of the 1878 Act, had a prototype built for his personal use and evaluation. It was of very similar design to the popular De Dion Quad being a two-seater with the driver seated upon a cycle seat at the rear and a passenger seat mounted between the two front wheels. The engine was a single-cylinder, air-cooled, 1¾hp De Dion unit of 239cc (66mm bore and 70mm stroke), with automatic inlet valve and mechanically operated side exhaust valve. It was mounted behind the rear axle and drive taken to the rear wheels by an exposed reduction-gear. The petrol-air mixture was supplied by a surface carburettor.

First Motorised Royal Enfields

The prototype Quadricycle created considerable interest and Eadie and Smith were eager to enter the market with a similar machine. The following year a second quadricycle was built under licence from De Dion Bouton. The following year the company began to produce motor tricycles and motor quadricycles built under licence from De Dion Bouton. Fitted with 2¼hp De Dion power units, which had square dimensions of 70mm bore and stroke giving a capacity of 269cc, they were first offered for sale on 1 September 1899. The De Dion-derived tricycles and quadricycles were built and sold in fair numbers for several years before being discontinued.

By 1900 motor car production had grown to be an important and expanding part of the company's activities, and in January 1900 Walter

Fernando Grew was appointed as manager of the motor department. Grew was a keen racing cyclist and had a background in railway carriage manufacturing and the cycle and wheel building industry.

Future Planning

By 1900 Eadie and Smith were firmly convinced that the future of the Enfield Cycle Company Ltd would rest firmly with powered vehicles and most probably with the motor car. At the time only De Dion based tricycles and quadricycles were in production, but plans were in hand to extend the range to include motor cars and possibly motorcycles. The ongoing work to develop a range of motor cars is described in Chapter 3. This would ultimately prove unsuccessful and the motor car side of the business was sold to Alldays & Onions in 1908. Meanwhile the motorcycle development was continuing and proving more successful. The early motorcycle developments over the years from 1900 to 1909 are described in Chapter 4. By 1910 the company had developed sound designs and become an established motorcycle manufacturer. The further motorcycle designs are described in detail on a year by year basis from chapter 5 onwards. Of course whilst all the work on powered vehicles was proceeding the Enfield cycle company remained a major manufacturer of pedal cycles. Before continuing with the history of the development of the powered vehicles a brief history of the Royal Enfield cycles is given in Chapter 2.

Albert Eadie

Albert Eadie, the 'Emperor of Redditch'.

Albert Eadie, who was born in 1863, was brought up in the jewellery quarter of Birmingham.. His father was an engraver-jeweller and it was expected that Albert would follow this trade. His interests, however, lay in engineering and at thirteen he left school to become an office clerk before taking a job as traveller for Perry & Co Ltd of Lancaster Street, Birmingham. Perry's was at the time a small manufacturing company, producing items such as pen nibs and pencil cases, but it did not take long for the enthusiastic young man to persuade the company to enter the bicycle components trade. He was rewarded by being appointed sales manager, responsible both for the buying-in of components for sale and acting as travelling salesman.

Most of the fittings sold by Perry's were supplied by D.Rudge & Co. Ltd, Coventry. Eadie became a frequent visitor to the firm and quickly gained respect for the engineering skills of Rudge's assistant manager, Robert Walker Smith. The two men combined to manage and subsequently take over the ailing George Townsend & Co Ltd.

Within a period of some fifteen years, his business skills had resulted in the growth of industry in the town to the extent that the town had doubled in size and he had become by far its largest employer of labour. Eadie would be recognised as the most successful salesman the country had ever known. He was also an engineering innovator and his development of transmission chain and the Eadie Coaster Hub contributed to the development of the bicycle.

Albert was an imposing larger than life character. Tall and well-built, he possessed the stature to stand out. He was recognised as much for his combined geniality and hospitality as much as for his many business skills. It is unsurprising that with his stature, flamboyant dress and benevolence as much as his importance to the prosperity of the town that he became known as the 'Emperor of Redditch'. Albert appears to have been proud of the title recognising it as a measure of the esteem in which he was held by the town and work colleagues. From 1894 he adopted 'Emperor', Redditch as his telegraphic address.

By 1907 Albert Eadie was in poor health and it was reported at the time that his weight was down to only 18 stone. It was time to reduce his commitments and at the annual meeting of the Enfield Cycle Company Ltd held in the new Hewell Road works on 7 October 1907 he resigned his post and announced his resignation from the board of directors. He retained his shareholdings in the Enfield Cycle Company Ltd but sold his other assets, including the Eadie Redditch factory, to BSA in 1907. He was appointed a director of BSA and subsequently became the managing director of the bicycle and motorcycle departments.

He suffered indifferent health during his last few years, which were spent in semi retirement and he passed away on 17 April 1931.

Robert Walker Smith

Unlike Albert Eadie, Robert Walker Smith came from an engineering background. He was born in 1857 near Stamford in Lincolnshire, but whilst still a child the family moved to Wolverhampton, when his father obtained a post of engineer in the Great Western Railway workshops. The young Robert gained his technical training at the Wolverhampton School of Arts and through an engineering apprenticeship in the same GWR workshops as his father.

Upon leaving the GWR works he gained appointment as assistant works manager at D Rudge & Co Ltd. Walter Phillips was general manager at this time and the two men were jointly responsible for the improvements to existing models and the development of new machines including the Olympia tandem tricycle and the Rudge triplet. Dan Rudge, who had begun building bicycles in Woverhampton in 1870, died in June 1880. His widow continued to run the company under the management of Walter Phillips until November of that year before agreeing its sale to George Woodcock, a wealthy Coventry solicitor. Woodcock relocated the company at Coventry and amalgamated it with the bankrupt businesses of Haynes & Jefferis and Smith, Starley & Co to form the Tangent & Coventry Cycle Co. Both Phillips and Smith transferred to Coventry with the new company.

Whilst at Coventry R.W.Smith was responsible for the manufacture in the Rudge factory of the parts and fittings which Perry & Co sold to the trade. It was this link with Perry that brought him into contact with the manager of Perry's cycle department, Albert Eadie. Eadie had been responsible for Perry entering the cycle trade and through his drive had built up

a substantial business. An ambitious man he could not ignore the possibilities of establishing his own business. The two men got on well together and their skills were complementary; Eadie was the extrovert salesman and skilled businessman whilst Smith was the talented engineer and manager. The two men found the opportunity they were seeking in the ailing George Townsend & Company and formed a partnership in 1891 for the management of the company and its subsequent acquisition.

George Townsend & Co was renamed the Eadie Manufacturing Co in 1892. The two men continued their active collaboration through the creation of seven new companies including the Eadie Chain Co. and the Enfield Cycle Co. The association between them lasted until 1907 when the Eadie Company was amalgamated with BSA. Both men were given seats on the BSA board.. R.W.Smith remained a director of BSA for two years before being compelled to resign owing to the pressures of ill-health and a heavy work load. He then became managing director of the Enfield Cycle Co Ltd, a position he would occupy until his death at the age of 75 on 6 April 1933.

R.W.Smith was a brilliantly inventive engineer and the design brain behind the early Royal Enfield designs and innovations. He was also a skilled businessman and instrumental in ensuring the profitability of the Enfield company through periods of both industrial growth and depression. Although described as being of a retiring disposition, he was well liked and generally known simply as RW, although in later years he was more affectionately referred to as 'The Guv'nor'. Despite the growth, under his management the company retained the attributes of a well run family business and he enjoyed the loyalty of staff members.

RW had four sons, Frank Walker, George H., Stanley Herbert and Gilbert D., all of whom followed their father into the business. The eldest son Frank shared with his father the management of the company and after an illustrious career in the services during the First World War returned to become joint managing director with his father.

RW had been a first-rate racing cyclist in his youth before becoming a motoring pioneer. In later years his sporting interests were catered for with billiard cue and golf club. In the early years of the Twentieth century he acquired an estate in Worcestershire and began to farm in a modern scientific manner. In later years he would spend much of his time enjoying the life of a country gentleman. He died in 1932.

CHAPTER TWO

Bicycles

Once R.W.Smith, together with Albert Eadie, had gained control of the ailing Townsend & Co, he wasted little time before developing a new range of bicycles. These discarded the earlier cross frame designs with their high wheels and solid tyres driven by block chains. Instead his new machines were front drivers (see Page 10) of modern design and featured small wheels fitted with pneumatic tyres. By the time the machines were in manufacture at Hunt End the company had begun trading as the Eadie Manufacturing Company. The new machines were marketed under the Enfield name, in recognition of the recently completed contract with the Royal Small Arms factory at Enfield which had helped to secure the future of the struggling company. Public announcement of the new Front Driver Enfield cycles was made at the Stanley Show in October 1892, where two machines were exhibited alongside six of the old safety models. The new machines were well received and proved popular. Sales at this time were handled by the Enfield Manufacturing Company Ltd, which was registered on 24 February 1893.
The Royal Enfield name was first used later the same year to market a superior version of the Enfield bicycle. The first appearance of Royal Enfield machines is believed to have been at the Stanley Show held during November 1893. The firm occupied Stand 79 and displayed twenty-four machines, ten of which were branded Royal Enfield. The exhibits consisted of safety bicycles, suitable for ladies and gentlemen, and included a Front Driver and a tandem. The latter had design flaws, such as a lack of adjustment of the chain, and was subject to press criticism. The design would be considerably improved before the following show.

Growth and Security

By the time of the 1894 Stanley Show the bicycle industry in the United Kingdom had entered a boom period and the Eadie group of companies was beginning to prosper. Nineteen safeties and two tandems were displayed on the Enfield stand. The Lady's model safety machines for 1895 featured a distinctive frame design. This had been introduced at the instigation of Albert Eadie, the marketing expert, who recognised the advantages of distinctive features making the Enfield products stand out from the competition and to be uniquely identifiable.

Enfield Cycle Company Ltd.

The booming cycle industry caused Eadie to reorganise his businesses in order to establish new larger premises at Lodge Hill, Redditch. The new works were equipped with new plant and machinery to improve productivity and increase cycle manufacture. We have seen in Chapter 1 how the continued demand for bicycles was the impetus to set up the New Enfield Cycle Company Ltd specifically for bicycle manufacture. The company was registered on 1 July 1986 and became the Enfield Cycle Company Ltd on 1 November 1897.

The bicycles displayed at the 1896 Stanley Show were fitted with the 'Eadie' Coaster rear hub brake, operated from the bottom bracket by back pedaling. The firm was reluctant to incorporate disc adjustment hubs, for chain tensioning, at the time of the show although these would be introduced during the following year.

The 1895 Lady's safety model. The model shows the distinctive frame, which Albert Eadie instigated to make the Royal Enfield easily recognised.

The Enfield stand at the 1897 Stanley Show displayed ten safeties, two tandems, a tricycle and a quad. Features of the new machines included a novel bottom bracket, which had the bearing parts in a separate sleeve, and a fulcrum for the front brake fitted to a bar projecting horizontally from the handlebar stem. The lady's dress guard cords were now wound around studs on the mudguards rather than passing through holes in the mudguard.

'Made like a Gun'

By 1898 all machines were identified as Royal Enfields and the motto 'Made like a Gun' was being used to summarise the quality of manufacture. They were supplied with R.W.Smith's Patent Disc-adjusting bracket and Enfield Lever Pattern brakes. The Light Roadster weighed in at 30 lb.

The cycling boom years of 1895 and 1896 had been followed by a slump in 1897. In a successful attempt to boost trade Albert Eadie arranged for several novelties to be incorporated into the 1899 Royal Enfield models displayed at the 1898 Stanley Show. Eadie's innovations included the incorporation of rubber inserts into the steering head tube to absorb vibration and the introduction for the first time on any machine of an eccentric chain adjustment at the crank axle. Adjustment disc plates were also introduced into the fork ends of some models. The band brake was promoted especially on the ladies models. The tricycle was particularly well liked because of the eccentric bracket chain adjustment.

Advertisement from 1901 for the Flexible bicycle.

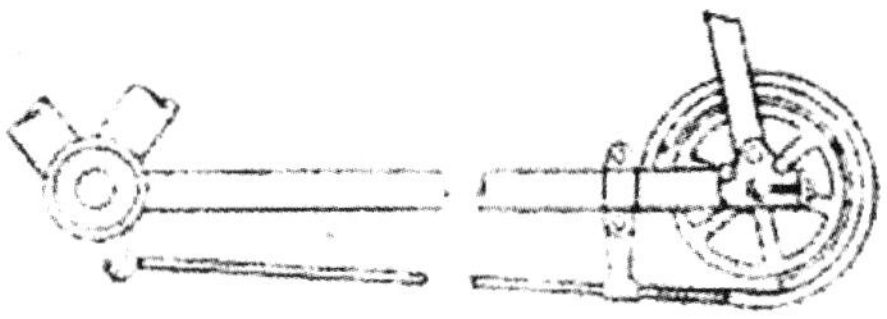

The band brake fitted to the Lady's models for 1899. The band consisted of a leather cord and was held within a split tube, which curved round the drum and formed a natural spring. A clutch on the crank axle operated the connection rod when the rider back-pedaled.

The Flexible Lady's Bicycle. It was described as providing 'The Acme of Comfort' and was the most expensive model in the Royal Enfield range.

The 'Flexible'

Fourteen safety bicycles, including eight ladies machines and two tandems were exhibited on the Enfield stand at the 1899 Stanley Show. A prominent feature of the display was a flexible spring-frame bicycle designed by Rucker and Philpott. The Flexible frame design was taken up by Enfield's and was subject to a major re-design before being offered for sale in 1901. It proved unsuccessful and a further re-design was carried out before it was again offered for sale in 1902. It was built with a compression stay of flat steel, the seat stays being divided and connected to the saddle lug by two flat steel scroll springs. Front fork blades were made from two flat steel springs, placed with broader surfaces parallel to the wheel hub. The longer blade, which was at the rear, was looped over at its lower end to provide an eyelet for the wheel spindle. The shorter forward blade extended downwards to a point just below the start of the curve in the rear blade. Its end was partly lapped round the rear blade for location. The same fork was offered on Royal Enfield motorcycles. The rear forks were of similar construction, whilst sprung handlebars and seat pillar were also fitted.

The Flexible was withdrawn in 1904 following on from the introduction of the 'Girder' models. However Flexible forks continued to be available to special order through to 1912.

Modèle Riche

The Royal Enfield bicycles had always been manufactured to the highest quality with the aim of being amongst the finest machines available. Indeed the company would go on to claim 'it occupies a place entirely apart from the usually accepted high-grade bicycle and is absolutely unequalled in the matter of refinement and distinctiveness'. Nevertheless by 1899 discerning wealthy customers were demanding that their cycles should be finished to a higher than normal standard and be supplied already equipped with the best fitments. They expected the frame to be enamelled black and to be hand polished before being handsomely lined, preferably using gold leaf. Enfield identified bicycles finished to this standard as Riche. The description Modèle Riche, or subsequently Riche, would be used to identify the top of the range Royal Enfield models for more than twenty years.

The 1901 Model Range

The Royal Enfield range already catered adequately for most users' requirements and was unchanged for 1901. It consisted of eight different designs of safeties, two tandems and two tricycles. Few improvements were required; the steering lock was changed from bolt through to head clip and a pull-up front stirrup brake became available on the ladies models. The top of the range solo models, either lady's or gentleman's, were priced at 15 guineas. A freewheel and rim brakes, which were optional on the touring versions of these models, were included within this price. Similarly the Road Racer was supplied with a freewheel and rear rim brake if required. The popular machines were the 'Ten Guinea' 'Model D' economy models. The Lady's and Roadster versions came equipped as standard with a front spoon brake; a front rim brake and Morrow freewheel cost an additional 10s. The 'Model D' Road Racer came without brakes and fitment of a front rim brake and Morrow freewheel cost an additional 15s. Special versions of the 'Model D machines were available for the first time and cost an additional 2 guineas. They came equipped with a freewheel, two rim brakes, an alternative handlebar and plated wheel rims. The

Illustrated in the 1901 Royal Enfield Catalogue were eight different safety models, two tandems and two tricycles. The back cover illustrated the Royal Enfield Motor Tricycle,

catalogue did not include illustrations of the Modèle Riche but noted that these were still available at a price of twenty guineas for customers desiring 'refined gold'.

The Tandem for a lady and gentleman cost £25 and came as standard with a two-speed hub gear. The gentleman's Tandem was priced at £21 as a Road Racer or at £24 as a Roadster with a two-speed hub gear. Chain guards for either machine cost an extra £1.

The tricycles were finished to Modèle Riche specification. The

1901 Model Range	
Lady's	£15 15s
Roadster	£15 15s
Light Roadster	£15 15s
Road Racer	£15 15s
Racer	£15 15s
Model D Lady's	£10 10s
Model D 'Special' Lady's	£12 12s
Model D Roadster	£10 10s
Model D 'Special' Roadster	£12 12s
Model D Road Racer	£10 10s
Model D 'Special' Road Racer	£12 12s
Modèle Riche	£21
Tandem for Lady and Gentleman	£25
Tandem for two Gentlemen	£21
Lady's Tricycle Modèle Riche	£25
Tricycle Modèle Riche	£24

1901 Lady's Bicycle

1901 Roadster

1901 Light Roadster

1901 'Model D' Lady's

1901 'Model D' Roadster

1901 'Model D' Road Racer

1901 Road Racer

1901 Racer

1901 Tandem for Lady and Gentleman

1901 Lady's Tricycle Model Riche

1901 Tricycle Modèle Riche

1901 Tandem for Two Gentlemen

lady's model was equipped with rear mudguards and available with or without a free wheel for £25. The basic gentleman's model was available for one pound less, but was similarly priced when equipped with rear mudguards and a leather tool carrier. Tool carriers were also available for most of the other models of the range. A leather case cost 10s 6d whilst a detachable metal case cost twice as much.

Girder Frame

In order to improve the rigidity of the simple diamond frame, several manufacturers had begun offering cycles with an additional diagonal bracing tube. These cross-tube designs significantly strengthened the upper part of the frame but did little to brace the frame at the lower bracket where the greatest forces were exerted. In the Enfield design a further tube was added to triangulate the frame and provide extra support at the crank bracket. The new design was considerably more expensive to produce but quickly demonstrated its worth through competition success; the company was soon able to claim 'Our Girder frame is the best hill-climber in the world'.

The new Girder frame appeared at the 1901 show and was initially made available on the Modèle Riche models. It would subsequently be widely fitted to all but the cheapest models of the range. A year or so following its launch the Girder frame was subjected to exhaustive testing by Lloyds. Whereas the Royal Enfield diamond frame had withstood a creditable stress up to 18cwt before collapse, the Girder frame of the same size and weight withstood a sustained strain of 27cwt without failure

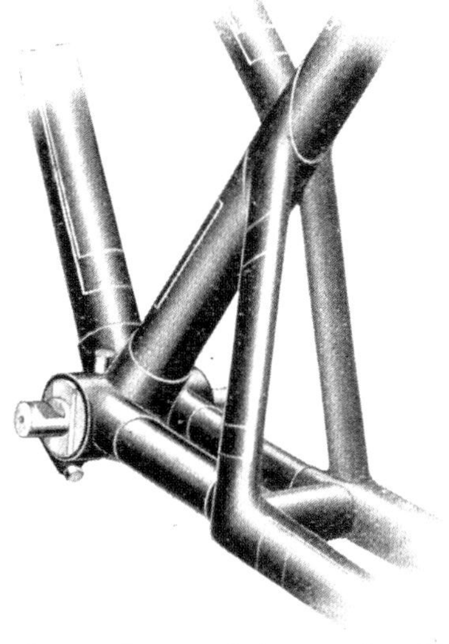

The lower frame of the Duplex Girder design which became available in 1907.

The Girder frame design was gradually made available on all except the cheapest bicycles of the range. An improved version of the Girder design was introduced at the 1907 Stanley Show. Known as the Duplex Girder frame, it replaced the single strut by a pair of tubes thereby further enhancing the lateral rigidity of the design. It remained in production on the top of the range models through till the late 1920's. The original frame with single bracing strut continued to be offered on the cheaper machines

1903 Model Year

The 1903 range consisted of the Modèle Riche Lady's and Roadster models at 18 guineas, the middle range Royal Enfield models at 15 guineas, the Model B machines at 12 guineas and the Model D at 10 guineas. The Royal Enfield models were available as a Lady's, Roadster, Lightweight Roadster, Road Racer or Path Racer. The Model B and Model D machines were available as Lady's, Roadster or Road Racer. Three Tandem models were listed. Both the Tandem for Lady and Gentleman and the Tandem for Two Gentlemen were priced at £24, whilst the Racer or Road Racer version cost £21. A detachable stay to make the Lady and Gentleman Tandem safe for use by two gentlemen was available for £1. The range was completed by the Lady's and Gent's Flexible models at £20; this would be the last year of their production.

The Special Lightweight Roadster model illustrates clearly the new Girder frame.

Mr Cooke's achievement was eclipsed some ten years later by Mr Gill who covered over 100,000 miles on his 1905 'Girder' model.

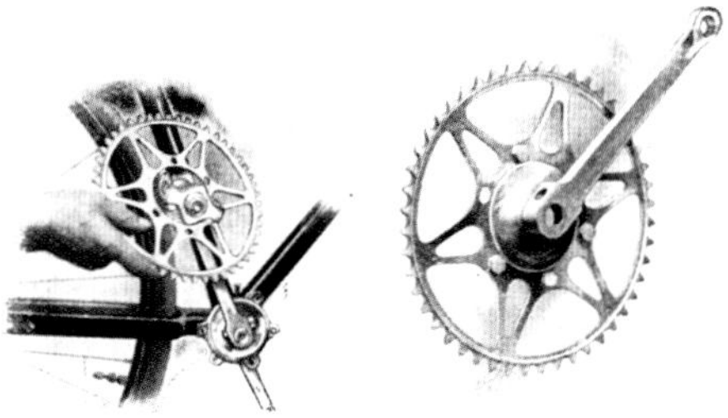

The chainwheel on the Flexible and early Girder models was distinctive in having six pear-shaped cut-outs. It secured to the crank in a novel manner so that it might be readily detached for the purpose of changing gearing without removal of the pedal and crank.

1903 'Flexible' Gent's Bicycle

THE
ROYAL ENFIELD
"FLEXIBLE"
GENT'S BICYCLE.

SPECIFICATION.

FRAME, 22in., 24in., or 26in. high—28in. to order (measurement is taken from the centre of crank axle to top of seat lug); WHEELS, equal 28in.; HUBS, dust-proof and oil-retaining; SPOKES, double-butted, anti-vibration tangent, heavily nickel-plated; RIMS, "Roman" aluminium; TYRES, Dunlop-Welch or Welch-Bartlett 1½in.; STEERING LOCK, Smith's Enfield patent; HANDLE-BAR, flat or upturned, 17in. wide (see page 11); HANDLES, special celluloid or Cooper's white felt; BRAKES AND FREE WHEEL, Enfield Bowden back rim brake with inverted hand lever and New Patent Hub (brake and free wheel combined); MUDGUARDS, metal detachable, beaded edges, flexible stays; CHAIN COVER, metal detachable, with celluloid panels; CHAIN, best quality ½in. pitch roller; CRANKS, detachable, 7in. long; STANDARD GEARING, 42 by 18 = 65in. (44 by 18 = 68in. can also be fitted); SADDLE, Brooks' B91 or B28, with plated springs; PEDALS, rat-trap or rubber, 3¾in. or 4½in. inside measurement; FINISH, Frame enamelled black, four coats and hand polished, exquisitely lined in fine burnished gold; Rims, black centres, with burnished gold lines; all usual parts heavily nickel-plated on copper. Lamp Bracket, Triangular Valise, Spanners, Oil Can in seat-pillar, Celluloid Inflator with special Clips.

PRICES.

As Standard Specification	£20 0 0
By Easy Payments, or part exchange . .	£24 12 0

AN EXPERT'S OPINION

OF THE

Royal Enfield

"MADE LIKE A GUN"

MODÈLE RICHE.

Though at one time – and with good reason, too, considering the rotten way in which the machines were made—featherweight bicycles were hounded out of the market, such great strides have been made in bicycle-building that these light-weight machines have clearly entered upon a new and what gives every promise of being a long lease of life.

If we can get a machine to stand the same pressure and produce the same results as a bicycle some 6 lbs. or 8 lbs. heavier, it stands to reason we're going to have the first-named, for to pull the extra poundage over a long mileage makes such a lot of difference. Makers have fully recognised this, and most of the big firms have laid themselves out to produce machines which are in no way hampered by unnecessary weight.

The Enfield Company, Redditch, in particular, have designed a featherweight on exceptionally sound and scientific lines, and I have just had the pleasure of putting one through an extended trial. My riding weight at present is 14 st. odd, and yet the machine of only 26½ lbs. weight has carried me like the proverbial bird over some scores of miles of all sorts of going.

I know a good deal of the practical side of cycling, and in my time I have ridden dozens of machines of one kind and another, but I must confess to a feeling of genuine pleasure and satisfaction that one of the lightest roadster machines that is made should prove such a wonderful weight-carrier. This power is secured by incorporating the Royal Enfield girder frame, which is so designed as to give strength to the whole frame, and to give perfect rigidity where it is most needed—in and around the crank bracket.

MY ROYAL ENFIELD is fitted with the new back-pedalling brake, which is wonderfully sensitive, the least backward pressure makes itself felt at once in retardation, and when anything like full pressure is applied, why, I can stop dead in the middle of slopes which, under less favourable conditions, might be calculated to try one's nerves a bit. It's a good thing.

You should secure our Free Art Booklet upon Royal Enfield Bicycles, from

£10 10s. to £20.

It is a work of art and can be obtained from any of our Agents.

The Enfield Cycle Co., Ltd., Redditch.

London Depot—25, Bartlett's Buildings, Holborn Circus, E.C.
Glasgow Depot—142, West Nile Street.

MENTION THIS PAPER.

The catalogues produced during the early years of the century drew attention to the well equipped modern factory involved in the production of the Royal Enfield bicycles. Whilst it was usual to include photographs of the machine and assembly shops, it was also considered important to demonstrate the value the company placed upon its workforce. 'That we are not unmindful of the comfort and amusement of our employees will be readily appreciated from a glance at our club house, which is devoted exclusively to the employees of the Company.'

1903 Modele Riche Lady's Bicycle. The illustration is from the catalogue issued by the Dutch Royal Enfield agent, J. Witmondt of Amsterdam.

This year saw the introduction of a patent drum hub brake. The plated fork crown was replaced with boxed crown (although the boxed crown with flexible forks would continue to be available to special order for several more years). Waisted seat stays replaced straight and inverted brake levers were adopted.

Fagan Two-speed Hub

John Fagan applied for a patent for an epicyclic two-speed hub in December 1901. The design was taken up and manufactured by Albert Eadie and featured on Royal Enfield models displayed at the 1903 Show. It was fitted as standard on the Modèle Riche machines for 1904 and would remain in production until 1906 and the availability of the Sturmey-Archer three-speed hub.

The 1904 Model Range

Although the market for cycles remained strong there were more companies keen to take a share and Enfield were forced to make substantial price reductions to remain competitive. The Modèle Riche machines were reduced in price by 2 guineas and were now made available, inclusive of the Fagan two-speed gear, for 16 guineas. A 3 guineas reduction

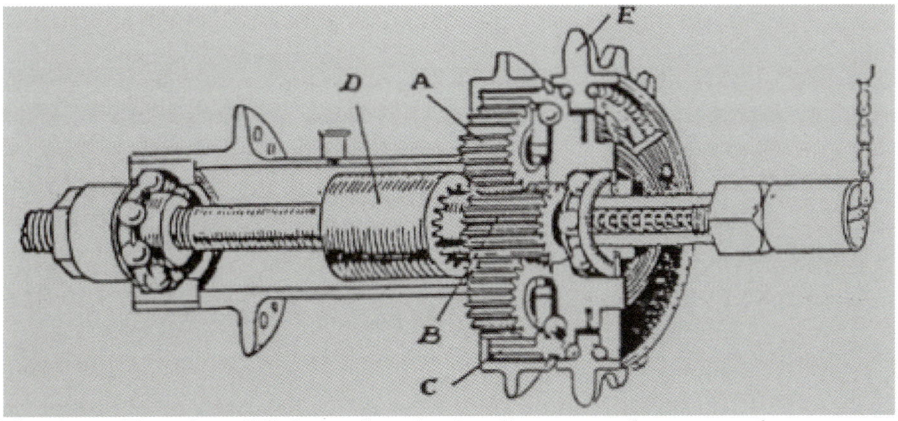

The Fagan Two-Speed Hub employed epicyclic gears and was an early attempt to provide gears on a bicycle.

was applied to the Royal Enfield models, which now cost 12 guineas. The Model B machines were reduced from 12 guineas to 10 guineas. This was the same price that had been asked in 1903 for the Model D, which was now omitted from the catalogue. The company perhaps realised that further price reductions were not possible at this price, noting: 'We venture to say that this bicycle is superior to any £12 12s, bicycle upon the market'. Both the 'Royal Enfield' and 'Model B' models were now fitted with a Royal Enfield coaster rear hub brake. The Modèle Riche Tandem remained available and was described in the catalogue, but was not illustrated. The Modèle Riche Roadster was offered as a lightweight machine or as a full Roadster for the heavyweight rider

1904 Model Range	
Modèle RicheLady's	£16 16s
Modèle RicheLightweight Roadster	£16 16s
Modèle Riche Road Racer	£16 16s
Lady's	£12 12s
Lightweight Roadster	£12 12s
Road Racer	£12 12s
Path Racer	£12 12s
Model B Lady's	£10 10s
Model B Roadster	£10 10s
Model B Road Racer	£10 10s

General Specification of 1904 Modèle Riche	
Price:	£16 16s 0d
Tyres	Dunlop 1½ in
Handlebars:	Silver finish with back celluloid grips.
Front brake:	Enfield rim with inverted lever.
Rear brake:	Enfield rim with inverted lever.
Gear:	Two-speed Fagan with freewheel upon both wheels.
Saddle:	Brooks B20.
Finish:	Frame, black enamel lined in 22 ct gold. Wheel rims nickel plated with black centre and bronze lining.
Accessories:	Spanners. Oil can in seat pillar, celluloid tyre inflator.
Options:	Dunlop-Bartlett (beaded edge) tyres. Enfield-Bowden rear wheel brake. Flat handlebars, B13 saddle, Rubber 3½in pedals.

1904 Modèle Riche Lady's Bicycle
Additional specification
Sizes: 20 in, 22 in or 24 in (26 in to order)
Tyres: Dunlop 1½in
Handlebars: Upturned.
Gears: 66 in high, 51 in low.
Gearcase: Enfield aluminium.
Pedals: Rubber, 3 1/8 wide.
Accessories: Saddle tool wallet, green waterproof dress guard.

1904 Modèle Riche Lightweight Roadster
Additional specification
Sizes: 22 in, 24 in or 26 in (28 in to order)
(Full Roadster available for heavy-weight riders)
Handlebars: Upturned.
Gears: 80 in high, 61 in low.
77 in high, 59 in low optional
Pedals: Rat-trap, 3 1/4 wide.
Optional: rat-trap 4 or 4½ in, rubber 3½, 4 or 4½ in
Accessories: Triangular tool wallet.
Aluminium gearcase (12s 6d)

1904 Modèle Riche Road Racer
Additional specification
Sizes: 22 in, 24 in or 26 in
Handlebars: Half-drop
Flat or upturned bars optional.
Gears: 84 in high, 64 in low.
91 in high, 69 in low
Pedals: Rat-trap, 3¾ wide with toe-clips.
Accessories: Triangular tool wallet.
Gearcase optional

LEFT: The 1904 Catalogue illustrated eleven different bicycles, three different motorcycles and a 10hp car.

The company already appreciated the importance of military contracts and during the year received valuable publicity with an order for 72 cycles destined for the First Volunteer Battalion of Royal Welsh Fusiliers.

General Specification of 1904 Mid-range Bicycles	
Price:	£12 12s 0d
Tyres:	Dunlop-Welch Roadster 1½ in
Front brake:	Enfield rim with inverted lever.
Rear brake:	Enfield patent Coaster hub.
Saddle:	Brooks B20.
Finish:	Frame, black enamel lined in green & bronze. Wheel rims nickel plated with black centre and bronze lining.
Accessories:	Spanners. Oil can & celluloid tyre inflator.

1904 Royal Enfield Lightweight Roadster

Additional specification

Sizes:	22 in, 24 in or 26 in (28 in to order)
Handlebars:	Flat with celluloid grips.
Gear:	74 in.
Pedals:	Rubber, 4 in wide.
Accessories:	Triangular tool wallet.
Options:	Rear rim brake operated by inverted lever, chaincase in metal (10s 6d) or aluminium(£1 1s), upturned handlebar, pedals 4½in rubber or 3¾ or 4½ in rat-trap.

1904 Royal Enfield Lady's Bicycle

Additional specification

Frame sizes:	20 in, 22 in or 24 in (26 in to order)
Handlebars:	Upturned with celluloid grips
Gear:	62 in.
Chaincase:	Enfield metal.
Pedals:	Rubber, 3 1/8 in wide.
Accessories:	Saddle tool wallet & waterproof dress guard.
Options:	Leather chaincase with Celluloid panels, rear rim brake operated by inverted lever, flat handlebars, 3½ in rubber pedals, aluminium chaincase (12s 6d extra)

1904 Royal Enfield Path Racer

Additional specification

Sizes:	22 in, 24 in or 26 in
Tyres:	Dunlop 1 1/8 in single tube sprint.
Wheels.	26 in wood rims
Handlebars:	Full drop.
Gears:	89 in. (48 x 14 chain wheels
Saddle:	Brooks B11
Pedals:	Rat-trap, 3½ in wide., with toe clips.
Accessories:	Triangular tool wallet, spanners, oil-can & celluloid tyre inflator..
Options to order.	28 in wheels, 1⅜ or 1½ Dunlop tyres, steel rims, half drop

ABOVE & RIGHT: A Royal Enfield advertisement dating from 1903.

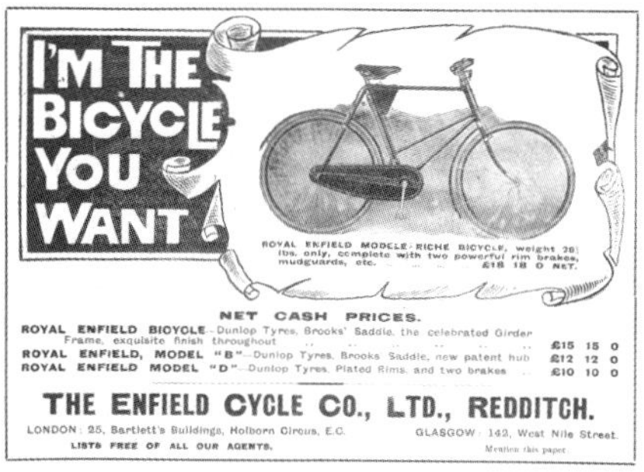

1904 Royal Enfield Road Racer

Additional specification

Sizes:	22 in, 24 in or 26 in
Handlebars:	Flat with celluloid grips.
Gear:	81 in.
Pedals:	Rat-trap, 3¼ in wide.
Accessories:	Triangular tool wallet.
Options:	Half dropped handlebar, rear-rim brake.

SHE IS TELLING HER

ABOUT THE

Royal Enfield Bicycles

We are giving away a beautiful Art Booklet which will tell **you** all about them, and which also contains many interesting photos.

Royal Enfield Flexible	£20 0 0 net
Royal Modele Riche -	18 18 0 ,,
Royal Enfield - -	15 15 0 ,,
Royal Model "B" -	12 12 0 ,,
Royal Model "D" -	10 10 0 ,,

ALSO UPON GRADUAL PAYMENT SYSTEM.

ENFIELD CYCLE CO., LTD.,
REDDITCH.

Mention this paper.

General Specification of 1904 Model B Bicycles

Price:	£10 10s 0d
Tyres:	Dunlop -Welch Roadster 1½ in
Front brake:	Enfield rim with inverted lever.
Rear brake:	Enfield patent coaster hub.
Saddle:	Brooks B21.
Finish:	Frame, black enamel lined in crimson & white. Wheel rims nickel plated.
Accessories:	Saddle tool wallet, spanners, oil can & celluloid tyre inflator.

1904 Model B Lady's Bicycle

Additional specification

Sizes:	20 in, 22 in or 24 in (26 in to order)
Handlebars:	Upturned with Celluloid grips.
Gear:	63 in.
Pedals:	Rubber, 3 1/8 wide.

1904 Model B Road Racer

Additional specification

Sizes:	22 in, 24 in or 26 in
Handlebars:	Flat
Gear:	80 in.
Pedals:	Rat-trap, 3 1/4 wide.

1904 Model B Roadster

Additional specification

Sizes:	22 in, 24 in or 26 in (28 in to order)
Handlebars:	Flat.
Gear:	71 in.
Pedals:	Rubber 4 in wide.

1904 Juvenile

Price:	£6 15s 0d, or £7 7s 0d with Coaster hub
Sizes:	18 in
Tyres:	Detachable.
Wheels:	24 in.
Handlebars:	Up-turned.
Front brake:	Enfield rim
Gear:	58 in.
Pedals:	Rubber 3 ½ in wide.
Finish.	Black, rims nickel plated
Accessories:	Saddle tool wallet, spanners, oilcan & inflator.

Steady Development: 1905 - 1909

By 1905 the Royal Enfield range of bicycles was firmly established and catered for most user tastes. Few additional changes to the basic models were required although detailed improvements would be added as these became available. The bicycle market was still extremely competitive and further price reductions were introduced. The price of the Modèle Riche machines remained unaltered but the 'Royal Enfield' and Model B models were reduced to 12 guineas and 8 guineas, respectively. The gap between the later two machines was filled by the Model A at 10 guineas. Wood wheel rims were made optionally available on the racing models.

The company was confident in its policy of price reduction and was well able to afford the further reductions proposed. It had a well-respected product range and previous price reductions had boosted sales and increased profits. The bicycle side of the business was at this time subsidising the car production, yet profits had increased from £8,258 in 1903 to £11,431 in 1904 and would reach £12,617 in 1905 and £18,663 in 1906.

The following year, 1906, marked the changeover from cable operated to roller-lever rod brakes. The 'Girder' Road Racer was offered with a two-speed coaster hub, whilst a Featherweight version of the Modèle Riche with a single-strut Girder frame and weighing just 27 lbs was made available. The Modèle Riche Lady's and Roadster models were offered with the recently introduced Sturmey-Archer three-speed gear. Within a few years this would dominate the market and the two-speed gears would go out of production.

The boom years of the cycle industry at the end of the nineteenth century had resulted in the creation of numerous firms who were keen to benefit from the sellers' market and enjoy the good profits being made by the cycle industry at that time. More manufacturers and a declining market quickly resulted in over-production. The well-established manufacturers, such as the Enfield Cycle Co, with good products and efficient production methods continued to thrive but many of the smaller, less efficient, newcomers were forced out of business. One such was the Coventry Cross Cycle Company and in 1906 this was acquired by the Enfield Cycle Company. The following year Enfields also acquired Ric Cycles.

By the time of the 1906 Show, Royal Enfield were able to display more than forty models. However, perhaps reflecting the emphasis on turning round the car side of the Enfield business, there were few innovations other than the adoption of the Sturmey-Archer 3-speed hub on the Modèle Riche. Prices of the machines ranged from £17 15s for the top of the range Riche Girder models, with Sturmey-Archer Three-speed Gear, down to £9 10s for the fixed gear Lady's Standard or Gentleman's Light Roadster Models. The Gentleman's Road Racer was similarly priced at £9 10s, although a lightweight Special Racer version, with fixed gear and finished in black enamel, but without the crimson and gold lining, was available for £8 17s 6d.

Royal Enfield had been targeting its bicycles at the wealthy but discerning customer, who wished to purchase a quality product at a reasonable price. By 1907, however, such buyers were beginning to purchase motorised bicycles and sales of pedal cycles were proving difficult. The 1907 sales brochure recorded:

> *In response to the repeated and urgent appeals of our Agents throughout the country, and Cyclists the world over, that we should place upon the market a machine somewhat lower in price than that at which Royal Enfields can be purchased, we have decided to set apart a special department in our factory for the manufacture of 'COVENTRY CROSS' Cycles.*
>
> *Hitherto we have felt that in the face of our inability to keep pace with the Royal Enfield demand, such action was impossible, and we have therefore acted up to our*

1907 Model Range

Model	Description	Gear	Price*
123	Lady's Riche Girder	63	£14 6s.
124	As Model 123	63 Enfield Coaster Hub	£14 11s.
140	As Model 123	70 x 53 Fagan Gear	£16
141	As Model 123	70 x 53 Enfield Coaster Hub	£16
144	As Model 123	Sturmey-Archer 3-speed Gear	£16 10s.
150	Lady's Riche Girder	70 x 53 Fagan Gear	£17 5s.
151	As Model 151	70 x 53 Enfield Coaster Hub	£17 5s.
160	As Model 151	Sturmey-Archer 3-speed	£17 15s.
125	Gent's Light Roadster Riche Girder	74	£14 6s.
126	As Model 125	74 Enfield Coaster Hub	£14 11s.
142	As Model 125	84 x 64 Fagan Gear	£16
143	As Model 125	84 x 64 Enfield Coaster Hub	£16
145	As Model 125	Sturmey-Archer 3-speed	£16 10s.
152	Gent's Light Roadster	84 x 64 Fagan Gear	£17 5s.
153	As Model 152	84 x 64 Enfield Coaster Hub	£17 5s.
161	As Model 152	Sturmey-Archer 3-speed	£17 15s.
105	Lady's Girder	63	£11 15s.
106	As Model 105	63 Enfield Coaster Hub	£12
120	As Model 105	70 x 53 Fagan Gear	£13 8s.
130	As Model 105	70 x 53 Enfield Coaster Hub	£13 15s.
132	As Model 105	Sturmey-Archer 3-speed	£14 5s.
107	Gent's Light Roadster 'Girder Type'	74	£11 15s.
108	As Model 107	74 Enfield Coaster Hub	£12
121	As Model 107	84 x 64 Fagan Gear	£13 8s.
131	As Model 107	84 x 64 Enfield Coaster Hub	£13 15s.
133	As Model 107	Sturmey-Archer 3-speed	£14 5s.
109	Girder Road Racer 'Club Model'	80	£11 15s.
1010	As Model 109	80 Enfield Coaster Hub	£12
122	As Model 109	80 x 64 Fagan Gear	£13 8s.
1011	Girder Path Racer	80	£11 15s.
80	Lady's Standard	63	£9 10s.
81	As Model 80	63 Enfield Coaster Hub	£9 15s.
100	As Model 80	70 x 53 Fagan Gear	£11 4s.
101	As Model 80	70 x 53 Enfield Coaster Hub	£11 10s.
97	As Model 80	Sturmey-Archer 3-speed	£12
82	Gent's Light Roadster	74	£9 10s.
83	As Model 82	74 Enfield Coaster Hub	£9 15s.
102	As Model 82	84 x 64 Fagan Gear	£11 4s.
103	As Model 82	84 x 64 Enfield Coaster Hub	£11 10s
98	As Model 82	Sturmey-Archer 3-speed	£12
84	Gent's Road Racer	80	£9 10s.
85	As Model 84	80 Enfield Coaster Hub	£9 15s.
104	As Model 84	84 x 64 Fagan Gear	£11 4s.
86	Special Racer	84	£8 17s 6d.
50	Juvenile Girl's	51	£6 10s.
51	As Model 50	51 Hyde Free-wheel	£7 5s.
52	Juvenile Boy's	51	£6 10s.
53	As Model 52	51 Hyde Free-wheel	£7 5s.
60	Lady's	63	£8
61	As Model 60	63 Eadie Coaster Hub	£9 15s
62	As Model 60	70 x 53 Fagan Gear	£9 15s
63	As Model 60	Eadie 2-speed Coaster	£10 5s.
64	As Model 60	Sturmey Archer 3-speed	£10 5s.
65	Gent's Light Roadster	74	£7 10s
66	As Model 65	74 Eadie Coaster Hub	£7 15s
67	As Model 65	84 x 64 Fagan Gear	£9 10s.
68	As Model 65	Eadie 2-speed Coaster Hub	£10
69	As Model 65	Sturmey-Archer 3-speed Gear	£10 10s.
70	Gent's Road Racer	80	£7 10s.
71	As Model 70	80 Eadie Coaster Hub	£7 15s.
72	As Model 70	84 x 64 Fagan Gear	£9 10s.

* The prices were subject to a 10% discount for cash payments.

opinion that to do one thing well was infinitely better than doing two things indifferently.

Moreover, we have always held that while the advantages and experience of the producer of the high-grade bicycle must be of infinite value in the production of the cheaper article, yet it is an absolute essential to carry on the manufacture of the two grades entirely and absolutely apart.

This essential can now be fulfilled and without in any way interfering with our turnover in Royal Enfields, we are enabled to offer the Public that for which they have for so long appealed a low-priced bicycle of absolute reliability; and we trust that its reception will be as hearty as we believe the appeals made to us to adopt this course have been sincere.

Telegraphic Address

The adjacent table lists the forty-eight different Royal Enfield models together with the thirteen cheaper Coventry Cross models, Models 60 - 72. All models, with the exception of the Juvenile machines, were available with a choice of frame size and altogether a total of 162 combinations of model/frame variation were available for the Royal Enfield models. The cheaper Coventry Cross branded models were available in 39 combinations. At this period the quoted prices were the total payable under the manufacturers easy payment scheme based upon twelve equal monthly payments; cash buyers were offered a 10% reduction.

The duplex girder frame was introduced on the top of the range Riche models during 1907.

1910 Model Range

By 1910 Royal Enfield was offering a range of twenty-seven different models including a Constabulary Roadster and a Tradesman's Carrier. Top of the range remained the Riche models with Duplex Girder frame. Described as the cycle for the connoisseur, they were finished to the highest standard in either black or green enamel with 22ct gold lining and came fully equipped with an

TELEGRAPHIC CODE.

Type	Model	Size	Code word
Riche Duplex Girder Type	Lady's Model 150	20 in.	Roller
		22 in.	Retrieve
		24 in.	Retinue
		26 in.	Requital
	Lady's Model 151	20 in.	Raiment
		22 in.	Ramble
		24 in.	Ramrod
		26 in.	Reaper
	Gent's Roadster Model 160	22 in.	Resemble
		24 in.	Reprieve
		26 in.	Ripple
		28 in.	Rocket
	Gent's Roadster Model 161	22 in.	Replete
		24 in.	Requisite
		26 in.	Rotund
		28 in.	Rupee
	Gent's Featherweight Model 162	22 in.	Revenue
		24 in.	Resilient
		26 in.	Resource
Duplex Girder Type	Lady's Model 123	20 in.	Rival
		22 in.	Richness
		24 in.	Refinement
		26 in.	Reliable
	Lady's Model 124	20 in.	Regal
		22 in.	Refulgence
		24 in.	Renown
		26 in.	Reserve
	Gent's Roadster Model 125	22 in.	Recompense
		24 in.	Reflect
		26 in.	Regent
		28 in.	Refresh
	Gent's Roadster Model 126	22 in.	Ravine
		24 in.	Rebuff
		26 in.	Recess
		28 in.	Reform
Standard Girder Type	Lady's Model 113	20 in.	Elevate
		22 in.	Eloquent
		24 in.	Eulogy
		26 in.	Embellish
	Lady's Model 114	20 in.	Easter
		22 in.	Elfin
		24 in.	Ember
		26 in.	Ermine
	Gent's Roadster Model 115	22 in.	Energy
		24 in.	Enjoyment
		26 in.	Eager
		28 in.	Enough
	Gent's Roadster Model 116	22 in.	Enhance
		24 in.	Evident
		26 in.	Encounter
		28 in.	Exclusive
	Gent's Roadster Model 117	22 in.	Earnest
		24 in.	Easiness
		26 in.	Ebony
	Gent's Featherweight Model 118	22 in.	Engine
		24 in.	Emotion
		26 in.	Entire
The Enfield	Lady's Model 90	20 in.	Bargain
		22 in.	Badger
		24 in.	Banner
	Lady's Model 91	20 in.	Banshee
		22 in.	Balance
		24 in.	Banjo
	Gent's Roadster Model 92	22 in.	Beacon
		24 in.	Bullion
		26 in.	Blanket
	Gent's Roadster Model 93	22 in.	Berlin
		24 in.	Bismuth
		26 in.	Bluebell
	Gent's Racer Model 94	22 in.	Blossom
		24 in.	Bobbin
		26 in.	Bracelet
	Speedman's Model 95	22 in.	Balloon
		24 in.	Bamboo
		26 in.	Barbacan
	Gent's Racer Model 96	22 in.	Buckle
		24 in.	Bridle
		26 in.	Biscuit
Tradesman's Carrier	Model 99	22 in.	Kindred
		24 in.	Kingly
		26 in.	Knight
Special Green	Lady's Model 100	20 in.	Binnacle
		22 in.	Blunder
		24 in.	Bowler
	Gent's Roadster Model 101	22 in.	Boxwood
		24 in.	Bounty
		26 in.	Brevet
		28 in.	Beaver
	Gent's Racer Model 102	22 in.	Barrel
		24 in.	Battery
		26 in.	Bayonet
Juveniles	Girl's Model 50	18 in.	Jasper
		20 in.	Jockey
	Boy's Model 51	18 in.	Jingle
		20 in.	Justice

The above table, which has been reproduced from the 1909 catalogue, gives the code words used to identify the various models when placing orders with the factory by telegraph. It identifies twenty-eight different models and a total of ninety-five different model/frame-size combinations. Models, such as 161, 123, 125, 113, 115, 117, 100, 101, 102, 90, 92 and 94 were also available with a Royal Enfield Coaster hub, for an extra 5s, or a Royal Enfield two-speed hub for an extra £1 5s.. In addition the Special Green Models 100 and 102 could be supplied with a BSA or Sturmey-Archer three-speed hub for an additional £1 15s. Options with the Tradesman's Carrier Model 99 comprised fitment of a coaster hub, 5s, and choice of enamel colour; 10s to paint the frame, forks and rims plus an additional 5s to enamel the bicycle all over including the bright parts. The total number of model/frame/equipment variants was in excess of two hundred.

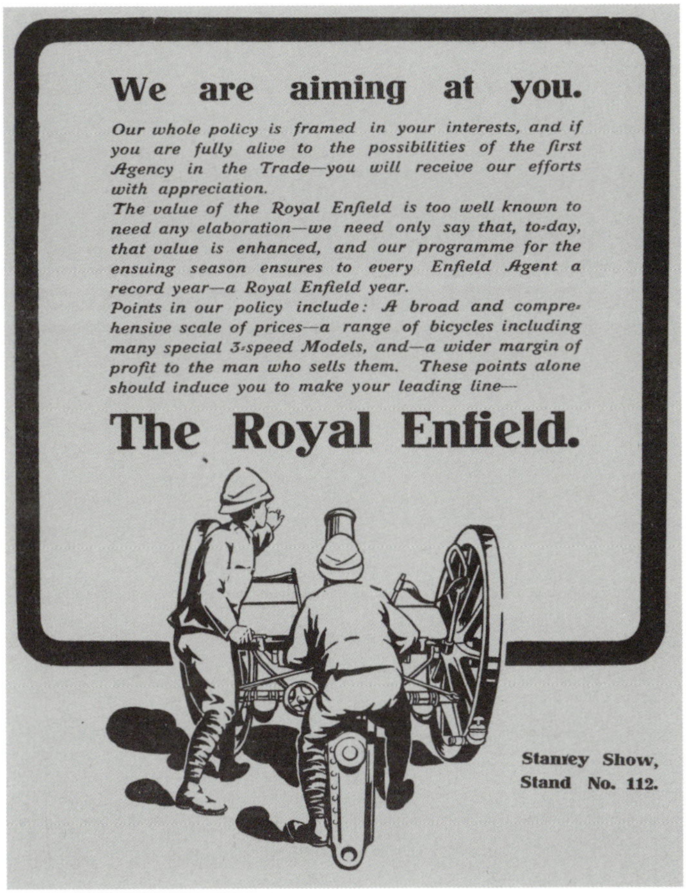

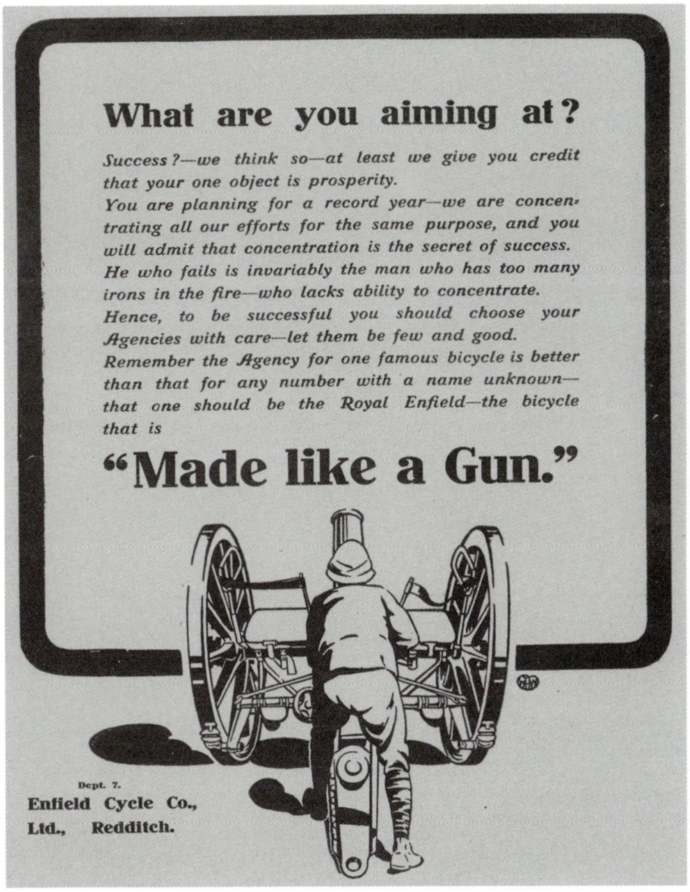

With the continued growth of business, the Enfield Cycle Co. needed to grow its network of agents. This two-page advertisement was placed in The Cycle & Motor Trades' Review in November 1908, at the time of the Stanley Show.

aluminium luggage carrier, lighting equipment and a bell. They were offered with either a BSA or Armstrong three-speed gear at 16 guineas or with single speed and free-wheel for 14 guineas.

The middle range machines were priced at 11 guineas for the All-weather Roadster, which dispensed with plated parts in preference for a more durable black celluloid finish for the handlebar, cranks and brake tubes, and 10 guineas for the remaining machines fitted with either Duplex Girder or Roadster frames. The economy ranges consisted of four single-speed with freewheel Duplex Girder framed machines at £8 15s each and three three-speed models, a Lady's model at £8 7s 6d, a Roadster and a Speedman's model both at £8 2s 6d. Special Green models, which had first been introduced for the 1909 season were listed and priced at £7 17s 6d for the Lady's model and at £7 12s 6d for the Roadster and Road Racer versions. These were single-speed machines although they could be supplied with a BSA or Armstrong three-speed gear for an extra £1 15s; alternatively a Royal Enfield Coaster hub could be supplied for 5s. The lowest priced machines comprised the Standard Lady's model at £6 17 6d, the Standard Roadster and Standard Road Racer at £6 12s 6d and the Standard Racer at £6 2s 6d. The remaining models of the range were the two Juvenile models, a Special Constabulary Roadster and the Tradesman's Carrier.

The innovations for the 1910 season consisted of an eccentric rear wheel spindle to simplify chain adjustment on the Riche models, an oil-bath chain case and a new design of

LEFT: The 1910 brochure illustrated the full Royal Enfield range of twenty-five different models of bicycle and two motorcycles.

The Royal Enfield Range of Bicycles for 1910

Model No.	Description	Frame Sizes (in)	Finish Gears	Enamel	Lining	Price
151	Riche Duplex Girder Lady's	20,22,24,26	47,62,81[1]	Black or green	22ct gold leaf	£16 16s.
150	Riche Duplex Girder Lady's	20,22,24,26	62	Black or green	22ct gold leaf	£14 14s.
161	Riche Duplex Girder Roadster	22,24,26,28	54,71,93	Black or green	22ct gold leaf	£16 16s.
160	Riche Duplex Girder Roadster	22,24,26,28	71	Black or green	22ct gold leaf	£14 14s.
136	All-weather Roadster	22,24,26,28	54,71,93	Black	22ct gold	£11 11s.
114	Three-speed Duplex Girder Lady's	20,22,24,26	47,62,81	Black	Crimson/green	£10 10s
116	Three-speed Duplex Girder Roadster	22,24,26,28	54,71,93	Black	Crimson/green	£10 10s.
130	Three-speed Tourist Lady's	20,22,24,26	47,62,81[1]	Black	22ct gold leaf	£10 10s
131	Three-speed Tourist Roadster	22,26,26,28	54,71,93	Black	22ct gold leaf	£10 10s.
120	Constabulary Roadster	24,26,28	72	Black	Crimson/gold	£9
113	Duplex Girder Lady's	20,22,24,26	62	Black	Crimson/green	£8 15s
115	Duplex Girder Roadster	22,24,26,28	71	Black	Crimson/green	£8 15s.
117	Duplex Girder Road Racer	22,24,26	80[2]	Black	Crimson/green	£8 15s.
118	Duplex Girder Featherweight	22.24.26	71	Black	Crimson/green	£8 15s.
91	Three-speed Lady's	20,22,24,26	47,62,81	Black	Bronze/crimson	£8 7s 6d
93	Three-speed Roadster	22,24,26	54,71,93	Black	Bronze/crimson	£8 2s 6d.
95	Speedman's Bicycle	22,24,26	63,84,110	Black	Bronze/crimson	£8 2s 6d.
100	Special Green Lady's	20,22,24,26	62	Green[3]	Gold	£7 17s 6d
101	Special Green Roadster	22,24,26	71	Green[3]	Gold	£7 12s 6d
102	Special Green Road Racer	22,24,26	80[2]	Green	Gold	£7 12s 6d
90	Standard Lady's	20,22,24,26	62	Black	Bronze/crimson	£6 17s 6d
92	Standard Roadster	22,24,26	71	Black	Bronze/crimson	£6 12s 6d
94	Standard Road Racer	22,24,26	80[2]	Black	Bronze/crimson	£6 12s 6d
96	Standard Racer	22,24,26	84	Black		£6 2s 6d.
50	Juvenile Girls	20	51	Black	Crimson/white	£5 15s.
51	Juvenile Boys	18,20	51	Black	Crimson/white	£5 12s 6d
99	Tradesman's Carrier	22,24,26	67	Tan		

1 Gears with 26in wheels fitted to the smallest frame model: 44, 57, 75

2 Standard wheel size is 26in, the gear with the optional 28in wheel is 84.

3 The Special Green cycles were also available in black with gold lining to special order.

chain wheel for the higher grade models which featured the Royal Enfield gun emblem. Three different patterns of brake mechanism were in use at this time. The Riche models had an exclusive design in which the rod for the front brake was concealed and the clips and guides for both front and rear brakes were brazed on forming an integral part of the handlebar and frame. The other two patterns were of a new design in which the rolling handlebar levers appeared as a single item. The design for the more expensive models used brazed on fittings whilst the cheaper models used clip fittings. The Tradesman's Carrier cycles were fitted with a patented brake balancing device.

A head transfer replaced the embossed metal head badge on all models except the top-of-the-range Riche models.

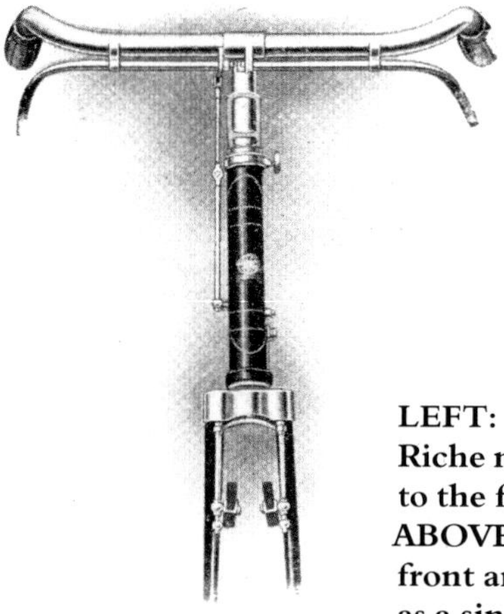

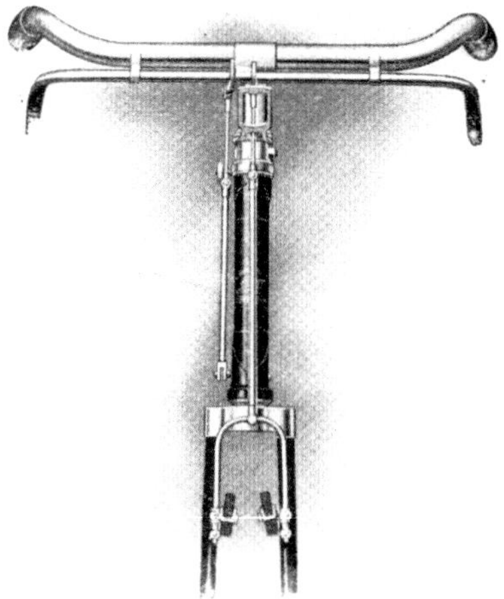

LEFT: The front brake actuation rod of the 1910 Riche models was concealed and the fittings brazed to the frame.
ABOVE: On the other models of the range, the front and rear brake levers were designed to appear as a single item.

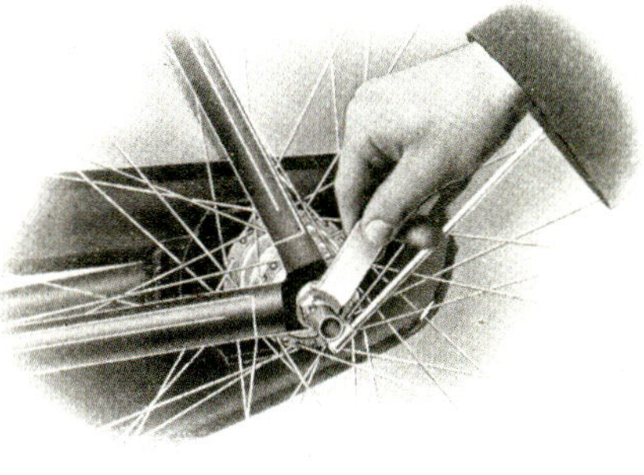

LEFT: The new design of chain wheel, introduced on the higher grade cycles in 1910, featured the field gun trade mark.

The patented chain adjustment on the 1910 Riche models consisted of an eccentric mounting of the rear spindle, which was adjusted by a peg spanner once the wheel nuts had been slackened.

1910 Riche Duplex Girder Lady's Model No. 151
(Model No.150 is a single-speed version of the above.

ROYAL ENFIELD

"MADE LIKE A GUN"

RICHE DUPLEX GIRDER LADY'S.

Model No. 151.

This Model is supplied with or without Royal Enfield Spring Front Fork to order and without extra charge.

FRAME Duplex girder (Registered), with eccentric chain adjustment, 20 in., 22 in., 24 in. and 26 in.

WHEELS 26 in. on 20 in. frames, other sizes 28 in.

TYRES Dunlop-Welch, $1\frac{1}{2}$ in.

HANDLEBAR Special upturned.

BRAKES Special Royal Enfield pattern.

THREE-SPEED GEAR B.S.A. or Armstrong. (Or tri-coaster and rear rim brake in conjunction with spring fork.)

GEAR 47, 62, 81, with 28 in. wheels; 44, 57, 75, with 26 in. wheels.

CRANKS $6\frac{1}{2}$ in.

GEAR CASE Royal Enfield detachable oil-bath.

CHAIN Hans-Renold, $\frac{1}{8}$ in. × $\frac{1}{2}$ in. pitch.

SADDLE Royal Enfield E 66. (Made by J. B. Brooks & Co.)

PEDALS Aluminium rubber-studded, $3\frac{1}{8}$ in.

FITTINGS Rear aluminium luggage carrier and reflex light, best quality lamp and bell; grips fitted with oilcan and repair outfit.

FINISH Black or green enamel, hand polished; lined 22ct. gold-leaf.

PRICE - - Net Cash £16 16 0

Easy Payments - - £18 13 0, or £1 11 1 per month.

Model 150 - - Net Cash £14 14 0

With Free Wheel and 2 hand-applied brakes.

Easy Payments - - £16 7 0, or £1 7 3 per month.

Catalogue description of the 1910 Royal Enfield Riche Duplex Girder Lady's Model 151.

1910 Riche Duplex Girder Roadster Model No. 161
(Model No. 160 is a single-speed version of the above)

1910 All- Weather Roadster Model No. 136.

The catalogue noted: 'Justifies its title to the letter. Effectively solves the "rust problem" – is easy to clean and easy to ride.' The machine had no plated parts. The handlebar, cranks and brake tubes were covered in black celluloid. The wheel rims were aluminium and the pedals were aluminium with rubber studding. The Brooks B75 saddle used black enamel springs.

1910 Three-speed Duplex Girder Lady's Model No. 114

1910 Three-speed Duplex Girder Roadster Model No. 116

1910 Three-speed Tourist Lady's Model No. 130

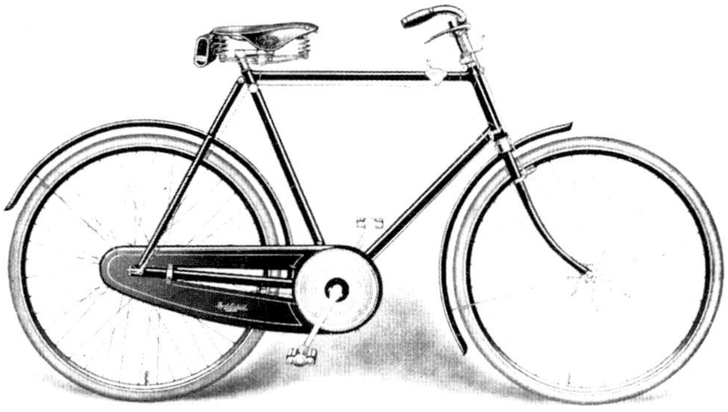

1910 Three-speed Tourist Roadster Model No. 131

1910 Duplex Girder Lady's Model No. 113

1910 Duplex Girder Roadster Model No. 115

1910 Duplex Girder Road Racer Model No. 117

1910 Duplex Girder Featherweight Model No. 118

ROYAL ENFIELD
MADE LIKE A GUN

CONSTABULARY ROADSTER.

Model No. 120.

FRAME	26 in. and 28 in., with extra rail, as illustrated; 24 in. can also be supplied, but without extra rail.
WHEELS	28 in. diameter.
TYRES	Dunlop Special Roadster, 1¾ in.
HANDLEBAR	Upturned.
BRAKES	Enfield pattern, with rolling levers.
FREE WHEEL	Ball-bearing, frictionless.
CHAIN	$\frac{3}{16}$ in. × ½ in. pitch.
GEAR	72.
CRANKS	7 in.
SADDLE	Brooks' B 90/3.
PEDALS	Rubber, 4½ in.
FINISH	Enamelled black; lined crimson and gold.

PRICE - - Net Cash £9 0 0

Easy Payments - - - - £10 0 0, or 16/8 per month.

With Royal Enfield Coaster Hub. Net Cash - £9 5 0

Easy Payments - - - - £10 5 0, or 17/1 per month.

With B.S.A. or Armstrong Three-Speed Gear.
Net Cash - £10 18 0

Easy Payments - - - £12 2 0, or £1 0 2 per month.

Catalogue description of the 1910 Royal Enfield Constabulary Roadster Model 120.

1910 Constabulary Roadster Model No. 120

1910 Three-speed Lady's Model No. 91

1910 Three-speed Roadster Model No. 93

1910 Speedman's Bicycle Model No. 95

1910 Special Green Road Racer Model No. 102

1910 Special Green Lady's Model No. 100

1910 Special Green Roadster Model No. 101

1910 Standard Lady's Model No. 90.

ROYAL ENFIELD

MADE LIKE A GUN

SPECIAL GREEN ROADSTER.

Model No. 101.

FRAME 22 in., 24 in. and 26 in.

WHEELS 28 in. diameter.

TYRES Dunlop-Welch, 1½ in.

HANDLEBAR Special upturned.

BRAKES Enfield pattern, with rolling levers.

FREE WHEEL Ball-bearing, frictionless.

CHAIN ⅛ in. × ½ in. pitch.

GEAR 71.

CRANKS 7 in.

SADDLE Brooks' B 18.

PEDALS Rubber, 4 in.

FINISH Enamelled green and lined gold. Can also be supplied in black enamel, gold lined, to order.

PRICE - - Net Cash £7 12 6

Easy Payments - - - £8 11 0, or 14/3 per month.

With Royal Enfield Coaster Hub. Net Cash - £7 17 6

Easy Payments - - - £8 16 0, or 14/8 per month

With B.S.A. or Armstrong Three-Speed Gear. Net Cash - £9 7 6

Easy Payments - - - £10 10 0, or 17/6 per month.

Catalogue description of the 1910 Royal Enfield Riche Special Green Roadster Model No. 101.

1910 Standard Roadster Model No. 92.

1910 Standard Road Racer Model No. 94.

ROYAL ENFIELD

MADE LIKE A GUN

TRADESMAN'S CARRIER.

Model No. 99.

FRAME 22 in., 24 in. and 26 in.
FRONT FORKS Of great strength.
TYRES Carrier strength, 28 in. × 1½ in.
HANDLEBAR Upturned.
BRAKES Royal Enfield pattern, very powerful.
FREE WHEEL Ball bearing, frictionless.
CHAIN Best quality roller, $\frac{3}{16}$ in. × ½ in.
GEAR 67.
CRANKS 7 in.
SADDLE E 23, with plated springs—extremely strong.
PEDALS Rubber.
CARRIER Oval or square wicker basket, supported by strong carrier of our own registered design.
ADVERTISING PLATE As illustrated, or full frame plate to order, lettered to the extent of 25 letters each side.
FINISH Best quality tan enamel.

PRICE - - Net Cash £7 17 6
Easy Payments - - - - - £8 16 0, or 14/8 per month.

EXTRAS

Coaster Hub and Front Rim Brake - - - -	5 0
Enamelling bicycle all over, including usual bright parts - - - - - - - - - - -	15 0

Catalogue description of the 1910 Royal Enfield Tradesman's Carrier Model No. 99.

1910 Standard Racer Model No. 96.

1910 Juvenile Girls Model No. 50.

1910 Juvenile Boys Model No. 51.

1910 Tradesman's Carrier Model No. 99.

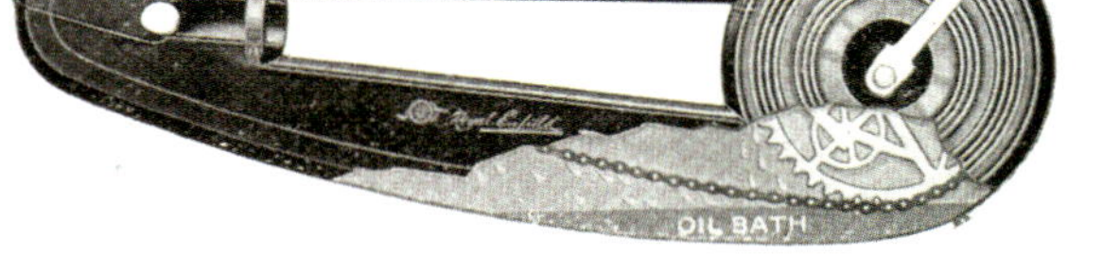

The Royal Enfield oil-bath chaincase.

Commercial Carriers

Almost from the outset a range of Enfield carrier tricycles, bicycles and delivery hand trucks, suitable for light commercial delivery purposes, had been offered alongside machines for personal transport. For 1911 a special catalogue was produced to describe the comprehensive range of machines and body styles available.

Two tricycle box carriers were available. The small No. 2 Carrier, illustrated, was fitted with a box measuring 27in long, 21in wide and 19½in deep with front opening doors. The No. 1 Carrier was fitted with larger 28in wheels and a box measuring 32in long, 24in wide and 19½in deep with front and top opening doors. The boxes were coach finished in either GPO red or olive green: lettering was extra.

A few of the prominent users of Royal Enfield Carriers.

H.M. Government, per G.P.O.
Metropolitan Railway Co.
Amalgamated Press Ltd., Lavington Street, S.E.
Selfridge & Co., Ltd., Oxford Street.
Dr. Jaeger's Sanitary Woollen System Co., Ltd., Regent Street, W.
Half-Tone Engraving Co., Farringdon Avenue.
Dictaphone Co., Oxford Street.
Tagent & Co., Great Portland Street.
Penkala Pencil Co., Charles Street, Hatton Garden.
Longs, Ltd., Queen Street, E.C.
Spillers & Nephew, London and Cardiff.
Turnbull & Asser, Outfitters, Jermyn Street, E.C.
Adams Bros. & Shardlow, Ltd., 59 Moor Lane, E.C. and Leicester.
Grandella & Co., Lisbon, Portugal.
J. Hart & Son, Butchers, Central Markets.
Waterlow Bros. & Layton, Upper Thames St., E.C.
J. & B. Dodsworth, 127 St. George's Street, E.
Surrey Davison & Co., Willesden, N.W.
Anglo-Engraving Co., Ltd., Farringdon Avenue, E.C.
Philip Morris & Co., Ltd., Bond Street, W.
Parry & Co., Ltd., Upper Thames Street, E.C.
Thomas & Sons, Maiden Lane, E.C.
Rofe & Co., Cranbrook, Kent.
Veg Meat Co., Compton Street, W.
Percival & Co., Blackfriars, S.E.
Faulks & Sons, Jermyn Street, W.
T. R. Roberts, Ltd., Islington, N.
Marshall's Dairies, Rupert Street, W.
Godfree, Felton & Co., Regent Street, W.
Abbey Press, Westminster.
British Bread Co., Westminster.
A. Sanderson & Son, Wall Papers, Berners Street, W.

ROYAL ENFIELD

Carrier Bicycle

Is not an ordinary bicycle with simply a carrier attached, but a specially constructed machine of enormous strength capable of withstanding a tremendous amount of hard wear and rough usage.

Cash Price £7 17 6

Progressive Scheme, £8 - 16 - 0, or 14/8 per month.

Coaster Hub and Front Rim Brake, 5/- extra.

Enamelling bicycle all over, including usual bright parts, 15/- extra.

SPECIFICATION.

Attention is drawn to the construction of the carrier (illustrated on opposite page) which is an integral part of the machine, and so cannot move out of position.

Frame.	22-ins., 24-ins. and 26-ins.
Front Forks.	Of great strength, supported by liner.
Wheels	28-ins. diameter, built specially strong.
Tyres.	Carrier strength, 28-ins. × 1½ ins.
Handlebar.	Flat or up-turned.
Brakes.	Actuating upon front and back rims.
Free Wheel	Ball-bearing, frictionless.
Chain.	Best quality roller, [illegible]-in. × [illegible]-in.
Gear.	67-ins.
Saddle.	E23 of best leather and with strong springs.
Pedals.	Rubber, G.P.O. Carrier type.
Carrier	Oval or square wicker basket, supported by strong carrier of our own registered design.
Advertising Plate.	As illustrated, or full frame plate to order, lettered free to the extent of 25 letters each side.
Finish	Best quality Tan Enamel smart and attractive.

MADE LIKE A GUN.

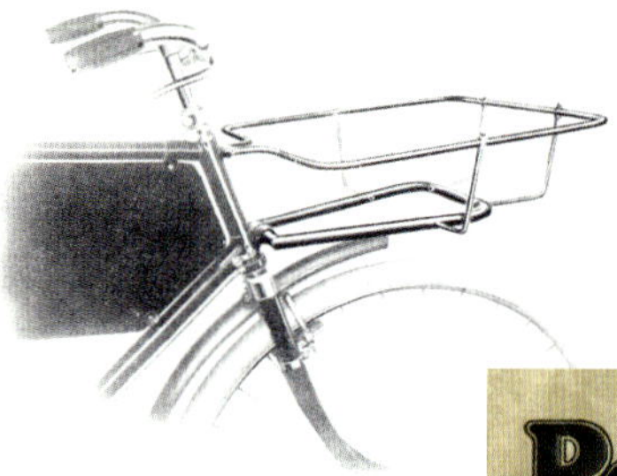

ROYAL ENFIELD

Carrier Bicycle.

ROYAL ENFIELD

Small—— Carrier Tricycle

No. 2

Is similar to that described on the previous page, though smaller throughout as the specification shows. In every respect it is an exceedingly serviceable vehicle.

Cash Price - **£14**

Progressive Payments, £5 deposit and £1 - 11 - 6 per month for 6 months.

Maintenance Scheme - 6 - per week, payable 3 months in advance.

SPECIFICATION.

(See Illustration on opposite page.)

BOX.

Built of best material. Corners strengthened with light angle plates. Door to open at end. Fitted with strong lock. Coach finished in G.P.O. red or olive green.

Size. 27-ins. long, 21-ins. wide, 19½-ins. deep (at corners).

Lettering (extra). German silver detachable letters, English gold leaf, or white. Lettered both sides and front to order.

CHASSIS.

Frame . .	Best quality weldless steel tubing.
Steering Head.	Registered design, ensuring easy and steady steering.
Wheels. . .	26 ins. × 1⅜-ins., ball bearing, oil-retaining dust-proof hubs.
Rims. . .	Westwood plated, centres black enamelled.
Tyres. . .	Warwick Carrier pneumatic, heavy non skid 1¾ in. tread.
Brakes . .	Compensating front rim brakes, operated by a single turn of grips on handlebar, and fitted with locking device of great service when descending long gradients and when carrier is left standing.
Mudguards. .	Motor car pattern, extra wide and very strong.
Chain . .	Best quality roller.
Cranks .	7-in., specially strong.
Saddle. .	Gough, G43.
Pedals.	Specially made for carrier work.

MADE LIKE A GUN.

ROYAL ENFIELD
MADE LIKE A GUN

Small Carrier Tricycle
No. 2.

Delivery Hand Truck No. 5

LEFT: A works photograph of the No. 1 Carrier Tricycle. The assistants are holding up a white sheet to provide an uncluttered background to allow the photograph to be more easily touched up for use in promotional material. The photograph dates from around 1912.

We are prepared to undertake the construction of any type of Carrier other than standard, and to gratuitously prepare drawings and estimates to customers' specifications. :: ::

The illustrations on these two pages are of Carriers which we have designed for special purposes. :: :: :: :: :: :: :: :: ::

We can manufacture both boxes and baskets for either Carrier tricycles, bicycles, or trucks of any design to suit customer's requirements. :: :: :: :: :: ::

Two pages reproduced from the 1911 Royal Enfield Catalogue of Delivery Carriers illustrating the range of body styles that could be made available. Promotional bodies could also be designed to meet the customers' requirements.

1912 Model Range

No.	Description	Gear	Price
151	Riche Duplex Girder Lady's	3-spd	£16 6s.
161	Riche Duplex Girder Roadster	3-spd	£16 16s.
137	Lady's All-weather Roadster	3-spd	£11 11s.
136	All-weather Roadster	3-spd	£11 11s.
130	Three-speed Tourist Lady's	3-spd	£10 10s.
131	Three-speed Tourist Roadster	3-spd	£10 10s.
120	Constabulary Roadster*		£9
120	Constabulary Roadster	3-spd	£10 2s 6d.
99	Tradesman's Carrier Cycle		£7 17s 6d.
113	Duplex Girder Lady's*		£8 15s.
114	As Model 113	3-spd	£9 17s 6d
115	Duplex Girder Gent's*		£8 15s.
116	As Model 115	3-spd	£9 17s 6d
97	Featherweight Lady's		£7 17s 6d.
98	Featherweight Roadster		£7 12s 6d.
100	Special Green Lady's Roadster*		£7 17s 6d
100	Special Green Lady's Roadster	3-spd	£9
101	Special Green Roadster*		£7 12s 6d.
101	Special Green Roadster	3-spd	£8 15s.
102	Special Green Road Racer*		£6 12s 6d.
102	Special Green Road Racer	3-spd	£8 15s
90	Standard Lady's Roadster*		£6 17s 6d.
91	As Model 90	3-spd	£8
92	Standard Roadster*		£6 12s 6d.
93	As Model 92	3-spd	£7 15s.
138	Lady's 'All-Black' Roadster		£7 2s 6d
139	'All-Black' Roadster		£6 17s 6d.
95	Speedman's Bicycle	3-spd	£8 2s 6d.
94	Standard Road Racer*		£6 12s 6d.
96	Standard Racer		£6 2s 6d.
50	Juvenile Girls		£5 15s.
51	Juvenile Boys		£5 12s 6d
200	Tandem		£19
201	Tandem	3-spd	£20 3s 6d

* Additional cost for Enfield Coaster Hub; 5s
Three-speed models models were fitted with either BSA or Armstrong hub gear.

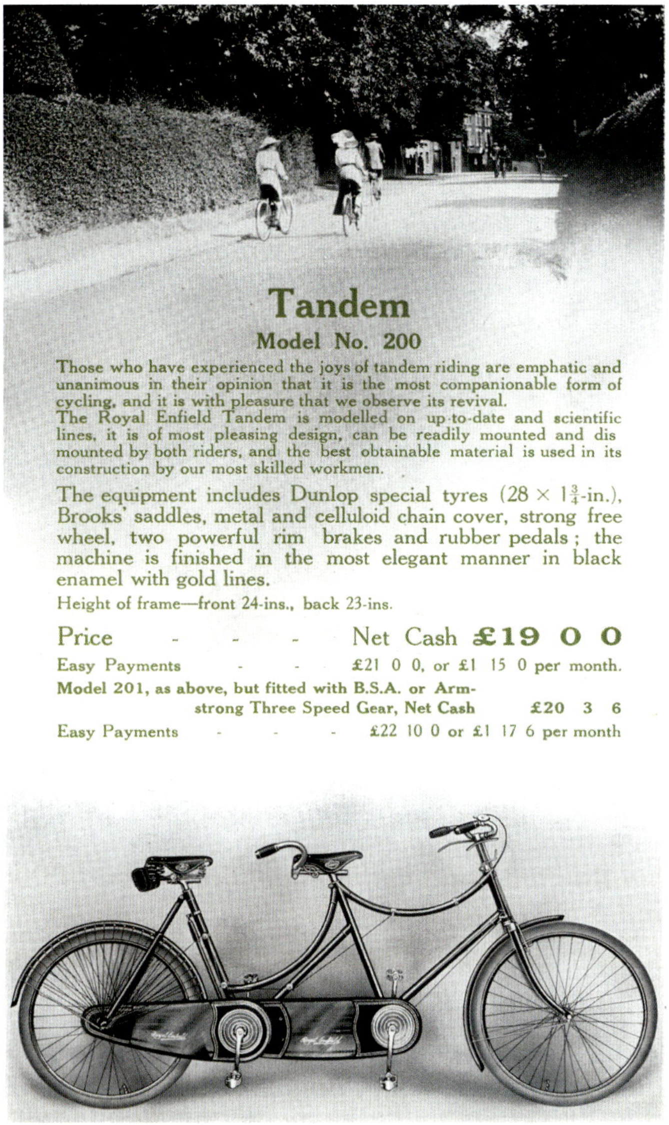
Tandem

Model No. 200

Those who have experienced the joys of tandem riding are emphatic and unanimous in their opinion that it is the most companionable form of cycling, and it is with pleasure that we observe its revival.
The Royal Enfield Tandem is modelled on up-to-date and scientific lines, it is of most pleasing design, can be readily mounted and dis mounted by both riders, and the best obtainable material is used in its construction by our most skilled workmen.

The equipment includes Dunlop special tyres (28 × 1¾-in.), Brooks' saddles, metal and celluloid chain cover, strong free wheel, two powerful rim brakes and rubber pedals ; the machine is finished in the most elegant manner in black enamel with gold lines.

Height of frame—front 24-ins., back 23-ins.

Price - - - Net Cash £19 0 0
Easy Payments - - £21 0 0, or £1 15 0 per month.
Model 201, as above, but fitted with B.S.A. or Armstrong Three Speed Gear, Net Cash £20 3 6
Easy Payments - - - £22 10 0 or £1 17 6 per month

Enfield reintroduced a tandem model for the 1912 season. The illustration is from the combined catalogue describing the company's bicycles and motorcycles.

Model Years 1912 - 1914

In 1912 handlebars were covered in celluloid, all-weather roadsters were offered and a tandem was re-introduced with pronounced downward curve to the gent's top tube. A constabulary girder-frame model proved very popular. The gear change lever moved to the handlebar for lady's machines. The three field-gun chainwheels were fitted to most models in 1913 and the field-gun was also on the lamp bracket replacing the previously plain design. There was a big publicity campaign for 1914, based on W. Gill's achievement of 100,000 miles on a 1905 'Girder'. Mr Gill's achievement was considerably in excess of the 10,000 miles achieved by Mr Cooke on his 1905 Girder Model, which was achieved without replacement of any bearings (see page 21). That Mr Gill was able to average approximately 11,000 miles per annum over nine years speaks highly of the improved reliability of cycles in general and Royal Enfield cycles in particular

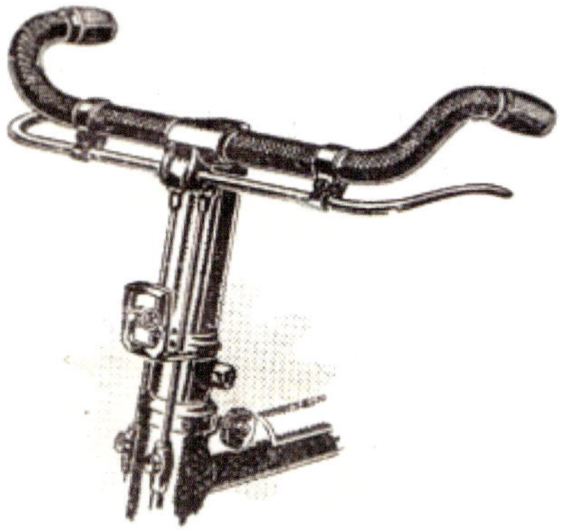

The celluloid covered handlebars fitted to the All-weather model.

The Enfield Cycle Shop, Redditch, circa 1914.

A rider poses with his Three-speed Tourist Roadster. His wing collar and tie, plus jacket with button-hole flower, was typical wear of the period for a social riding occasion.

Wartime Production

The production of both bicycles and motorcycles was maintained during the early stages of the war. During this period the Enfield factories were gradually being moved over to war production, but it was not until 17 August 1915 that the factories were requisitioned and turned over solely to war production. Even then the manufacture of motorcycles was continued to meet orders from foreign governments. Bicycle production was also continued as machines were in demand by the military and essential to provide mobility for the civilian workforce.

Three new 'Popular' models, priced at £5 10s, were announced in December 1914. They comprised both Lady's and Gent's roadsters and a road-racer. A special military bicycle was produced in 1915 and priced at 8 guineas. Future wartime production would concentrate on this model. The model range was rationalised during the war and the more expensive Racer Model discontinued

Combined catalogues, listing both Royal Enfield bicycles and motorcycles, continued to be produced throughout the war and were surprisingly lavish considering the wartime restrictions. The bicycle range was slightly reduced in number, but still consisted of twenty-one machines. The main omissions from the range were the Tandem and Juvenile models, although Racer models were included. Despite the restrictions imposed during the war on the manufacture of civilian machines, the factory continued to

Despite the war, the 1916 Royal Enfield combined catalogue of bicycles and motorcycles was a lavish production with an embossed cover.

Wartime Catalogue Models

No.	Description
151	Riche Duplex Girder Lady's
161	Riche Duplex Girder Roadster
137	All-weather Lady's
136	All-weather Gent's Roadster
130	Three-speed Tourist Lady's
131	Three-speed Tourist Roadster
120	Constabulary Roadster
99	Tradesman's Carrier Cycle
113	Duplex Girder Lady's
115	Duplex Girder Gent's
100	Special Green Lady's
101	Special Green Roadster
90	Standard Lady's Roadster
92	Standard Roadster
138	'All-Black' Lady's
139	'All-Black' Roadster
95	Speedman's Bicycle
94	Standard Road Racer
86	'Enfield' Lady's
87	'Enfield' Gent's Roadster
88	'Enfield' Road Racer

A works photograph of a Duplex Girder machine adapted to carry a rifle. The photograph is thought to date just prior to the war. Wartime production models would carry the rifle diagonally across the frame.

The illustration from the 1916 catalogue shows a fully equipped Roadster and the position for mounting the rifle adopted on production machines.

The wartime catalogue of cycles and motorcycles produced towards the end of the war. It illustrated twenty different bicycles, two solo motorcycles and three motorcycle combinations.

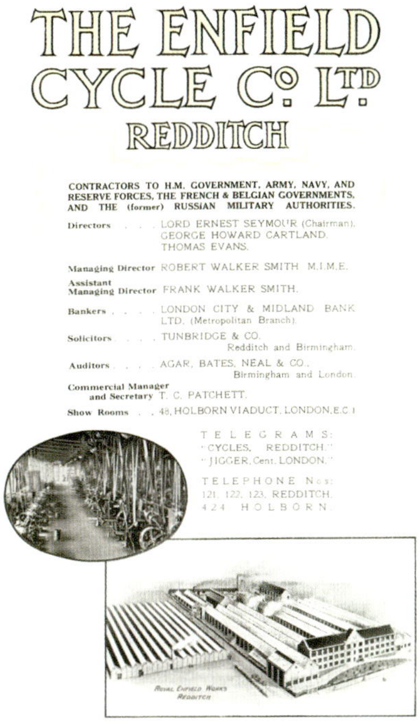

THE ENFIELD CYCLE Co. Ltd. REDDITCH

CONTRACTORS TO H.M. GOVERNMENT, ARMY, NAVY, AND RESERVE FORCES, THE FRENCH & BELGIAN GOVERNMENTS, AND THE (former) RUSSIAN MILITARY AUTHORITIES.

Directors . . . . LORD ERNEST SEYMOUR (Chairman), GEORGE HOWARD CARTLAND, THOMAS EVANS.

Managing Director ROBERT WALKER SMITH M.I.M.E.

Assistant Managing Director FRANK WALKER SMITH.

Bankers . . . . LONDON CITY & MIDLAND BANK LTD. (Metropolitan Branch).

Solicitors . . . . TUNBRIDGE & CO. Redditch and Birmingham.

Auditors . . . . AGAR, BATES, NEAL & CO., Birmingham and London.

Commercial Manager and Secretary T. C. PATCHETT.

Show Rooms . . 48, HOLBORN VIADUCT, LONDON, E.C.1

TELEGRAMS: "CYCLES, REDDITCH." "JIGGER, Cent. LONDON."

TELEPHONE Nos: 121, 122, 123, REDDITCH. 424 HOLBORN.

The frontispiece of this wartime catalogue recorded that the company was contractor to H. M. Government. Army, Navy and reserve forces, the French and Belgiam governments and the Russian Imperial Military Authorities, although by 1918 and the time of this catalogue the latter was referred to as the (former) Russian Military Authorities.

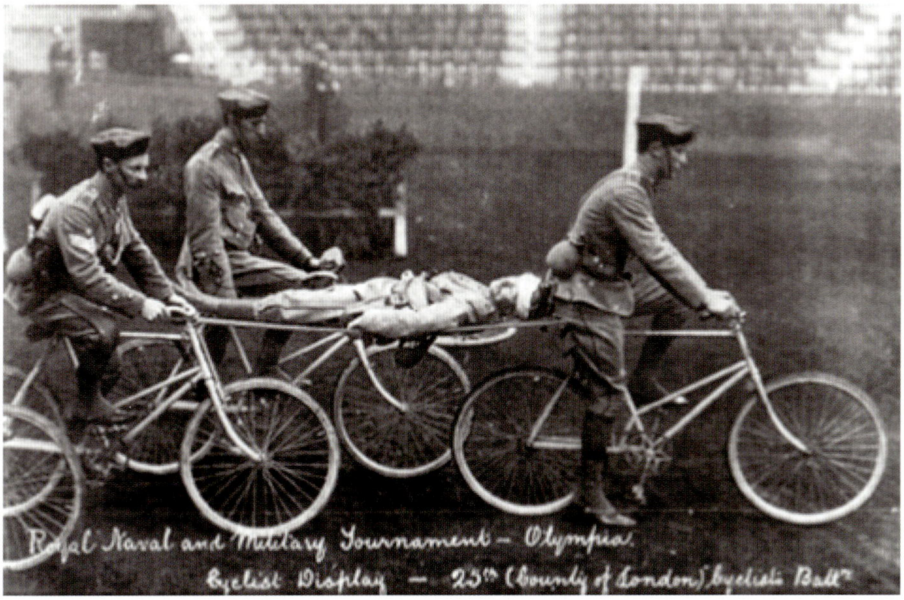

The Display Team of the 25th (County of London) Cyclist Battalion.

produce luxury machines, such as the Riche Duplex models, and purely sporting or race machines as well as the cheaper, more utilitarian, models. The two Juvenile machines, Models No. 50 and 51 were not listed. The range remained largely unchanged although was subject to gradual price increases.

Post-war Production

The signing of the Armistice on 11 November 1918 resulted in the immediate cancellation of all government contracts and manufacturing companies were forced to reorganise quickly to resume civilian production. The Enfield Cycle Company was more fortunate than other similar companies in that it had been able to maintain the production of bicycles and the limited production of the 6 hp motorcycle model throughout the period of hostilities. Consequently, it was in a better position than many of its competitors in being able to revert back to peace-time production.

The 1921 Sales Brochure.

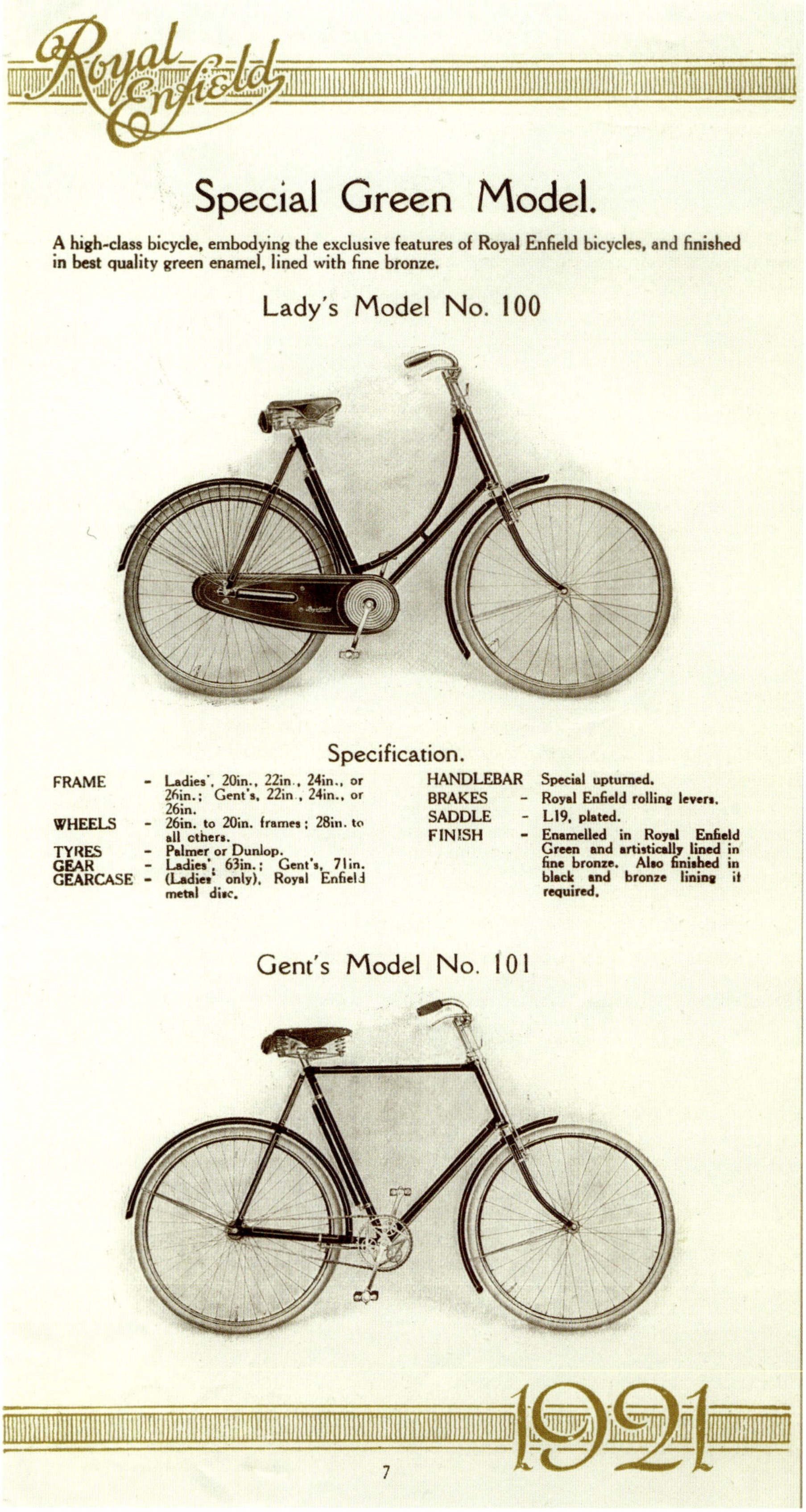

A page from the 1921 Bicycles Sales Brochure. The illustrations are of the Special Green Models, which like other models in the range were little changed from before the start of the war.

This reproduction of an aerial photograph of the Enfield Cycle Co. Ltd was included in the 1922 Bicycle catalogue. The works extended over approximately 18 acres.

1921 and 1922 Model Range

No.	Description	Price
151	Riche Duplex Girder Lady's	£20 0s.
161	Riche Duplex Girder Gent's	£20
137	All-Weather Lady's	£16 10s.
136	All-Weather Gent's	£16
130	Three-speed Tourist Lady's	£16 10s.
131	Three-speed Tourist Gent's	£16
113	Duplex Girder Lady's	£15
115	Duplex Girder Gent's	£14 10s
100	Special Green Lady's	£13 10s.
101	Special Green Gent's	£12 10s.
90	Standard Roadster Lady's	£12 10s.
92	Standard Roadster Gent's	£12
138	'All-Black' Lady's	£12 10s.
139	'All-Black' Gent's	£12
95	Speedman's	£13
94	Speedman's Road Racer	£12

1925 Model Range

No.	Description	Gear	Price
151	Riche Duplex Girder Lady's	3-spd	£16
161	Riche Duplex Girder Roadster	3-spd	£16
137	Lady's All-weather Roadster	3-spd	£13 10s.
136	All-weather Roadster	3-spd	£13
130	Three-speed Tourist Lady's	3-spd	£13 10s.
131	Three-speed Tourist Roadster	3-spd	£13
90	Standard Lady's Roadster*		£10
92	Standard Roadster*		£9 10s.
138	All-black Roadster Lady's*		£10
139	All-black Roadster*		£9 10s.
95	Speedman's Bicycle	3-spd	£10 5s.
94	Standard Road Racer*		£9 10s.
99	Tradesman's Carrier Cycle		£12 10s.
59	Tradesman's Carrier (medium load)		£10
101	Lady's	3-spd	£10
100	Gent's	3-spd	£9 10s
55	Lady's*		£8 2s 6d.
50	Gent's*		£7 10s
52	Road Racer		£7 10s.
* Sturmey-Archer 3-speed gear additional			£ 1 1s.

The return home of ex-service men with gratuities to spend and a newly gained appreciation of mechanical transport created an unprecedented demand for bicycles and powered transport. However, Royal Enfield and the other manufacturers struggled to satisfy this post-war demand owing to the shortages of raw materials and in the supply of components, such as tyres. Production gradually returned to normal during 1919 and the industry ejoyed a boom period. This would be brought to an end through increasing materials and labour costs which caused consequential price rises. By the end of 1920 supply problems had been resolved, but prices had risen considerably compared with pre-war figures.

There was a rationalisation of the range during the immediate post-war period. The Constabulary Roadster Model No. 120, the 'Enfield' Lady's, Gent's Roadster and Road Racer, Models Nos. 86, 87 and 88 respectively of the wartime catalogue, were discontinued as was Model No. 86 and the Juvenile Models

By the 1924 Season the Special Green Models Nos. 100 (Lady's) and 101(Gent's) had been dropped from the range. The additions were a new Tradesman's Carrier, Model No. 59, intended for medium loads, and three new economy models. These were a Road Racer, Model No. 52, and a Gent's Model No. 50, both priced at £6 18s 6d, and a Lady's Model No. 55 priced at £7 8s 6d. Models Nos. 50 and 56 were single-speed and a Sturmey-Archer 3-speed gear was available for an additional 21s.

The Models Nos. 100 and 101 were re-introduced for the 1925 Season although simply identified as Lady's or Gent's models. Dropped from the range were the single speed Duplex Girder Models Nos. 113 and 115.

Perhaps reflecting the more difficult market conditions during the mid-twenties, the 1925 Sales Brochure gave the hire-purchase payment terms over six or twelve months.

LEFT: In the mid-nineteen twenties the Royal Enfield sales brochures finally shook off wartime austerity and appeared with colourful depictions of the joys of cycling. The brochure illustrated dates from 1925.

BELOW: The illustrations, reproduced from the 1925 brochure, are of the four new models first introduced for the 1924 season.

Lady's Model No. 55

Price - £8 2 6 net cash

or **£1 15s. 0d.** down and 6 monthly instalments of **£1 2s. 6d.**, or 12 monthly instalments of **12s. 0d.**

Specification :

FRAME ...	Lady's, 20 in., 22 in., or 24 in.; Gent's, 22 in., 24 in., or 26 in.	**GEAR** ...	Lady's, 63 in.; Gent's, 71 in.
WHEELS ...	26 in. to 20 in. frames; 28 in. to all others.	**GEARCASE**	(Lady's only) metal.
TYRES ...	Warwick.	**BRAKES** ...	Rolling lever type.
HANDLEBAR	Upturned.	**SADDLE** ...	Good three-coil spring saddle.
		FINISH ...	Enamelled black and lined.

Sturmey-Archer 3-Speed Gear, 21s. extra.

Gent's Model No. 50

Price - £7 10 0 net cash

or **£1 13s. 0d.** down and 6 monthly instalments of **£1 1s. 0d.**, or 12 monthly instalments of **11s. 0d.**

TRADESMAN's CARRIER Model No. 59

(For Medium Loads)

Price - £10 0 0 net cash

or **£2 6s. 0d.** down and 6 monthly instalments of **£1 7s. 6d.**, or 12 monthly instalments of **14s. 6d.**

Specification :

FRAME ...	22 in. or 24 in.	**BASKET** ...	Square, wicker.
TYRES ...	Clipper Carrier.	**SADDLE** ...	Three-coil spring saddle.
HANDLEBAR	Upturned.	**FINISH** ...	Enamelled in black.
GEAR ...	68 in.		
BRAKES ...	Rolling lever.		

The price in each case includes lettering on the advertising plate (up to 25 letters each side) gratis.

ROAD RACER Model No. 52

Price - £7 10 0 net cash

or **£1 13s. 0d.** down and 6 monthly instalments of **£1 1s. 0d.**, or 12 monthly instalments of **11s. 0d.**

Specification :

FRAME ...	22 in. or 24 in.	**BRAKES** ...	Rolling lever type.
WHEELS ...	26 in.	**SADDLE** ...	Good quality saddle.
TYRES ...	Warwick.	**FINISH** ...	Enamelled in best black enamel and lined.
GEAR ...	80 in.		
HANDLEBAR	North Road upturned.		

The 1930 Model Range

The market for bicycles during the late nineteen twenties was very competitive and Royal Enfield, as other manufacturers, had to cut prices to maintain sales. Inevitably there became a utilitarian feel about the range. The Riche Duplex Girder machines were dropped during 1929 and were absent from the 1930 brochure. When last available they were priced at £12 19s 6d for the Gent's Model 161 and £13 9s 6d for the Lady's Model 151. By comparison the Tourist Models, also with Sturmey-Archer Three-speed Gears were priced at £7 19s 6d for the Model 100 and £8 9s 6d for the Lady's Model 100. The cheapest single-speed machines were available for £5 19s 6d. By 1930 single-speed machines were available from £4 19s 6d

The 1929 Riche Duplex Girder machines were the first to boast to have all plated components Chromium Plated. This became the standard finish for the plated components for all machines from 1930 and the cover of the sales brochure proudly proclaimed 'CHROMIUM PLATED BICYCLES'. The cover of the 1930 catalogue also informed that the machines were

'Guaranteed for all time' although a similar guarantee had been in place since 1924. However, the guarantee excluded tyres, rims, chains, saddles, variable gears, coaster hubs, enamelling and any parts not made of metal, which had a guarantee furnished by the suppliers, plus all defects caused by wear and tear, misuse or neglect.

1930 Model Range

No.	Description
40	Popular Gentleman's
45	Popular Lady's
47	Road Racer (Low bottom bracket)
52	Road Racer (Sloping top tube)
50	Standard Roadster Gent's
55	Standard Roadster Lady's
74	Club Model Gent's
76	Club Model Gent's
75	Club Model Gent's
70	Club Model Lady's
100	Three Speed Tourist Gent's
101	Three Speed Tourist Lady's
102	All-weather Model Gent's
103	All-weather Model Lady's
161	Riche Duplex Girder Gent's
151	Riche Duplex Girder Lady's

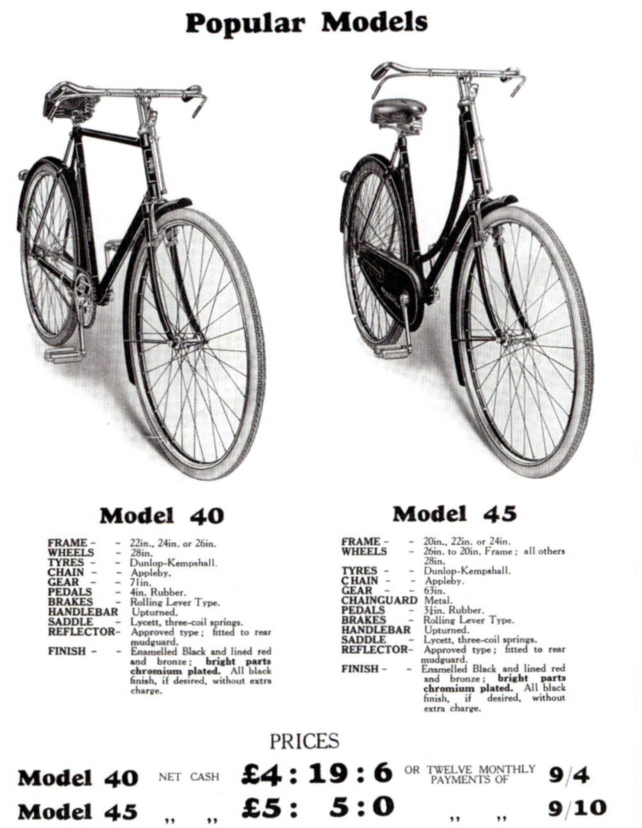

Standard Roadster

Model 50

FRAME - - 22in., 24in. or 26in.
WHEELS - 28in.
TYRES - - Dunlop.
CHAIN - - Hans Renold.
GEAR - - 71in.
PEDALS - 4in. Rubber.
BRAKES - Rolling Lever Type.
HANDLEBAR Upturned, solid brake lugs.
SADDLE - Waterproof Top, made by J. B. Brooks & Co., Ltd.
REFLECTOR- Approved type; fitted to rear mudguard.
FINISH - - Enamelled Black and lined red and bronze; **bright parts chromium plated.** All black finish, if desired, without extra charge.

Model 55

FRAME - - 20in., 22in. or 24in.
WHEELS - 26in. to 20in. Frame; all others 28in.
TYRES - - Dunlop.
CHAIN - - Hans Renold.
GEAR - - 63in.
GEARCASE - Metal.
PEDALS - 3½in. Rubber.
BRAKES - Rolling Lever Type.
HANDLEBAR Upturned, solid brake lugs.
SADDLE - Waterproof Top, made by J. B. Brooks & Co., Ltd.
REFLECTOR- Approved type; fitted to rear mudguard.
FINISH - - Enamelled Black and lined red and bronze; **bright parts chromium plated.** All black finish, if desired, without extra charge.

PRICES

Model 50	NET CASH	**£5 : 19 : 6**	OR TWELVE MONTHLY PAYMENTS OF	**11/3**
Model 55	,, ,,	**£6 : 10 : 0**	,, ,,	**12/3**

Club Models

Model 74

FRAME - - 21in.
WHEELS - 26×1¼in. Drop-out Type, Endrick Patent Steel Rims.
TYRES - - Dunlop-Kempshall Road Racing
CHAIN - - Appleby.
GEARS - - Two Fixed Cogs, giving gears of 66in. and 70in.
PEDALS - Steel Rat-trap.
BRAKES - Calliper Brake to rear wheel.
HANDLEBAR Marsh Racing Pattern.
SADDLE - Lycett Racing.
MUDGUARDS Lightweight steel.
REFLECTOR- Approved type; fitted to rear mudguard.
FINISH - - **Enamelled Maroon; bright parts chromium plated.**

Model 76

FRAME - - 20in., 22in. or 24in.
WHEELS - 26in.×1¼in. Drop-out Type, Endrick Patent Steel Rims.
TYRES - - Dunlop Road Racing.
CHAIN - - Hans Renold.
GEARS - - One Fixed Wheel 70in., one Free Wheel 66in.
PEDALS - Light Steel Rat-trap.
BRAKE - - Calliper Brake to rear wheel.
HANDLEBAR Marsh Type, with forward extension lug, cellulose finish.
SADDLE - Made by J. B. Brooks & Co., Ltd.
MUDGUARDS Bluemel's celluloid.
REFLECTOR- Approved type; fitted to rear mudguard.
FINISH - **Enamelled Black, with bright blue head; handlebar black cellulose finish; bright parts chromium plated.**

PRICES

Model 74	NET CASH	**£5 : 5 : 0**	OR TWELVE MONTHLY PAYMENTS OF	**9/10**
Model 76	,, ,,	**£6 : 15 : 0**	,, ,,	**12/8**

Club Models

Model 75

FRAME - - 20in., 22in. or 24in.
WHEELS - 26in.×1¼in. Drop-out Type, Endrick Patent Steel Rims.
TYRES - - Dunlop Road Racing.
CHAIN - - Hans Renold.
GEARS - - One Fixed Wheel 70in., one Free Wheel 66in.
PEDALS - Light Steel Rat-trap.
BRAKES - Two Calliper Brakes.
HANDLEBAR Marsh Type, with forward extension lug.
SADDLE - Made by J. B. Brooks & Co., Ltd.
MUDGUARDS Lightweight, steel.
REFLECTOR- Approved type; fitted to rear mudguard.
FINISH - - **Frame Enamelled Dark Green, Mudguards Light Green. Bright parts chromium plated.**

Model 70

FRAME - - 20in. or 22in.
WHEELS - 26in.×1¼in. Drop-out Type, Endrick Patent Steel Rims.
TYRES - - Dunlop Road Racing.
CHAIN - - Hans Renold.
GEARS - - One Fixed Wheel 70in., one Free Wheel 66in.
PEDALS - Light Steel Rat-trap.
BRAKES - Two Calliper Brakes.
HANDLEBAR Enfield Half-drop.
SADDLE - Made by J. B. Brooks & Co., Ltd.
MUDGUARDS Lightweight, steel.
REFLECTOR- Approved type; fitted to rear, mudguard.
FINISH - - **Frame Enamelled Dark Green, Mudguards Light Green. Bright parts chromium plated.**

PRICES

Model 75	NET CASH	**£6 : 15 : 0**	OR TWELVE MONTHLY PAYMENTS OF	**12/8**
Model 70	,, ,,	**£7 : 0 : 0**	,, ,,	**13/2**

Three-speed Tourist

Model 100

FRAME - - 22in., 24in. or 26in.
WHEELS - 28in.
TYRES - - Dunlop.
CHAIN - - Hans Renold.
GEAR - - Sturmey-Archer 3-speed.
GEARCASE - Detachable Oil-bath.
PEDALS - 4in. Rubber.
BRAKES - Rolling Lever Type.
HANDLEBAR Upturned, solid brake lugs.
SADDLE - Made by J. B. Brooks & Co., Ltd.
REFLECTOR- Approved type; fitted to rear mudguard.
FINISH - - Enamelled Black and lined red and bronze; **bright parts chromium plated.** All black finish, if desired, without extra charge.

Model 101

FRAME - - 20in., 22in., 24in. or 26in.
WHEELS - 26in. to 20in. Frame; 28in. all others.
TYRES - - Dunlop.
CHAIN - - Hans Renold.
GEAR - - Sturmey-Archer 3-speed.
GEARCASE - Detachable Oil-bath.
PEDALS - 3½in. Rubber.
BRAKES - Rolling Lever Type.
HANDLEBAR Upturned, solid brake lugs.
SADDLE - Made by J. B. Brooks & Co., Ltd.
REFLECTOR- Approved type; fitted to rear mudguard.
FINISH - - Enamelled Black and lined red and bronze; **bright parts chromium plated.** All black finish, if desired, without extra charge.

PRICES

Model 100	NET CASH	**£7 : 19 : 6**	OR TWELVE MONTHLY PAYMENTS OF	**15/-**
Model 101	,, ,,	**£8 : 9 : 6**	,, ,,	**16/-**

Utility Model

(With low Bottom Bracket)

Tradesman's Carrier

Model 51

FRAME - -	22in. or 24in.
WHEELS -	26in.
TYRES - -	Dunlop.
CHAIN - -	Hans Renold.
GEAR - -	71in.
GEARCASE -	Detachable Oil-bath.
PEDALS -	4in. Rubber.
BRAKES -	Rolling Lever Type.
HANDLEBAR	North Road Upturned, solid brake lugs.
SADDLE -	Waterproof Top, made by J. B. Brooks & Co., Ltd.
MUDGUARDS	Mudguards valanced.
REFLECTOR-	Approved type; fitted to rear mudguard.
FINISH - -	**All Black Finish. Handlebars, Chain Wheel and Cranks finished cellulose. Rims chromium plated.**

Model 58

FRAME - -	22in.
TYRES - -	Dunlop.
CHAIN - -	Hans Renold.
GEAR - -	66in.
PEDALS -	4in. Rubber.
BRAKES -	Rolling Lever Type.
HANDLEBAR	Upturned.
SADDLE -	Three-coil Spring Saddle.
CARRIER -	Strong, tubular Carrier.
BASKET -	Wicker.
REFLECTOR-	Approved type; fitted to rear mudguard.
FINISH - -	**All Black Finish.**

PRICES

Model 51	NET CASH	**£6 : 15 : 0**	OR TWELVE MONTHLY PAYMENTS OF	**12/8**
Model 58	,, ,,	**£8 : 10 : 0**	,, ,,	**16/-**

Juveniles

Model 20

FRAME - -	18in.
WHEELS -	24in.
TYRES - -	Dunlop-Kempshall.
CHAIN - -	Appleby.
GEAR - -	53in.
PEDALS -	Rubber.
BRAKES -	Rolling Lever Type.
HANDLEBAR	Upturned.
SADDLE -	Lycett.
REFLECTOR-	Approved type; fitted to rear mudguard.
FINISH - -	Enamelled Black and lined red and bronze; **bright parts chromium plated.**

Model 22

FRAME - -	18in.
WHEELS -	24in.
TYRES - -	Dunlop-Kempshall.
CHAIN - -	Appleby.
GEAR - -	53in.
PEDALS -	Rubber.
BRAKES -	Rolling Lever Type.
HANDLEBAR	Upturned.
SADDLE -	Lycett.
REFLECTOR-	Approved type; fitted to rear mudguard.
FINISH - -	Enamelled Black and lined red and bronze; **bright parts chromium plated.**

PRICES

Model 20	NET CASH	**£4 : 15 : 0**	OR TWELVE MONTHLY PAYMENTS OF	**9/-**
Model 22	,, ,,	**£4 : 15 : 0**	,, ,,	**9/-**

Extras on Various Models.

	s	d.		s.	d.	
Eadie Coaster in lieu of Rear Rim Brake	12	6	cash or	1	2	per month.
Sturmey-Archer Three-speed Gear, top tube control ...	20	0	,, ,,	1	11	,, ,,
,, ,, ,, ,, ,, handlebar ,, ...	22	0	,, ,,	2	1	,, ,,
Detachable Oil Bath Gear-case (Model 40)	15	0	,, ,,	1	5	,, ,,
,, ,, ,, ,, (other Models) ...	21	0	,, ,,	2	0	,, ,,

All-weather Models

Model 102

FRAME - -	22in., 24in. or 26in.
WHEELS -	28in.
TYRES - -	Dunlop.
CHAIN - -	Hans Renold.
GEAR - -	Sturmey-Archer 3-speed.
GEARCASE -	Detachable Oil-bath.
PEDALS -	4in. Rubber.
BRAKES -	Rolling Lever Type.
HANDLEBAR	Upturned, solid brake lugs.
SADDLE -	Made by J. B. Brooks & Co., Ltd.
MUDGUARDS	Mudguards valanced.
REFLECTOR-	Approved type; fitted to rear mudguard.
FINISH - -	Enamelled Black and lined red and bronze. **Rims chromium plated.** Handlebars, Chain Wheel and Cranks black cellulose finish.

Model 103

FRAME - -	20in., 22in., 24in. or 26in.
WHEELS -	26in. to 20in. Frame; 28in. all others.
TYRES - -	Dunlop.
CHAIN - -	Hans Renold.
GEAR - -	Sturmey-Archer 3-speed.
GEARCASE -	Detachable Oil-bath.
PEDALS -	3¼in. Rubber.
BRAKES -	Rolling Lever Type.
HANDLEBAR	Upturned, solid brake lugs.
SADDLE -	Made by J. B. Brooks & Co., Ltd.
MUDGUARDS	Mudguards valanced.
REFLECTOR-	Approved type; fitted to rear mudguard.
FINISH - -	Enamelled Black and lined red and bronze. **Rims chromium plated.** Handlebars, Chain Wheel and Cranks black cellulose finish.

All-weather Models

Model 102

FRAME - -	22in., 24in. or 26in.
WHEELS -	28in.
TYRES - -	Dunlop.
CHAIN - -	Hans Renold.
GEAR - -	Sturmey-Archer 3-speed.
GEARCASE -	Detachable Oil-bath.
PEDALS -	4in. Rubber.
BRAKES -	Rolling Lever Type.
HANDLEBAR	Upturned, solid brake lugs.
SADDLE -	Made by J. B. Brooks & Co., Ltd.
MUDGUARDS	Mudguards valanced.
REFLECTOR-	Approved type; fitted to rear mudguard.
FINISH - -	Enamelled Black and lined red and bronze. **Rims chromium plated.** Handlebars, Chain Wheel and Cranks black cellulose finish.

Model 103

FRAME - -	20in., 22in., 24in. or 26in.
WHEELS -	26in. to 20in. Frame; 28in. all others.
TYRES - -	Dunlop.
CHAIN - -	Hans Renold.
GEAR - -	Sturmey-Archer 3-speed.
GEARCASE -	Detachable Oil-bath.
PEDALS -	3¼in. Rubber.
BRAKES -	Rolling Lever Type.
HANDLEBAR	Upturned, solid brake lugs.
SADDLE -	Made by J. B. Brooks & Co., Ltd.
MUDGUARDS	Mudguards valanced.
REFLECTOR-	Approved type; fitted to rear mudguard.
FINISH - -	Enamelled Black and lined red and bronze. **Rims chromium plated.** Handlebars, Chain Wheel and Cranks black cellulose finish.

ABOVE: The tag attached to new bicycles around 1930 to advise prospective purchasers of the guarantee applying to new Royal Enfield models.

TOP RIGHT: An extract from the sales brochure extolling the virtues of the chromium plated bicycles.

RIGHT: The illustration from a Royal Enfield factory photographs album, which dates from 1913, is an apt link between the present chapter describing the company's bicycles and the subsequent chapters describing the motorcars and motorcycles. Sadly Enfield cars were no longer available by this time, but Royal Enfield continued to offer a range of two-wheeled pedal or powered machines to satisfy any owner.

Three and Four Wheelers

From the outset Albert Eadie was keen that the New Enfield Cycle Company Ltd, which had been formed in July 1896 by the combination of the Enfield Manufacturing and Eadie Manufacturing Companies, should become involved in the rapidly developing market for powered vehicles. A prototype quadricycle, built in 1897 by R.W.Smith for his own use, has already been mentioned in Chapter 1. The machine had proved successful and created considerable interest. Both Eadie and Smith realised the new business opportunity represented by motorised transport and were eager to enter the market with a similar machine of their own manufacture. However, it was first necessary to gain more experience and a second quadricycle was built in 1898, under licence from De Dion Bouton, for further evaluation

Enfield Quadricycles and Tricycles

In late 1899 the company began to manufacture and market motor quadricycles and subsequently motor tricycles built under licence from De Dion Bouton. The quadricycles were fitted with 2¼ hp De Dion power units, which had square dimensions of 70 mm bore and stroke giving a capacity of 269 cc. The tricycles were fitted with the smaller 1¾ hp De Dion engine of 239 cc (66 mm bore and 70mm stroke). Enfield manufactured the entire machines except for the engine, carburettor and ignition, which were of the current De Dion design and manufacture, and the Dunlop tyres. David Doyle was foreman in charge of motor quadricycle and motor tricycle

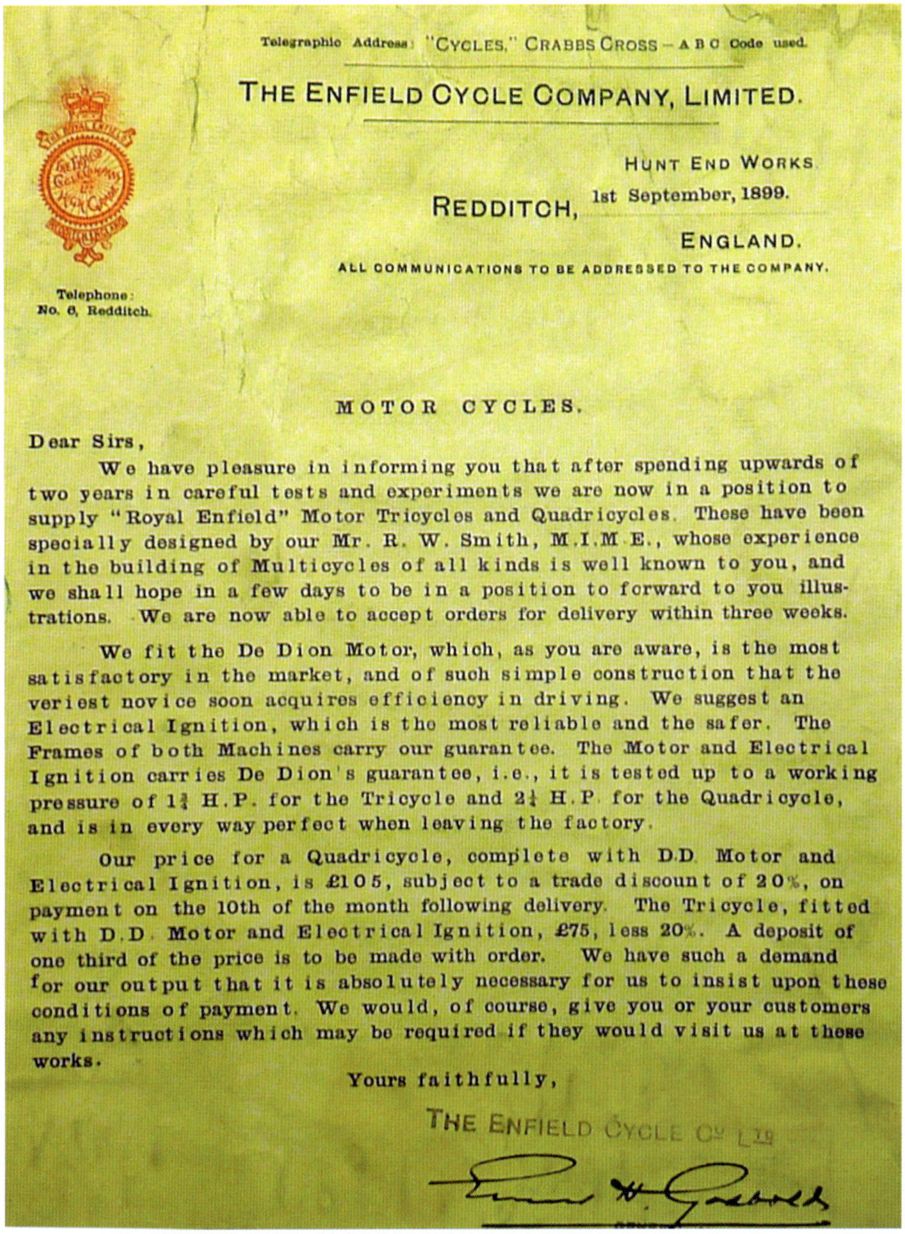

Telegraphic Address: "CYCLES," CRABBS CROSS – A B C Code used.

THE ENFIELD CYCLE COMPANY, LIMITED.

HUNT END WORKS

REDDITCH, 1st September, 1899.

ENGLAND.

ALL COMMUNICATIONS TO BE ADDRESSED TO THE COMPANY.

Telephone:
No. 6, Redditch.

MOTOR CYCLES.

Dear Sirs,

We have pleasure in informing you that after spending upwards of two years in careful tests and experiments we are now in a position to supply "Royal Enfield" Motor Tricycles and Quadricycles. These have been specially designed by our Mr. R. W. Smith, M.I.M.E., whose experience in the building of Multicycles of all kinds is well known to you, and we shall hope in a few days to be in a position to forward to you illustrations. We are now able to accept orders for delivery within three weeks.

We fit the De Dion Motor, which, as you are aware, is the most satisfactory in the market, and of such simple construction that the veriest novice soon acquires efficiency in driving. We suggest an Electrical Ignition, which is the most reliable and the safer. The Frames of both Machines carry our guarantee. The Motor and Electrical Ignition carries De Dion's guarantee, i.e., it is tested up to a working pressure of 1¾ H.P. for the Tricycle and 2¼ H.P. for the Quadricycle, and is in every way perfect when leaving the factory.

Our price for a Quadricycle, complete with D.D. Motor and Electrical Ignition, is £105, subject to a trade discount of 20%, on payment on the 10th of the month following delivery. The Tricycle, fitted with D.D. Motor and Electrical Ignition, £75, less 20%. A deposit of one third of the price is to be made with order. We have such a demand for our output that it is absolutely necessary for us to insist upon these conditions of payment. We would, of course, give you or your customers any instructions which may be required if they would visit us at these works.

Yours faithfully,

THE ENFIELD CYCLE Co LTD

The letter from Ernest Godbold, dated 1st September 1899, sent to all Enfield agents to inform them of the new Royal Enfield Tricycles and Quadricycles.

assembly. He would subsequently be given the same responsibility for motor cycle assembly.

The Autocar magazine announced the Enfield Motor Quadricycle in September 1899. 'For some months

Two views of the Royal Enfield Motor Quadricycle announced in September 1899. The machine was fitted with a 2¼hp De Dion engine and priced at £105.

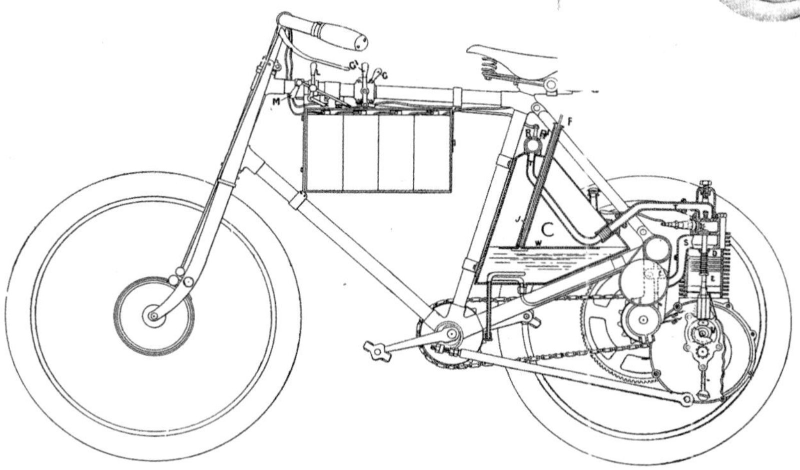

Drawing illustrating the layout of the Royal Enfield Tricycle, which was announced in September 1899 at the same time as the Quadricycle. The machine would be made available with either a 1¾hp or 2¼hp De Dion engine and was priced at £75.

The motor tricycle of 1900 with 2½hp De Dion engine. The machine illustrated is fitted with the two-speed gear introduced in 1900 for the 1901 season.

the Enfield Cycle Co. Ltd of Hunt End, Redditch, who enjoy an enviable reputation for the design and quality of their man-propelled bicycles, have been turning their attention to the production of motor cycles also. They have now completed designs, and have for some time been supplying a very handsome form of quadricycle.'

Eadie Motor Quadricycle and Tricycle

The Royal Enfield quadricycles and tricycles, marketed by the Enfield Cycle Company Ltd, were also offered to the trade by the Eadie Manufacturing Company Limited. The machines were of identical specification and offered at the same price. However, befitting a cycle components manufacturer, the Eadie Motor Quadricycles and Tricycles were also available in component form. The frames came enamelled and plated and ready for the motor to be attached. They were supplied complete with all components except engine, saddle, control levers, mudguards, wheels and a metal case to enclose the gear wheels. The price of the quadricycle frame was £45 and the tricycle frame £22 10s.

The De Dion-derived tricycles and quadricycles were built and sold in fair numbers for several years before being discontinued in 1903.

Appointment of Walter Grew

By 1900 motorcar production had grown to be an important and expanding part of the company's activities, and in January 1900 it was decided to appoint a manager with specific responsibility for the motorcar department. The man chosen was Walter Ferdinando Grew, who was a keen racing cyclist and had a background in railway carriage manufacturing and the cycle and wheel building industries. He remained with the company until the middle of 1903.

IF you keep a horse it will cost you 15s. per week for its food, and the hire of a man to groom it will average 25s. Here is a dead charge of £2 per week to begin with, without considering cost of stabling and the illness and accidents which horseflesh is heir to. For this expenditure you can be conveyed on the average 120 miles per week.

You can cover this distance in one day on an Eadie Motor Vehicle for an expenditure of 5s. in petrol. You can manage the vehicle yourself, store it in a cycle shed, and can get over the ground more quickly and comfortably.

The merits of the quadricycles and tricycles as spelt out in the Eadie Motor Vehicles booklet.

The Thousand Miles Trial of 1900

In April 1900 The Automobile Club of Great Britain and Ireland organised a strenuous road trial to increase public awareness of the motorcar and to demonstrate that the powered vehicle was now a practical form of transport. The trial started on Monday 23 April at Grosvenor Place, Hyde Park Corner and encompassed 1,108½ miles before returning to Whitehall on Saturday 12 May. The route passed through Bristol, Birmingham, Manchester, Kendal, Carlisle, Edinburgh, Newcastle, Leeds, Sheffield and Nottingham with one-day public exhibitions in many of the cities. The legal speed limit at the time was 12mph, reduced to 8mph through towns, and the utmost care was taken to comply with the law. However. the

The advertisement placed in The Autocar following Iliffe's success in winning a silver medal in the One Thousand Mile Trial.

performance of the competing vehicles was tested by a speed trial in Welbeck Park and three compulsory hill climbs.

Sixty-five motor vehicles lined-up for the start of the trial at 7 am. Amongst the private entrants in Section Two was Edward M Iliffe, who was later to become the first Baron Iliffe, owner of the publishing firm Iliffe,

Side and rear views of the De Dion engine of the type used in the Royal Enfield Tricycles and Quadricycles.

which published the Autocar magazine and three years later was to publish The Motor Cycle. He was driving his 2¾hp De Dion powered Enfield Quadricycle, which carried number A28. He was accompanied by W.F.Grew of Enfield's. Iliffe experienced his share of mechanical problems and breakdowns but was one of the forty-nine entrants completing the trial. The Gold Medal awarded for the most meritorious performance went to C.S.Rolls driving his 12 hp Panhard. Iliffe with his quadricycle received one of the ten Silver Medals presented.

R.W.Smith's Accident

On the afternoon of 7 August 1900 R.W.Smith was driving a 3½hp De Dion Bouton car from Stratford towards Alcester when he experienced steering problems whilst descending Red Hill. The car swerved sideways and overturned. The passengers were thrown clear and escaped with only minor injury, but Smith was pinned beneath the car. He suffered a double fracture of his left upper arm, damage to the right fore-arm and contusions about his face. He required thirty stitches for his injuries and was away from the factories for three months.

The 1901 Model Range

A three model range of motor tricycles and quadricycles was announced at the Stanley Cycle Show in London in November 1900 for the 1901 season. The basic tricycle and quadricyle models were continued with little change, although both were now powered by a larger, 2¾hp, De Dion engine and the surface carburettor replaced by one of the Longuemare type. The quadricycle also benefitted from improved springing of the front seat, the seat now being carried on leaf springs and the back rest supported by independent springs. Both models were fitted with cycle type pedals and chain drive to the rear wheels. The machines would be started by the rider pedaling away from a standstill, using an engine decompression device to reduce the engine resistance, until sufficient speed was obtained for the engine to fire

Royal Enfield Auto-Quad or Tandem Car

The third machine, known as the Auto-Quad or Tandem Car, had been considerably developed from the basic quadricycle. The 2¾hp engine had been fitted with a water-cooled cylinder head to improve cooling when the engine was working under load. The cooling water was circulated by thermo-syphon through a pair of interconnected finned copper radiators, each containing 3-gallons of water, located one on each side of the rider. A more comfortable front passenger seat was fitted and a seat of similar upholstered design was provided for the rider to replace the more usual saddle. The petrol tank was located under the seat. Three separate band brakes were applied to the rear wheels; two directly to the drums on the hubs of the wheels and the third indirectly through the rear axle.

The Royal Enfield Auto-Quad or Tandem Car was a more refined version of the Quadricycle. With water-cooled engine and two-speed gearbox it had a specification equal to many of the contemporary cars but still sat the passengers ahead of the driver.

Specification of the Auto-Quad

Framework:	Tubular
Wheels:	26 inch
Tyres:	Clipper 2½in Motor Cycle
Motor:	2¾hp De Dion with water cooled combustion head and valve jacket
Cooling:	Thermo-syphon, two 3 gallon water tanks.
Carburettor:	Float feed, spray.
Ignition:	Coil, dry cells or batteries carried under front seat
Petrol tank:	Under seat, 2 gallons capacity (sufficient for 130 miles).
Gears:	Two-speed Enfield patent.
Seating:	Luxuriantly sprung and upholstered.
Finish:	Frame enamelled black, bright parts plated, seats natural and varnished.
Accessories:	Two mackintosh cloth-lined rugs, motor horn, spanners, oil can, etc.
Price:	£140

The Enfield Two-speed Gear

All Enfield tricycles and quadricycles prior to the 1900 Stanley Show had been single-gear machines and the further innovation of the Auto-Quad was the introduction of the new Enfield two-speed gear.

The merits of having more than one gear were already well appreciated in respect of improved hill climbing capabilities and the capacity to cope with the extra load of carrying a passenger. Various designs were attempted but few in the early days of motoring were entirely successful. The design of two speed gear fitted to the Auto-Quad, which had been designed and patented by R.W.Smith, was one of the best of the early designs being simple to use and robust.

The two-speed gear mechanism was contained in an aluminium oil-tight case attached to the back cross member of the chassis frame and connected to the engine by a coupling. The gear was of the sliding cone type with the four gear wheels, two for each gear, being always in mesh. The cone was double ended and could be moved either left of right to engage either a low-speed or a high-speed clutch. Thus one set of wheels was locked to the shaft to provide drive whilst the other pair of gears ran free. In the central position of the cone both gears ran free to provide a free-engine position. The cone was actuated through a chain and worm gearing by a handle convenient to the driver's right hand.

The 1903 Model Range

A motor tricycle and two motorised quadricycles were displayed at the 1902 Stanley Show . The tricycle was fitted with a lengthened frame to reduce the effects of road vibration and to improve steering. The 2¾hp engine was now water-cooled and direct drive to the rear axle was retained. The quadricycles, fitted with either 3½hp or 2¾hp water-cooled De Dion engines and Bozier two-speed gear operated by a lever at the driver's side, were largely unchanged from the 1902 models. They were priced at about 120 guineas according to power.

The 6hp Ader

The Royal Enfield quadricycles and tricycles proved popular. Based upon the proven construction methods of the bicycle industry, they were lightweight, yet robust and offered excellent economy. However, even at the height of their popularity it was clear that the future of personal transport resided with the purpose-built motorcar rather than the adapted bicycle. By 1900 Enfields were already looking to enter into motorcar manufacture but were hampered by not having a suitable design available. As an interim step it was decided to buy one in from another manufacturer. The French were at this time dominant amongst the early pioneers of the motorcar and English manufacturers looked to France for motoring design and innovation. In January 1901, the Automobile Club de France, together with other motor and cycle organisations, staged an exhibition in the Grand Palais in Paris. The occasion provided the opportunity for the Société Industrielle des Téléphones (S.I.T.), Paris, which had been experimenting with car design for several years and had produced their first car in 1899, to reveal their new 6hp car. It was was designed by Clément Ader, who was in charge of motorcar development at the firm, and identified variously as an Ader, after the designer, or as a S.I.T. Clément Ader had been born in 1841 and was the telephone engineer credited for having installed the first telephone lines in France. He was a pioneer aviator and claimed to have made a flight in an aircraft of his own design as early as 1890; he also designed several unsuccessful steam powered aircraft during the 1890s. The Ader was again exhibited at the International Motor Exhibition held at the Agricultural Hall, London, in May 1901. Both four-seater tonneau and spider coach work were exhibited and the cars on this

A testament to the quality and reliability of the early Royal Enfield products was provided by this motor tricycle, which provided reliable transport for Dr Rock of Leeds for almost fifteen years.

Motor Cycling. *ADVERTISEMENTS.* August 20th, 1902.

THE

"MADE LIKE A GUN."

"ROYAL ENFIELD" MOTOR QUADRICYCLE.

With Spray Carburetter and Water-Cooling.

SPECIFICATION:

FRAMEWORK.—Best weldless steel tube of special gauge.

WHEELS.—26in., with high tension tangent spokes.

TYRES.—Clipper Motor Cycle, pneumatic, 2½in.

CRANKS.—Specially toughened, 6¾in. throw, with clutch to allow machine to run independently.

HUBS AND BEARINGS.—Dustproof and oil-retaining.

BRAKES.—Powerful band brake on balance-geared rear axle, double action spoon brakes on both rear wheels, applied by depressing the driver's footrest, this latter to be used in case of emergency only.

MOTOR.—De Dion 2¾ or 3½ h.p., water-cooled.

IGNITION.—Electrical ignition, dry battery or accumulator, induction coil, well insulated wiring, latest De Dion system single wire to switch handle, new coil carried in end of tank.

MUDGUARDS.—Steel, special pattern, fitted with solid steel-stamped stays, practically unbreakable.

SEATING.—Front seat self-colour dark wood, varnished, upholstering, in rain and dustproof material, the whole entirely insulated from the frame by C springs, in addition to the springs between front axle and frame. Waterproof cloth-lined apron.

CHAIN.—Albert-Eadie, 5-16in. block.

CHAIN CASE.—"Enfield" metal detachable, protecting chain from dirt and dust.

GEAR CASE.—Detachable metal with grease cap.

PEDALS.—Rubber, specially strong, extra wide.

TANKS.—Spare petrol tank in frame. Water tank over engine. Lubricating oil tank with pump for feeding crank chamber.

RADIATOR.—Copper coil carried under left half of rear axle.

TYRE PUMP CLIPS.—Rivetted to back of seat convenient for carrying pump, easily detachable, no straps, etc., held by spring only.

VALISE.—Special large, with De Dion and other spanners, oilcan, oil measure, etc.

DIMENSIONS.—Width 3ft. 5in.; height 5ft.; length 7ft. 6in.

Price, 2¾ h.p. Water-Cooled Engine - - - **105** Guineas net.
" 3½ " " " - - - **120** " "

EXTRA: Two-speed gear, worked by friction clutch, ***15*** Guineas net.

1¾ H.P. Engine.	**THE "ROYAL ENFIELD" MOTOR BICYCLE.** Full Particulars Post Free.	**50 Gs.** Net Cash.

The ENFIELD CYCLE CO., Ltd., Redditch, Eng.

occasion were identified as S.I.T.s. A deputation from Enfield's had inspected the Ader in Paris and after further examination in London, Eadie and Smith made the decision to market the Ader under the Royal Enfield brand name. An order was placed with the Société Industrielle des Téléphones for fifty cars. Walter Grew and two other drivers were sent to Paris to collect the first three cars, which were driven back to the Hunt End works. Two months later details of the car were released to the press with the implication that it was a new Royal Enfield machine in production at the Enfield factory with delivery expected in the New Year.

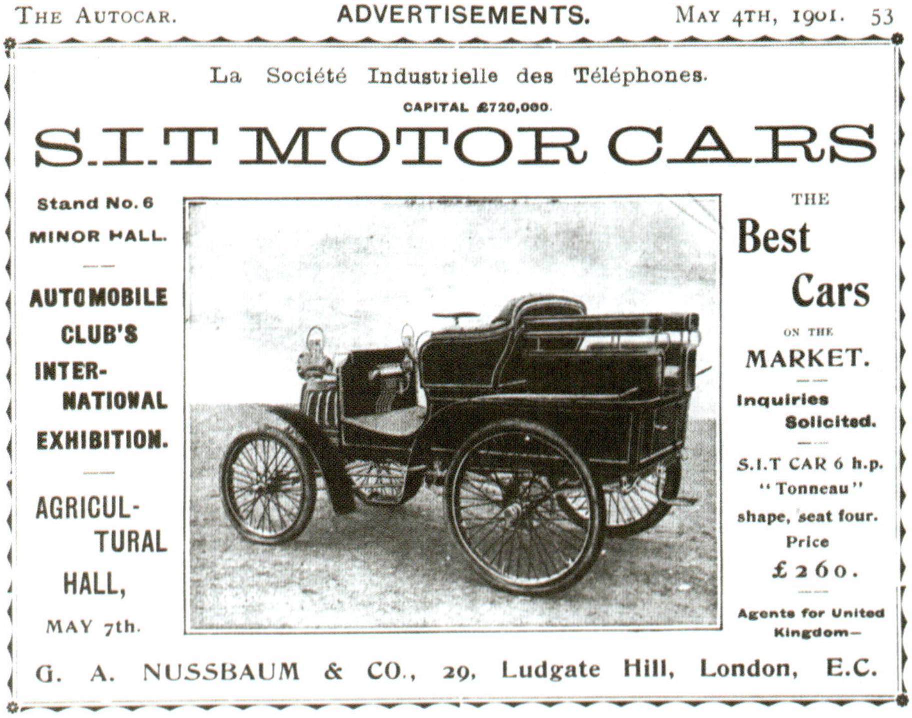

TOP RIGHT: This advert for the S.I.T. appeared in The Autocar near the time of the 1901 International Motor Exhibition in London.

RIGHT: The engine of the Ader was a 4-stroke V-twin of 6hp rating. The cylinders were set at 90 degrees and the capacity was 880cc (79.4mm bore, 88.9mm stroke). The flywheels were within the crankcase. The engine was water cooled using a condensing system. The carburation relied upon saturated wicks held in brass cages through which warm air was drawn through an outer pipe surrounding the exhaust pipe. An electric ignition system was used with separate coils for each cylinder. The usable revolution range was from 1,200 to 1,500rpm.

LEFT: An Ader Light Car of the type purchased by Enfields and to be marketed as an interim model.

Glasgow Reliability Trial

The Enfield Company entered one of the Ader-built Royal Enfield cars in the Glasgow Reliability Trial held in conjunction with the Glasgow International Exhibition and organised by the Automobile Club of Great Britain and Ireland. The trial took place from Monday 2nd to Friday 6th September 1901. The cars set out from the city each day and followed a fixed route before returning in the evening. Marks were awarded for the journey and for the ascent of the various test hills on the route. Over fifty cars took part and the Royal Enfield was allocated number 41. It ran in Section 1 for motor vehicles entered by their manufacturers or agents.

The Royal Enfield performed well on the first day of 116½ miles and gained 289 points out of the maximum of 300. However on the second day of the trial a connecting rod broke and the car was out of the trial. This was an unfortunate set back for the Enfield company as the motoring public assessed the worth of the available motorcars by their performance in the major reliability trials of the day. Lacking success in the trial, it is most likely that sales of the car would have been poor. It was determined that the connecting rods were a weak feature of the design and the outstanding order for the remaining 47 cars was cancelled.

Motorcar Design Studies

In 1901 the Enfield Company employed the French engineer Jules Lambert Joseph Gobiet to assist in the design of powered two-wheelers and the first Gobiet-designed motorcycle was exhibited at the Stanley Show in November 1901. With the motorcycle in production Gobiet, with R.W.Smith, was able to turn his attention to the design of a suitable car for quantity production by Royal Enfield. The two men were joined by Samuel Gallimore, who had been foreman of the gear-set manufacture department at the Wolseley Tool and Motor Company at Alderley Park, Birmingham. David Doyle was also amongst the design staff at this time.

A French-built car was obtained for study and assessment at the factory. It was referred to as the 'Yellow Car' and is known to have had a tonneau body and to have been fitted with a 8hp single cylinder, water-cooled, De Dion engine at the rear with rear axle drive, but the make is uncertain.

Catalogue illustrations of the Ader built Royal Enfield Light Car

Move to the British Mills

Two cars were completed at the Hunt End factory during 1902 before Mr Grew and the car building department moved to rented premises at the British Mills in Redditch, joining other parts of the company that had already moved to this site. At this time the British Mills was the largest factory space in the town having frontage onto Prospect Hill and extending back the length of Albert Street. The factory was equipped with a large steam powered beam engine to power the machinery although this was not utilised by Enfield's as all machining work continued to be carried out two miles away at the Hunt End works.

Staff Changes

There is an indication that the Enfield motorcar project was not proceeding smoothly at this time as a number of senior staff members sought employment elsewhere. The general manager, Ernest Godbold, left in July 1902 to take up the same position with the Premier Cycle Company Ltd. in Coventry. He was replaced by John Davis, who had been company secretary since 1899. Jules Gobiet left the company in 1903 and took employment in November that year with the Premier Cycle Company. Gobiet had been in charge of design although he had not been credited as the designer of any Enfield car. Walter Grew, manager of the motor department, also left in June 1903 to take employment with Iliffe's as editor of the new journal 'The Motor Cycle'. He was replaced as manager of the motor car and motor cycle department by Trevor Rapson, who joined Enfield's from the Wolseley Tool and Motor Co. Ltd. Rapson brought with him Edward McLaughlin, who was placed in charge of the motor car building section.

The impact of Rapson was dramatic. As sole designer/manager he was able in a very short space of time to design and build a 10hp light car the equal in price and quality with any on the market at that time. Despite his contribution to the Company Rapson did not remain in post very long as a replacement works manager was appointed in October 1904. The new man was W.M.Jenkins, who was formerly general manager of Garlick's Motor and Cycle Supply in Cape Town.

The 1904 Models

Enfield cars were exhibited for the first time at the Stanley Show, held in November 1903 at the Agricultural Hall, Islington. Three cars and two motorcycles were exhibited. Two cars were 10hp machines, with rear entrance and tonneau bodies, and the third was a 6hp machine. The Motor was able to report, 'The Enfield Cycle Co., Ltd., have now equipped an entirely separate factory for the production of motorcars and motor-bicycles, and from this it will be seen that the Enfield Company do not look upon motor-bicycles as a "side line." In placing cars upon the market this company is bringing to its aid the experience gained during the last few years in the building of motor quadricycles, tricycles and bicycles.'

The 10hp model was to be fitted with a vertical twin engine of Enfield manufacture, but the engine was not ready in time for the show. Instead the display machines were fitted with bought-in engines. An example of the new Enfield twin-cylinder engine was mounted on a stand for display alongside the cars. It was not in running order, but it was hoped that visitors to the stand would assume that similar, working, engines were fitted to the display machines. However, it would appear that the ruse proved unsuccessful as no orders were placed at the show.

The 1904 10hp Model

The 10hp machine used a vertical twin-cylinder engine of 1930cc (98.4mm bore x 127mm stroke) engine manufactured throughout by the Enfield company. The capacity would very quickly be increased to 2059cc by increasing the bore from 3 7/8 to 4 inches. The induction valves were mechanically operated. The engine was governed to operate at 850 revolutions per minute. Speed was controlled by the air supply to the carburettor and by a pedal, which acted to throw the governor out of action. The chassis was a tubular design and the engine was attached, using aluminium brackets extending through its crank chamber, to a tubular under-frame slung from transverse members at the front of the chassis. A foot pedal operated a band-brake on the gearbox shaft and a hand lever operated either an internal or an external brake on each rear wheel. The use of either the foot pedal or the brake

The Royal Enfield 10hp was a well-designed light car, typical of its period. The bodywork was well designed and upholstered. Three oil lamps were fitted as standard: two at the rear of the bonnet and one at the back of the car. The cars were priced at £300.

Specification of 1904 Royal Enfield 10hp Model B

Engine:	Side-valve vertical twin Enfield
Capacity:	1932cc (increased to 2059cc)
Bore & stroke:	98.4mm (increased to 101.6mm) x 127mm
Ignition:	High tension
Carburettor:	Longuemare
Cooling:	Water, thermo-syphon, fan assisted. A gear driven pump was subsequently added.
Gearbox:	3 forward gears and reverse
Chassis:	Tubular
Suspension:	Semi-elliptic springs front and rear
Transmission:	Cardan shaft
Brakes:	Transmission brake at rear of gearbox operated by pedal Rear wheel drum brakes operated by hand lever
Wheels:	Artillery 30in diameter, 3½in pneumatic tyres
Weight:	15 cwt
Dimensions:	Wheelbase 7ft 4in, track 4ft 5in Length 10ft 5in, width 5ft overall
Price:	£325 with leather upholstery and Modèle Riche finish

lever resulted in withdrawing the clutch and disengagement of drive.

The machine was available with Modèle Riche finish, comprising leather upholstery with spring cushions, or in standard finish and upholstered in Pegamoid.

The 1904 6hp Model

The 6hp Royal Enfield was fitted with a front mounted single-cylinder De Dion engine. Engine speed was regulated by the air supply to the carburettor and by an exhaust valve lifter operated by a lever on the steering column. A gilled tube radiator was fitted and the water circulation was by a friction-driven pump. The chassis was of tubular design and supported on half-elliptic leaf springs. Three speeds and reverse were provided with drive taken by Cardan shaft to a bevel gear rear axle. The braking arrangement was the same as the 10hp model. The car at the Stanley Show was fitted with artillery wheels and large diameter Continental tyres, but the production machines would be offered with the choice of either artillery or wire wheels. The 6hp car on display was well finished in white, picked out in red, and had a comfortable phaeton seat with a large toolbox at the rear. Fitted with artillery wheels, it was priced at £185.

The 6hp, although marketed as a Royal Enfield machine, was not of Enfield manufacture. The show machine and all subsequent cars were bought-in complete. The manufacturer was never revealed by Enfield's although is thought of have been of Continental origin.

The 6hp Royal Enfield Car, Although marketed as a Royal Enfield, the car was not manufactured by the company.

Specification of the 1904 Royal Enfield 6hp Model C

Engine:	Single cylinder De Dion
Capacity:	653cc
Bore & stroke:	90.5mm x 101.6mm
Ignition:	High tension trembler coil
Carburettor:	De Dion
Cooling:	Water, pump circulation
Gearbox:	3 forward gears and reverse
Chassis:	Tubular
Suspension:	Semi-elliptic springs front and rear
Transmission:	Cardan shaft
Brakes:	Transmission brake at rear of gear box operated by pedal Rear wheel drum brakes operated by hand lever
Wheels:	Artillery 28in diameter, 3¼in pneumatic tyres or 28in wire wheel with 2½in tyres
Upholstery:	Leather, with spring cushions
Weight:	10 cwt
Dimensions:	Wheelbase 5ft 11in, track 4ft Length 8ft 9in, width 4ft 10in overall
Price:	£185 with artillery wheels, £175 with wire wheels

The 6hp Royal Enfield Car displayed at the Stanley Cycles and Motor Show was finished in white, picked out in red.

The 1904 8hp Royal Enfield

A surprise entry in the Reliability Trial organised by the Automobile Club of Great Britain and Ireland at Hereford from Monday 29 August to Saturday 3 September 1904 was an 8hp Royal Enfield. The car was fitted with a twin cylinder monobloc engine of 1261cc (88.9 x 101.6 mm) producing 8hp at 1,200rpm. Ignition was by coil and battery. A Longuemare carburettor was fitted and the speed was controlled by a governor. The engine was water cooled and circulation relied upon a friction driven water pump. The chassis frame was of tubular construction with brazed joints. The wheelbase was 7ft and the track 3ft 10in, the unladen weight was recorded as 1204 pounds.

The 8hp Royal Enfield which took part in the 1904 Small Car Trials.

The machine entered in the trial was fitted with two-seater bodywork and was driven by George Serrell. The stated selling price of the vehicle was £200, which placed it in class D, for machines costing over £175 but less than £200. The car suffered numerous problems during the trial, including seven involuntary stops on the first morning, and was finally retired on the third day with a broken oil lubrication pipe. Following its poor performance in the trial the car was never marketed.

Return to Hunt End

Splitting the motor car manufacture between British Mills and Hunt End had not proved entirely satisfactory and it was decided to bring the production back to Hunt End. Following negotiations with Mr W.H.Chambers, licensee of the Red Lion public house, the company was able to secure the purchase of a piece of land between the Red Lion and the works. The new Autocar manufacturing facilities were completed and operational by the Spring of 1904. It is thought some 20 to 30 cars were manufactured at the British Mills.

Vehicle manufacture at this time comprised the manufacture and assembly of the rolling chassis. The engines were of Enfield design and whilst most were manufactured at the works a number were made by an outside supplier using the Enfield drawings. The rear axles were all bought-in. Before the advent of the Royal Enfield sidecar combinations, the company did not have a coach building department. As a consequence the vehicle bodies were bought in from from external suppliers. They arrived at the works complete and ready to be fitted to the chassis.

1905 Model Range

Amongst the first tasks carried out by the new works manager, W.M.Jenkins, following his appointment in October 1904 was the design of a new car radiator. This was of almost circular section and with a matching bonnet resulted in the front of the cars having the appearance of a railway locomotive boiler The new radiator, which was of honeycomb design, was made in the tinsmiths' shop of the cycle works. It would be used on all Royal Enfield cars from 1906 to 1908. He also made a number of significant design improvements, resulting in a revised 10hp twin-cylinder model and brought out a new larger four-cylinder car designated the 20hp. The new 10hp car was revealed at the end of November 1904. It featured for the first time a pressed steel chassis frame in place of the earlier type of brazed tubular component. The water pump was discarded in favour of a thermo-syphon cooling system. The bodywork now featured a metal dash which curved towards the driver to enclose the ignition coils and oil drip feed lubricators.

The 6hp car was discontinued for 1905 to allow the factory to concentrate on the Jenkins designed vehicles.

ENFIELD AUTOCAR TALK

A brochure with the above title was produced for the 1905 season and addressed to 'Intending Purchasers of Moderate Priced Autocars'. It began: 'The Enfield Autocar Co., in putting upon the market the entirely new Models for 1905, which are fully illustrated and described in their Catalogues, are confident that the design is good, the quality is excellent, and the price is cheap. KINDLY NOTE THESE MODELS ARE ONLY CHEAP IN PRICE, NOT IN QUALITY OR APPEARANCE. The Enfield Autocars are not a new departure, but constructed on the generally accepted designs known to the automobile industry.'

The Enfield Cycle Co models introduced for 1905 were entirely new. Two different chassis were produced and offered either as a range of five different models or in chassis form, for customers wishing to fit their own choice of body. The bodies were of the Company's own design. The Model 20, previously identified as the 10hp or Model B was considered unique. It was designed to seat two and to have a large storage capacity useful for touring. However, if required to carry four persons it could be rapidly converted into the Model 21 by fitting the tonneau attachment. The van

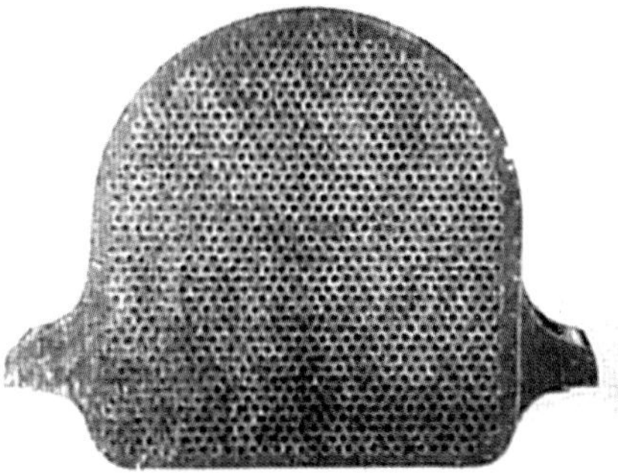

The new works manager, Mr Jenkins, lost little time in before creating a more modern design of radiator to be used on all Royal Enfield cars.

attachment could be fitted as easily as the tonneau to convert to the Model 22. It was claimed the purchaser of a Model 20 could drive his car to business and in less than five minutes convert it from a pleasure to a business vehicle with ample room for samples or light merchandise.

The Enfield Autocar Model 30 was a roomy touring car seating five passengers and capable of carrying all the luggage required for touring. The fifth machine, the Model 40, was fitted with Landaulette bodywork. The Company's sales literature informed the prospective purchaser: 'As the self-propelled vehicle is becoming very popular, and more horse-driven vehicles are being abandoned for the Autocar, we have brought out the Landaulette, known as the Model 40, which will be a popular car for getting about in extremely cold or wet weather.'

The 20hp chassis was a scaled up version of the 10hp model. In particular the engine, a four cylinder side-valve used pairs of blocks and the same 4in bore and 5in stroke as the 10hp model

Commercials

Details were released in December 1904 of a new van, which the company intended to market. It was based upon the 10hp chassis, but fitted with 32in wheels with solid tyres to carry a load of 15 cwt. It appears two were made and used for journeys to the Stanley Show in London in December 1904. The indicated price was £300, but it found no buyers.

Production Methods

The Enfield Cycle Co Ltd was by this time a recognised motorcar manufacturing company. Its vehicles were exhibited at the main motoring exhibitions and it had been a member of the Society of Motor Manufacturers and Traders since August 1904. However, at this stage the company was more a designer and assembler of vehicles than a manufacturer. The emphasis on bicycle manufacture

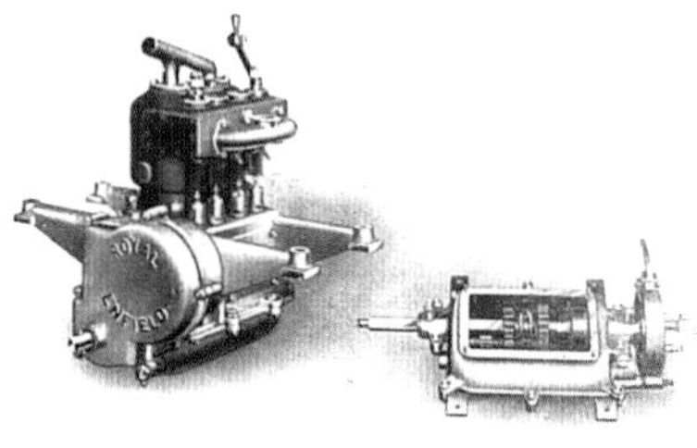

Model 20. Engine and Gear Box.

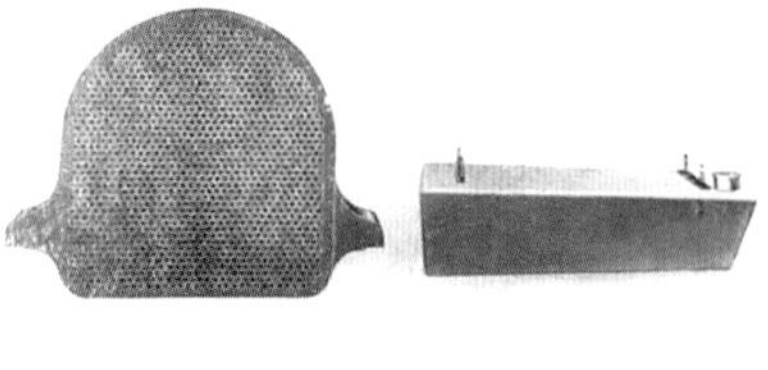

Model 20. Radiator and Petrol Tank.

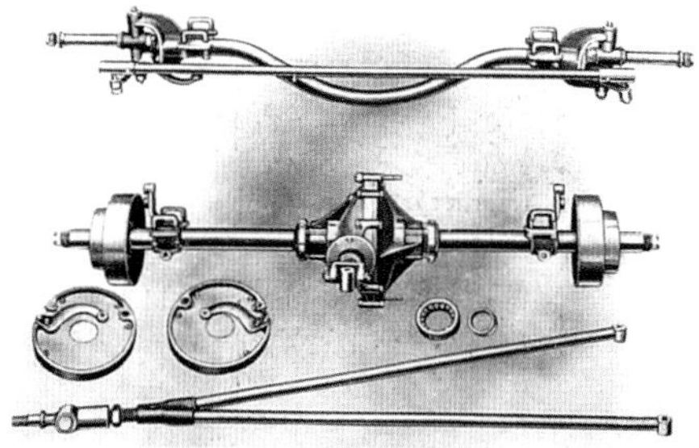

Model 20. Front and Rear Axles, with Expanding Brakes; also Radius Rod.

Model 20. Dash.

A page from the sales catalogue entry for the 1905 Royal Enfield 10hp Model 20 illustrating the Royal Enfield-produced two-cylinder engine and gearbox. Also shown are the front and rear axles and the new design of radiator , which resulted in a curved bonnet . The extended dash provided weather protection for the ignition coils and lubricators.

SPECIFICATION OF "ROYAL ENFIELD" 10-H.P. AUTOCARS.

Model 20. Price, 200 guineas.

ENGINE.—Twin-cylinder; bore, 4in.; stroke, 5in.; normal speed, 1,000 r.p.m.; accelerated, 1,500 r.p.m.; camshaft and timing gear enclosed; ball thrust on crankshaft; gudgeon pin held in position by ring round piston.

GEARING.—Of approved design; three-speed, forward and reverse, direct drive on top speed; running in oil; all gears actuated by one lever; drive, newest type flexible shaft and bevel gears.

FRONT AND REAR AXLE.—Of special design; made of strong material; live axle on balls; semi-elliptic springs, very resilient.

RADIATOR.—Resting on channel iron; strictly Enfield design; containing over 1,000 tubes; exceptionally large cooling surface.

PETROL TANK.—Fitted under seat, with capacity for 5 gallons.

DASH.—With curved sides to form protection for coil, accumulator, oiler, switch and commutator; also two cabinets for holding tools, etc.

FRAME.—The very best design of pressed steel, with all attachments hot rivetted.

BRAKES.—Powerful metal to metal band brake on brake drum, actuated by foot pedal; pair of powerful expanding internal brakes, operated by hand lever, acting in drums on rear wheels.

SILENCER.—Very small, but efficient.

WHEELS.—750 by 85; strong, but of light appearance; oak spokes, hickory felloe; fitted with Pattern 2 Clincher, Continental or Dunlop tyres.

BODY.—Enfield design; two-seated; large storage space at back. So designed that tonneau or van attachment can be fitted in a few minutes.

DIMENSIONS.—Wheel base, 7ft.; tread, 3ft. 11½in.

Model 21. Price, 225 guineas.
Specification—see Model 20.
Body: Detachable Tonneau making seating capacity for 4.

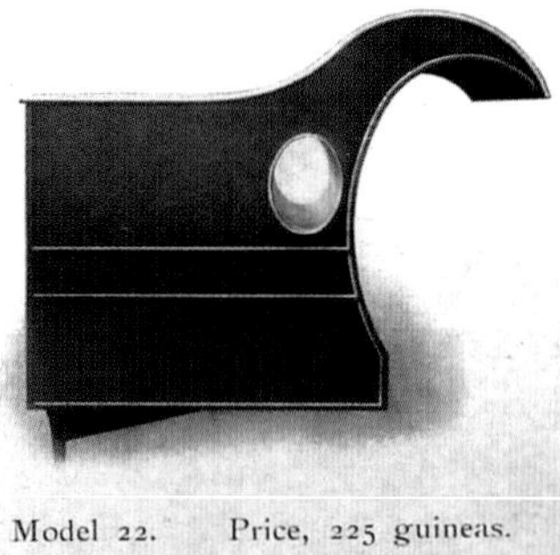

Model 22. Price, 225 guineas.
Specification—see Model 20.
Body: Detachable Van, for delivery of light merchandise.

A page from the 1905 sales catalogue giving the specification of the new 10hp Model 20. The basic Model 20 was a two-seater and priced at £200, but for an extra £25 additional bodywork could be purchased to convert it, within a few minutes, into either a four-seater tonneau, Model 21, or a van, Model 22.

For 1905 the Royal Enfield 10hp Autocar gained a new pressed steel chassis frame and a more stylish scuttle. Two-seater, five-seater and van variants were available, the two seater being identified as the Model 20

The 1905 five-seater tonneau bodywork fitted to the 10hp chassis was identified as the Model 21.

Specification of 1905 Royal Enfield 10hp

Engine:	Side-valve twin-cylinder
Capacity:	2059cc
Bore & stroke:	101.6 x 127mm
Ignition:	High tension trembler coil
Carburettor:	Longuemare
Cooling:	Water, thermo-syphon, vanes on flywheel act as cooling fan.
Clutch:	Leather-faced cone
Gearbox:	3 forward gears and reverse
Chassis:	Pressed steel
Suspension:	Semi-elliptic springs front and rear
Steering:	Rack and pinion type
Transmission:	Cardan shaft
Brakes:	Metal to metal transmission brake at rear of gearbox operated by pedal coupled with clutch pedal. Rear wheel internal drum brakes operated by hand lever
Wheels:	Artillery 30in diameter, 3½in pneumatic tyres
Weight:	15 cwt
Dimensions:	Wheelbase 7ft 4½in, track 3ft 11½in

A brochure was produced to promote the light delivery van conversion of the 10hp car. The van was identified as the Model 22.

meant the company lacked heavy duty press and foundry facilities and most major components were bought-in. The tubular chassis frames of the earlier motorcars had been manufactured in-house but the pressed steel frame, introduced from 1905, could not be produced at the works. The major components of the engines, including the cylinder blocks, crankshafts, camshafts and aluminium crankcases were bought-in. The blocks were pressure tested for blow holes at the works before being passed, together with the other components, to outside engineering firms for machining. Other minor components were produced in the company's machine shop. The early designs of radiators were bought-in, but after 1905, as has already been noted, the round design of Mr Jenkins was produced in the tinsmiths' shop. Nickel plating was carried out in the cycle works.

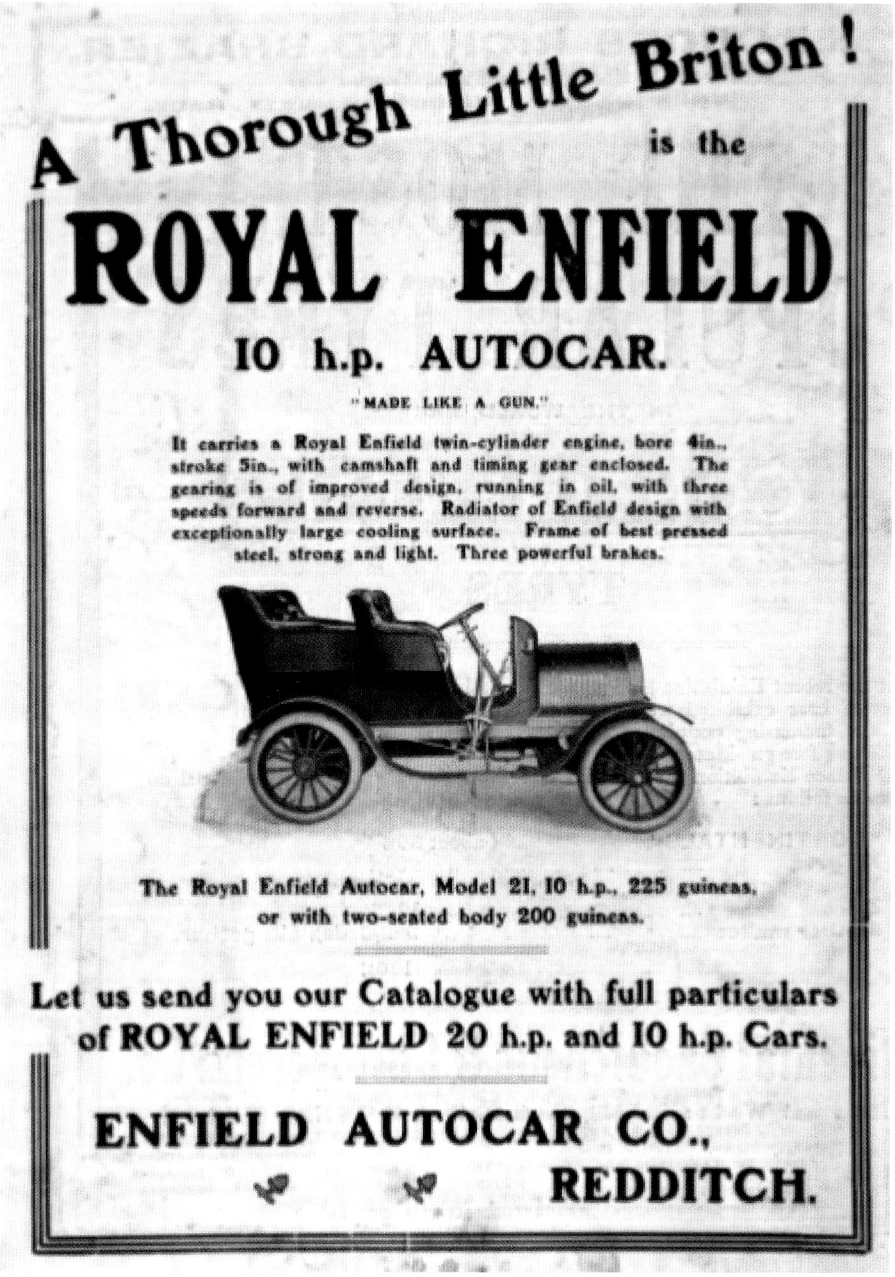

Advertisement from February 1905 promoting the 10hp Autocar.

New 1906 Model Range

The appointment of Mr Jenkins appears to have been very brief and he was soon replaced as works manager by William H.Guillon, who had had considerable design experience in France and was formerly works manager of Fabrique Nationale, Liège, Belgium. His assistant was A.Louis Hadley, whose fluency in French must have been of great assistance to the French-speaking engineer. Guillon's first task during the period from September to November 1905 was to design two entirely new models, designated the 15-20hp and the 24-30hp. Despite the very short timescale examples were completed in time to be exhibited on Stand 17 at the Olympia Show in November 1905. Both vehicles were to be sold as rolling chassis to be bodied to the customers requirements. The larger car was exhibited in chassis form whilst the smaller car was exhibited with five-seater double Phaeton bodywork. Both cars received favourable comment although the larger 24-30 model proved the more popular and the decision was made to proceed with the 30hp model only: the 15-20 model would not enter production.

The company did, however, continue to market a light commercial, which was announced in January 1906. This was based upon the earlier twin-cylinder 10hp machine. It was offered with either solid or pneumatic tyres and weighed 13cwt; it was priced at £300. With the arrival of Mr Guillon, the earlier automobile designs due to Mr Jenkins appear to have been discontinued.

RIGHT: The 1905 Royal Enfield 20hp with Side-entrance Double Phaeton bodywork. The 20hp chassis, with its four-cylinder engine, was a scaled-up version of the 10hp twin-cylinder model. Production was largely of the 10hp models.

1906 30hp Model Y

The car production side of the company was losing money at this time and there was an urgent requirement to promote sales of the new 30hp model. As soon as the 24-30 model was fully completed, Mr Guillon undertook a series of promotional tours during January and February 1906 to demonstrate the car at various exhibitions and to agents around the country. A journalist from The Autocar accompanied Mr Guillon on a journey from Redditch to Edinburgh. They departed in the dark at 5.30 in the morning. Whilst they were confident the car could complete the journey in the day, Mr Guillon had no intention of so doing. He commented, 'This is a touring car and we will run it under touring conditions'. An overnight stop was accordingly taken in Newcastle. The journey was not without its problems and they were compelled to stop on several occasions. However, the journalist considered the reasons for the halts to be trivial and to be expected on an absolutely new type of car completed in such a short space of time. He considered the performance to be really excellent.

Specification of 1906 Royal Enfield 24-30 Model Y

Engine:	Side-valve T-head four-cylinder, cast singly
Capacity:	6107cc
Bore & stoke:	120 x 135mm
RAC Rating:	35.7hp
Ignition:	Battery and coil ignition
Main bearings:	Three ball races
Cooling:	Water, pump
Clutch:	Leather-lined cone clutch.
Gearbox:	3 forward gears and reverse
Chassis:	Pressed steel, with sub-frame for engine and gearbox
Suspension:	Semi-elliptic springs front and rear
Steering:	Worm and nut type
Transmission:	Cardan shaft
Brakes:	Pedal operated transmission brake at rear of gear box. Hand lever operated contracting brakes on rear wheels.
Tyres:	Front 870 x 100mm, rear 880 x 120mm.
Price:	Only available in chassis form £210

New Redditch Factory

The directors of the Enfield Cycle Company were able to report a net profit of £12,616 at the ninth annual meeting of the company held on 20 October 1905. A dividend of 7% was paid on the preference shares and 6% on the ordinary shares. This was the best performance of any year since the boom and had resulted from the growth in bicycle sales; the motor car department, despite an investment of over £20,000 was still struggling and had recorded a loss. However, the directors expected improved results in the coming year. They were confident that the reorganised car department under its new manager, Mr Guillon, would in future operate profitably. As a consequence the decision was made to buy land for another factory in order to expand production. A plot of land near the railway line in the centre of Redditch and with frontage onto Hewell Road, which in total comprised just less than six acres, was purchased from Lord Windsor.

The 1906 30hp Enfield designed by Mr W. Guillon.

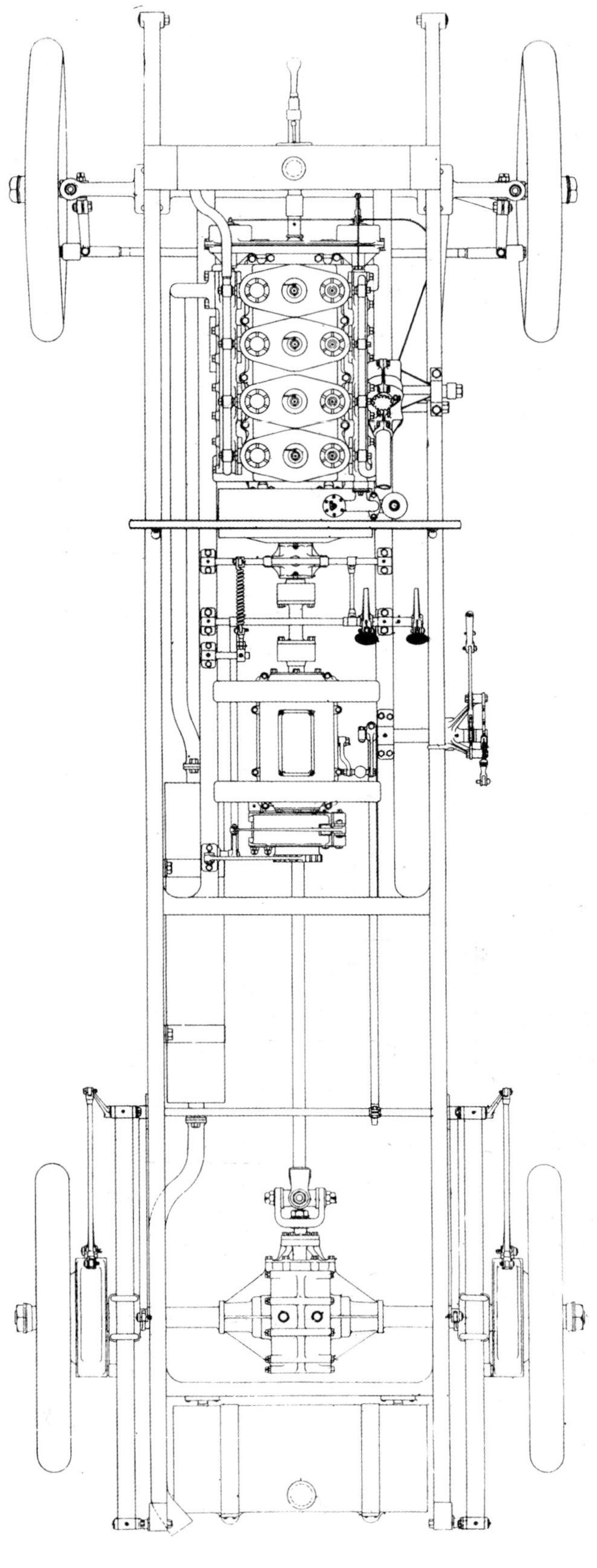

The 30 hp Enfield car, deigned by Mr W.Guillon, followed general design practices for the period and was of high quality. The four cylinders were separately cast with the exhaust valves located on the left hand side of the engine.

The right hand side of the engine showing the inlet manifold and carburettor. The engine developed 30hp at 900 revolutions per minute.

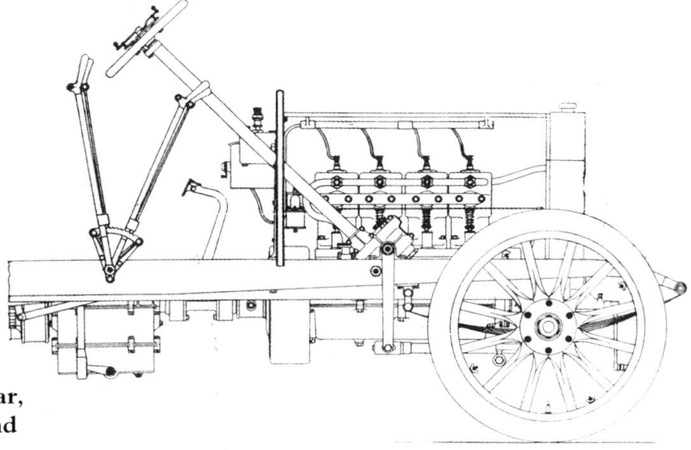

ABOVE AND RIGHT: The chassis of the Enfield 30hp car, which was designed by Mr Guillon. was of high quality and conventional for the period.

Specification of 1906 Royal Enfield 15-20 Model X

Engine:	Side-valve T-head four-cylinder, cast singly
Capacity:	3927cc
Bore & stroke:	100 x 125mm
Ignition:	Battery and coil ignition
Main bearings:	Plain
Cooling:	Water, pump
Clutch:	Cone clutch.
Gearbox:	3 forward gears and reverse
Chassis:	Pressed steel, with sub-frame for engine and gearbox
Suspension:	Semi-elliptic springs front and rear
Steering:	Worm and nut type
Transmission:	Cardan shaft
Brakes:	Pedal operated transmission brake at rear of gear box. Hand lever operated contracting brakes on rear wheels.
Tyres:	Front & rear 810 x 90mm
Price:	Chassis only £368
	Five seater, double Phaeton with side entrance £420

Early Financial Concerns

Since incorporation, the Enfield Cycle Co. Ltd. had consistently made a profit. The dividend paid on ordinary shares had been 10% for the first three years of the firm's existence and 5% for the subsequent five years. Whilst this was entirely satisfactory there was concern that the Motor Car Department was making a loss. In the Directors' Report for the year ended 31 August 1904 it was noted 'The Company have (sic) spent a considerable sum during the past year in developing the motor industry, and your Directors have every confidence that this branch of the business will eventually become profitable'. The report for the following year was no more encouraging: 'The Motor Department has not realised your Directors' anticipations, but they have good reason to expect improved results in the coming year.'

The Enfield Autocar Company Ltd

The manufacture of cycles, motorcars and motorcycles had been carried out by the single company, the Enfield Cycle Co Ltd until 1906. In that year the decision was made that the motorcar side of the business should stand on its own feet as an independent organisation. A special meeting of shareholders was called on 5 March 1906 at the Grand Hotel in Birmingham at which the directors put forward a proposal to form a separate company for the manufacture of motor cars. They further proposed that once the new works at Hewell Road, Redditch, were completed the Hunt End Works should be turned over wholly to the new company for car production. It was considered that this would provide the necessary working space for the car company to operate efficiently. The proposal was not well received by the shareholders who neither wished the company to be divided nor desired the expenditure on a new factory. Nevertheless it was agreed.

It would appear that the directors already had plans for the creation of the new company, whatever had been the outcome of the meeting, for within two days the Enfield Autocar Company was announced and shares offered to the public.

The Enfield Autocar Company Limited was structured with a capital of £150,000. This comprised 100,000 shares of £1, of which 15,000 were retained by the Enfield Cycle Company Ltd, with the remaining 50,000 shares held in reserve. The directors were listed as Albert Eadie (chairman), Lord Seymour (chairman of The Enfield Cycle Company), George Cartland, Thomas Evans, Robert W. Smith and Edward H. Lancaster. All were also directors of the Enfield Cycle Company, with the exception of Edward Henry Lancaster, who had been managing director of the Clement Motor Company and was to join the board and be appointed managing director of the new company after allotment. The share issue was over subscribed, although the financial pundits were not convinced of the viability of the new venture. The Investors' Review noted: 'No estimate of profits is ventured upon and certainly the history of the parent concern does not seem to hold out prospects of more than a moderate success.' whilst the The Statist was more direct: 'The issue is a speculative venture.'

A Memorandum of Agreement, dated 6th March 1906, agreed for the Enfield Autocar Company Limited to purchase, from the Enfield Cycle Company Limited, as a 'going concern of the goodwill trade name and mark of 'Enfield' in relation to motor cars and motor engines only and the freehold land and works with the engines fixed machinery plant and fixtures loose tools and fittings thereon' together with 'stock in trade manufactured and partly manufactured of and relating to motor cars, motors and motor engines'. The indenture for the sale was finally signed on 19 August 1907. Both companies retained the right to the trade marks and the 'Enfield ' trade name. The vendor company agreed no further involvement in the manufacture or sale of motor cars and motor engines other than applied to bicycles, tricycles and quadricycles, whilst the purchasing company agreed to refrain from the manufacture or sale of cycles.

This rather imprecise statement of the demarcation between the companies was acceptable at the beginning when the companies still retained management bonds. It did not, however, allow for the developments in the motor industry and, as we will see, would be the cause of future disagreement.

Mr Guillon remained as works manager of the new company. However, at the end of August 1906, having been with Enfields for less than a year, he left to join the newly formed motor car division of the Birmingham Small Arms Company Ltd at Sparkbrook, Birmingham. His replacement was by Charles F.Pougher, who joined the company from BSA at Small Heath and had worked previously at Hans Renold Ltd.. Whilst at BSA he had been involved in the development of variable speed gears for bicycles.

The new Hewell Road factory had been laid out for large scale bicycle manufacture and upon completion in 1906 the entire production of bicycles was moved to Redditch. The Enfield Cycle Company Limited finally vacated Givry Works at Hunt End in 1907.

Market Downturn and Further Sales Concerns

The outlook for the new Enfield Autocar Company Ltd initially looked promising with numerous orders being placed for the company's vehicles. The 24-30hp car had entered manufacture and at least ten examples had been produced by May 1906 and although a one model policy had been adopted it was to be available with five different body styles and as a 'one tonner' motor lorry. However, within six months of its formation the company was again facing financial difficulties due to a slump in demand brought about by surplus manufacturing capacity within the industry and a reduction in the market due to rising unemployment. Urgent steps were required to reverse the situation and Lancaster, assisted by Guillon, designed a new 15hp four-cylinder car to be sold alongside a redesigned version of the 24-30hp, which was now to be identified as the 25hp.

In September 1906, the Motor magazine announced the new models. It noted: 'They (Enfield Autocar Co Ltd) were too late to take any prominent part in the 1906 trade, but they have made their arrangements to be well in the fore-front in the coming season.' The magazine borrowed a 15hp model for a trip covering various roads in the Midlands and reported: 'Throughout the trip the motor ran delightfully smoothly. And we only dropped from the top gear once, notwithstanding that we had five healthy passengers aboard. It was a splendid test of a new car, and the Enfield came through with flying colours.'

The Motor was also able to report that the organisation of the factory at Hunt End was nearing completion and a coach building department had been established.

1906 Olympia Motor Show

The new 15hp and modified 25hp cars were completed in time to be exhibited at the motor show held at Olympia during October 1906. The new 15hp model was a handsome car and well received. It was priced at a modest 350 guineas (£367 50s) and orders for 150 cars were placed within the first few days of the show. This was a splendid outcome and a cause for celebration at the Enfield Autocar Co dinner held during the period of the show. After the loyal toasts, the chairman for the evening, Albert Eadie, proposed the toast of the company coupled with the name of Mr Lancaster. He was lavish in his praise for Mr Lancaster. He had been confident that Mr Lancaster would pull the company through and he was now certain that he had succeeded. The 15hp Enfield model was a car to satisfy the English market and the company could now expect a share of the £3,000,000 worth of trade currently enjoyed by motor cars imported from the Continent. The 'allied' companies at Redditch were at this time employing 2,700 people and making good profits. Furthermore the new works of the Enfield Autocar Company Ltd was capable of producing 400 cars a year.

The new style of radiator, introduced by E.W.Lancaster on the new 15hp and 25hp models introduced for the 1907 season. The Enfield trademark gun featured prominently on the radiator.

Specification of 1907 Enfield 15hp Model

Engine:	Side-valve T-head four-cylinder, cast in pairs
Capacity:	3261cc
Bore & stroke:	95 x 115mm
Ignition:	Battery and coil ignition
Main bearings:	Plain
Cooling:	Water, pump
Clutch:	Leather-faced cone clutch.
Gearbox:	3 forward gears and reverse
Chassis:	Pressed steel, tubular sub-frame for engine and gearbox
Suspension:	Semi-elliptic springs front and rear
Steering:	Worm and nut type
Transmission:	Cardan shaft
Brakes:	Pedal operated transmission brake at rear of gearbox. Hand lever operated contracting brakes on rear wheels.
Tyres:	Front & rear 815 x 105mm
Price:	£350

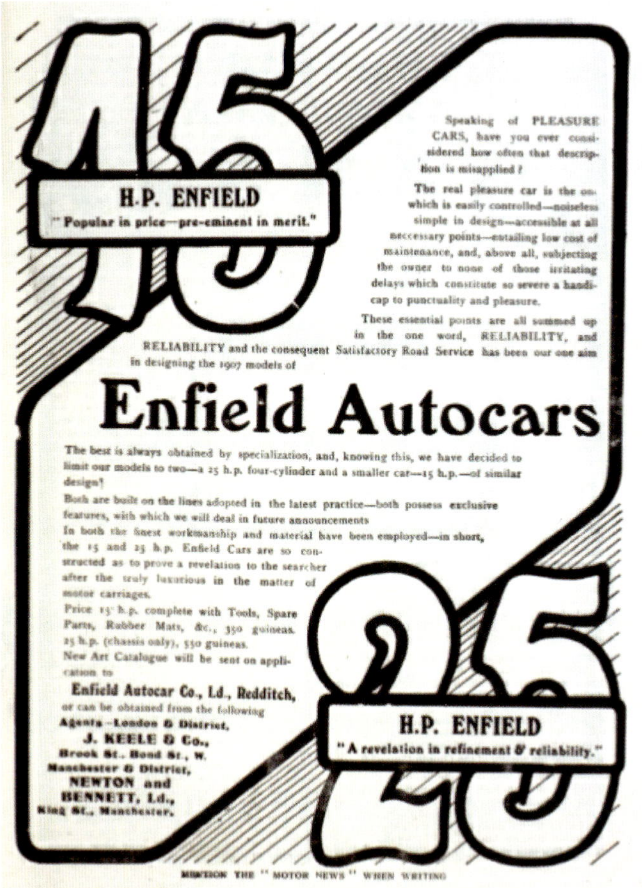

Two advertisements from December 1906 promoting the new 15hp and 25hp Enfield Models.

The 15hp car developed in time for the 1906 Olympia Motor Show,

Unfortunately it proved impossible to satisfy the demand owing to difficulty recruiting additional skilled labour and in obtaining the necessary machinery and raw materials. It would be February before the first orders were supplied and several more months before all production problems were resolved.

On-going Financial Concerns

In the financial report for the period ending 31st July 1907, the company reported a loss £19,264. Blame for the loss was apportioned both to the delays in getting the 15hp model into production and the cost of producing the 25hp model. Late in 1907, Charles Yeomans Hopkins joined the Enfield Cycle Company Ltd. He had briefly worked for the Eadie Manufacturing Company Ltd before moving on to Joseph Lucas Ltd and, subsequently, the Austin Motor Company Ltd where he was the company's first works superintendent. His role with the Enfield Autocar Company Ltd was to increase car production and ascertain whether profitability could be achieved and creditors reimbursed, or the works should be closed. He decided to adopt a one-model policy for the 1908 season based upon the popular 15hp (95 x 115mm) model.

The Enfield 15hp Model tested by The Autocar July 1907.

1907 Enfield 15hp Model from November 1907.

1907 Olympia Show

The Enfield exhibits at the Motor Show, held at Olympia in November 1907, consisted of three complete cars and one finished chassis. All were 15hp models, now identified as the 18-22hp, with a number of improvements incorporated to provide a little more power. The 'Standard Car' with detachable canopy was priced at £407 10s. It was described as an ideal car for use in town and country. The 'Single Landaulette' was priced at £480. The cheapest model, the 'Tulip Side-entrance Car', was priced at £367 10s and cost only £26 5s more than the 'Finished Chassis'.

Whilst the Enfield 18-22hp was unquestionably a good car and competitively priced for its market, sales continued to fall. Without additional capital, which would allow him to improve manufacturing efficiency and profitability, Hopkins had no alternative option than to recommend closure of the company.

Receivership

A few days before Christmas 1907 Thomas Evans presided over a meeting between the major shareholders and the directors. Evans had taken over the chairmanship of the Enfield Autocar Company Ltd following the acquisition of the Eadie Manufacturing Company Ltd by the Birmingham Small Arms Company Ltd and the resignation of Albert Eadie to join the BSA board. Eadie offered to invest £8,000 in the company if the shareholders would put forward the remainder of the working capital required. In the event, the investors were reluctant to allocate fresh capital to the motor industry and there was no option other than voluntary liquidation.

Edward Lancaster, the managing director, lost his position on the board and Mr Hopkins was appointed to manage the company on behalf of the official receiver.

Alldays & Onions

In April 1908 it was finally decided to put the motor car company up for sale.

An auction of the company's assets was held on the 6th April 1908 at the Grand Hotel in Birmingham. However no buyer was found prepared to offer the reserve price of £18,000 and the Enfield Autocar Co was

The Enfield 18-24hp Model. The illustration dates from March 1909.

An advert from July 1908.

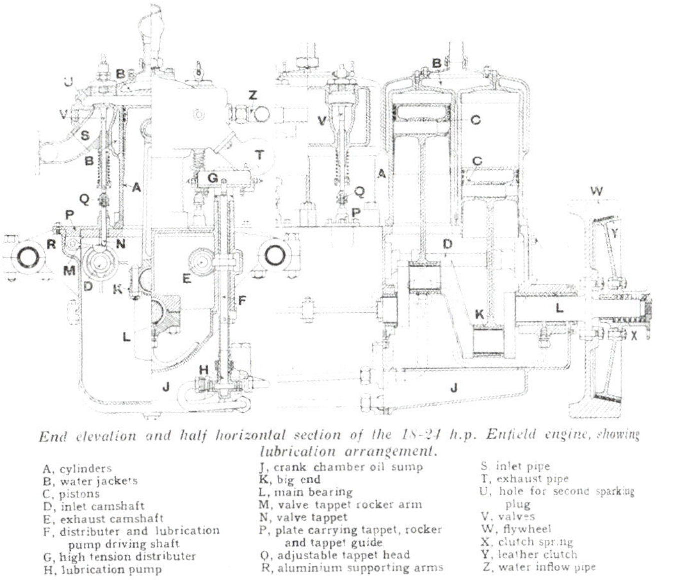

General arrangement of the 15hp (18-24hp) Enfield engine showing the lubrication system.

An 18-22hp model Enfield fitted with a high sided body similar to that exhibited at the Olympia Show in November 1909.

withdrawn from auction. A few days later it was sold to the Birmingham firm of Alldays & Onions for an undisclosed sum. The new owners immediately floated a new company under the name the Enfield Autocar Company Ltd to retain business continuity and moved production to the Alldays Matchless Works in Birmingham. The old company works at Redditch would remain empty for a considerable period.

The New Enfield Models

The new Alldays owned company would initially maintain the Hopkins single model policy and concentrated production on the Lancaster designed 15hp model. This appears to have been variously identified as the 18-22hp Enfield, as it had been when produced by Enfield's, or as the 18-24hp Enfield. The model was now fitted with a gate-change four-speed gearbox with third speed being direct. The braking system was also redesigned. The price, complete with five-seater bodywork was £367 10s, whilst the chassis only price was £304 10s. For the 1909 season the company offered two additional models. The first of these was a two-cylinder 1630cc lightweight, the Model 10-12 Popular, which employed the same bore and stroke (95 x 115mm) as the 18-22hp and would remain in production for four years. It appears to have been based upon the Alldays 10/12 model which had been first produced in 1906.

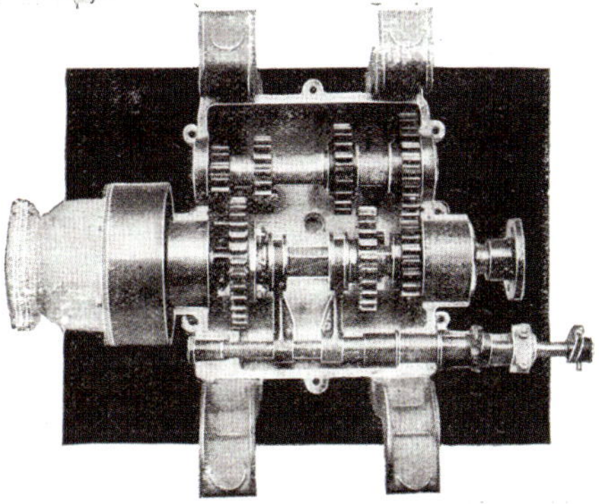

RIGHT UPPER: Underside view of the 18-12 Enfield engine showing the ample size of the bearings.
RIGHT LOWER: Gearbox of the 18-22 Enfield showing brake drum and universal joint casing.

Alldays & Onions

The firm of Alldays and Onions was created with the amalgamation in 1885 of William Allday & Sons with J. C Onions. Both firms were already well established by this date. There is record of a bellows maker named Onions working in Dudley in 1625 and John Onions is known to have established a bellows making business by 1650. The company Alldays & Co was founded in 1788 for the manufacture of bellows and portable forges. At the time of the merger the combined factory premises covered approximately eight acres. The company's name was changed in 1889 to Alldays & Onions Pneumatic Engineering Co to identify more clearly the main business interests. However, by this time the company had already begun to take an interest in the cycle industry and bicycles under the Alldays name were marketed as early as 1896. In July of that year the company submitted plans for a new factory in Fallows Road to be purpose built for the manufacture of cycles; it was to become known as the Matchless Works.

Competition in the cycle industry at this time was intense and Alldays & Onions, like Enfield, began to take an interest in the infant motor industry. Their first machine, a tricycle powered by a De Dion single-cylinder engine was produced in 1898, but is thought not to have gone into production. This was followed by a tiller steered two-seater quadricycle fitted with a 2¾hp De Dion engine and two-speed Champion clutch, which was known as the Traveller. It was later available with a 3½hp De Dion engine and offered in both private and commercial form. The first Alldays light car, a rear-engined flat-twin 7hp four-seater, entered production in 1903. Although this proved successful, it was the 10/12 1.6 litre vertical twin side-valve, introduced in 1905 that really established the company. A light van using the same chassis was added in 1906 and the model remained in production through to 1913. A 16hp four-cylinder model was also added to the range in 1906. Following the acquisition of the Enfield Autocar Co in 1908 the range was rationalised and most models were sold under both the Alldays and Enfield brand names. A 30/35hp six was included from 1911 to 1914 and a 990cc air-cooled V-twin Midget cyclecar added in 1913. The pre-Great War range also included 12/14, 16/20 and 25/30hp models with four-cylinder engines with paired cast cylinders, which were available as passenger vehicles and commercials. The company's trucks and ambulances saw service during the war. Rationalisation post-war saw the introduction of the Enfield-Allday range of cars.

The first Alldays motorcycle, a single cylinder machine, was produced in 1901. By 1903 the company was marketing a 2¼hp machine under the Alldays-Matchless brand name. The machine proved successful in the sporting trials of the day and achieved good sales. It was joined in 1905 by 2.5hp and 3.5hp single-cylinder models. An entirely new 2.5hp machine was added in the range in 1908. During this period motorcycle production had been scaled back to concentrate on car manufacture, but following extensions to the factory production was increased and new models added. Very soon an extensive range of side-valve singles of 499cc and 539cc, and V-twins of 798cc and 988cc were in production. A 292cc single-cylinder two-stroke was offered in 1915 and marketed under the Allon name. It was available postwar and was produced in limited numbers up to 1924. A brief attempt made in 1927 to revive production using proprietary engines proved unsuccessful.

Single-cylinder 2¼hp Alldays illustration from the advertisement published at the time of the 1903 Stanley Show in London.

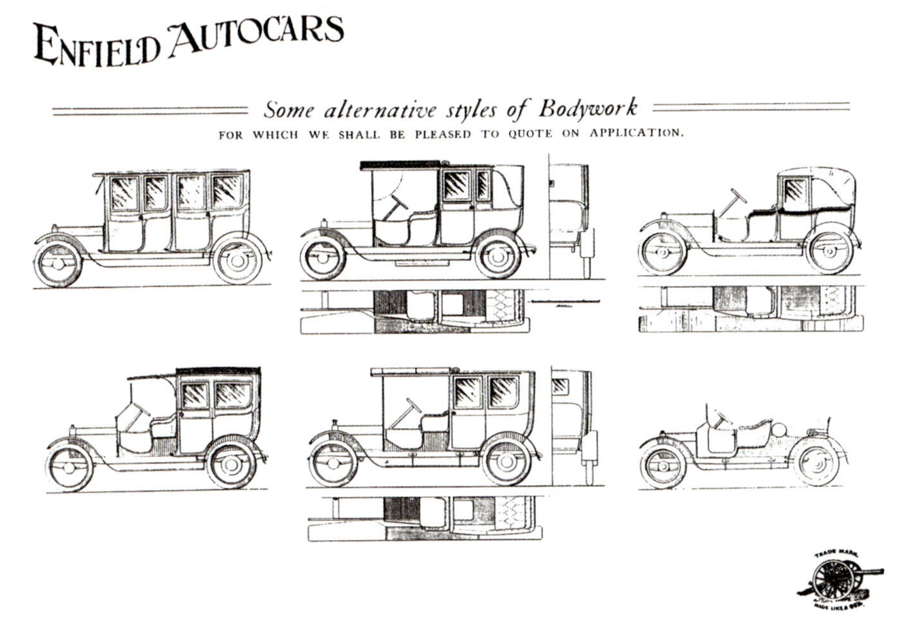

An illustration from the Enfield Autocars catalogue for 1910 showing the different body styles available on the Enfield chassis. A similar illustration was included in the Alldays catalogue.

The acquisition of the Enfield Autocar Company Ltd gave Alldays & Onions the right to the use of the Enfield name on motor cars and the use of the 'Made Like A Gun' trademark, but the Royal Enfield name, remained in the ownership of the Enfield Cycle Company. The trademark is shown on the adjacent illustration of the available body styles.

Fitted with a standard type of two-seater body it was priced at £222 12s. The second new model was the 30/35 Enfield, which was powered by a 6107cc (120 x 135mm) four-cylinder engine. This was derived from the 1906 24-30 Model designed by Mr Guillon and had no equivalent in the Alldays range; it would be offered for only two years. It was priced at £367 10s as a chassis for customers' own choice of body or for £441 with five-seater side-entrance bodywork fitted. It was also available with shooting brake bodywork designed for country use.

With the 30/35 Model Alldays & Onions were promoting the Enfield, rather than the Alldays model, as the brand leader aimed at the wealthier customer seeking a more luxurious car. This policy was continued until production ceased in 1916 and future Enfield models were largely more expensive versions of the equivalent Alldays models. The exceptions to this generalisation were the 4890cc six-cylinder Alldays 30/35 of 1911-14 and the various lightweights.

The five Enfield cars operated by Mr J.C.H.Martin, of The Garage, Lewes, in 1910. All are 18-22 hp models, one being a 1907 type, one 1908, one 1909 and two 1910.

The Enfield Autorette of 1913 was described as a two-seater Runabout built like a Car to give Car service and to provide all the pleasures of motoring at a minimum cost.

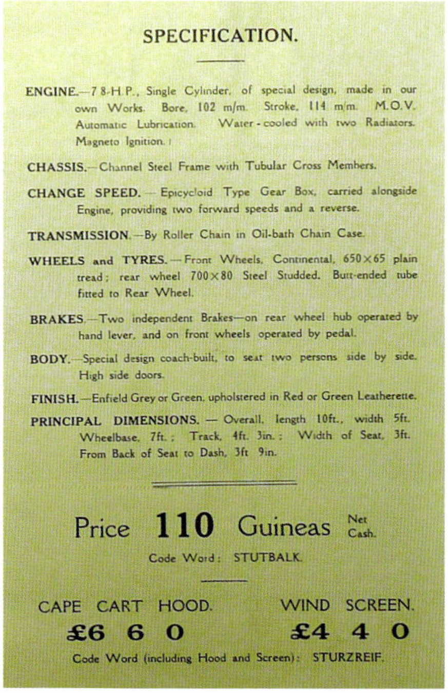

SPECIFICATION.

ENGINE.—7 8-H.P., Single Cylinder, of special design, made in our own Works. Bore, 102 m/m. Stroke, 114 m/m. M.O.V. Automatic Lubrication. Water-cooled with two Radiators. Magneto Ignition.

CHASSIS.—Channel Steel Frame with Tubular Cross Members.

CHANGE SPEED.—Epicycloid Type Gear Box, carried alongside Engine, providing two forward speeds and a reverse.

TRANSMISSION.—By Roller Chain in Oil-bath Chain Case.

WHEELS and TYRES.—Front Wheels, Continental, 650×65 plain tread; rear wheel 700×80 Steel Studded. Butt-ended tube fitted to Rear Wheel.

BRAKES.—Two independent Brakes—on rear wheel hub operated by hand lever, and on front wheels operated by pedal.

BODY.—Special design coach-built, to seat two persons side by side. High side doors.

FINISH.—Enfield Grey or Green, upholstered in Red or Green Leatherette.

PRINCIPAL DIMENSIONS.—Overall, length 10ft., width 5ft. Wheelbase, 7ft.; Track, 4ft. 3in.; Width of Seat, 3ft. From Back of Seat to Dash, 3ft 9in.

Price **110** Guineas Net Cash.

Code Word: STUTBALK.

CAPE CART HOOD. **£6 6 0**

WIND SCREEN. **£4 4 0**

Code Word (including Hood and Screen): STURZREIF.

An extract from the brochure produced by Alldays for the Enfield. Autorette.

The Enfield Autorette

By the second decade of the twentieth century, the motorcar had become a soundly engineered and reliable machine, but it was also extremely expensive. Motorcycles on the other hand were much more affordable, but were a less sociable form of transport. It was about this time that manufacturers conceived the idea of combining motorcycle construction methods and powerful motorcycle engines to construct cheap lightweight cars offering economical transport for two and excellent performance. These new machines quickly gained popularity and were known collectively as cyclecars. In 1912 the Enfield Autocar Co. Ltd., Birmingham, added to their range a three-wheeler light car, or cyclecar, designed by Charles Edward Simms and identified as the Enfield Autorette.

The Enfield Autorette was built on a car-type channel steel chassis frame with tubular cross members and fitted with a coach built car-type body designed to carry two people. It was powered by a water-cooled single cylinder engine of 7.8hp (102mm x 114mm). An epicycloid type gearbox provided two forward speeds and a reverse. Overall dimensions were length 10ft and width 5ft. The wheelbase was 7ft and the track 4ft 3in. It was priced at £115 10s with the cape cart hood and windscreen costing an additional 6 and 4 guineas, respectively.

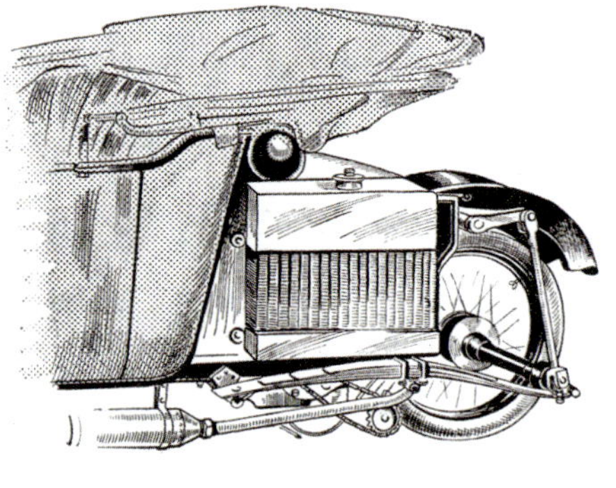

The engine of the Autorette was mounted behind the passenger seats. The rear suspension was by cantilever springs.

Autorette: Motorcar or Motorcycle?

Under the demarcation of manufacturing rights agreed with the Enfield Autocar Co. Ltd, at the time of its formation in March 1906, the Enfield Cycle Co. Ltd understood it had retained manufacturing rights for all lightweight motorcycle-derived machines. In February 1912 Mr Davis and Mr Walker Smith of the Enfield Cycle Company complained to William Allday of the Enfield Autocar Company concerning the manufacture and marketing of a tricycle, namely the Enfield Autorette. William Allday's written response was to assert that the Autorette, although a three-wheeled vehicle, was in fact a small car and not a tricycle. He went on: 'I would also point out that the Directors (of the Enfield Cycle Company) are advised that having regard to the terms of the Assignment of Goodwill of the 19 August 1907, although your Company are entitled to manufacture Motor Engines for use in connection with bicycles, tricycles and quadricycles, as made and understood at date of Agreement, they are not entitled to put them on the market under the name of "Enfield".' He concluded: 'It must however be clearly understood that they (the directors of the Enfield Autocar Company) are not prepared to forego their right to make a light runabout car such as the Enfield Autorette nor can they admit that your Company have the right to build a similar vehicle'.

The Enfield Cycle Company sought legal advice, which was obtained from David Fulton at Temple. His view following 'a fair

Two views of the experimental cycle car produced by the Enfield Cycle Co. Ltd early in 1913.

reading of the assignment' was that the Enfield Cycle Company had in their minds 'a light wire spoke wheel vehicle of low power with a saddle and single seat'. The Purchasers, correspondingly, had in their minds 'only a vehicle answering in weight, seating accom-modation and power to what was then and now known as a motor car'. He concluded that the addition of a car type body, windscreen and steering column to the skeletal frame of the motor tricycle was a logical advance and that the Autorette was therefore 'of a category reserved to themselves by the Enfield Cycle Company'. He was supported in this view by the recent decision of the R.A.C. to delegate to the Auto-Cycle Union the control of motor cycling, including the lightweight motorcycle-based 'bastard' type of machine.

Whilst Mr Fulton was convinced the Enfield Autorette was within the class which the Enfield Cycle Co reserved for themselves under the assignment, he noted should the matter come to litigation the court would not be bound by such evidence and rulings. He noted that 'neither of the parties are inclined to be unreasonable' and advised seeking the services of an arbitrator.

The archived records do not record the outcome of discussions to resolve the disagreement between the two companies, although subsequent letters suggest that these favoured the Enfield Autocar Company.

Royal Enfield Cyclecar

Perhaps to assert their rights for the manufacture of motorcycle derived vehicles, the Enfield Cycle Co Ltd designed and built an experimental cyclecar. The publication of photo-graphs of the machine, in the Cycle Car magazine, of 21 May 1913, brought a quick response from William Allday, managing director of the Enfield Autocar Co. Ltd. He wrote to R.W.Smith noting that any intention to put the Cyclecar on the market 'we should regard as a distinct infringement of this Company's rights'. Again archived records do not record the outcome of discussions between the companies. However the Enfield Cycle Company did not proceed with its Cyclecar suggesting the decision was not in the Company's favour. Also the Enfield Cycle Company, perhaps by way of retaliation, began a trade war with Alldays over the sale of tradesmen's pedal tricycles. On 7th July 1913 Wiliam Allday wrote to R.W.Smith to express his concern: 'Now, before you went away, we were speaking about the shocking way in which you were cutting the Carrier Tricycle trade. We have lost a good number of orders lately, and of course, my London people do not like it. You are under-selling us to the extent of 30/- per Carrier Tricycle, and in addition to this, you are paying carriage and not charging for crates; and further you are allowing £1 more per box than we allow.

'If you can afford to sell these goods at these prices (and I question whether you can) we shall have to do the same. We can certainly produce them quite as cheaply as you can, but we are in business to make money. Life is uncertain, and it is no use waiting for the coming year to make it. In my opinion there is no time like the present.'

The low prices charged by the Enfield Cycle Co. Ltd were also affecting Melens, the third major manufacturer of tradesmens' goods, who had proposed a fixed trade price agreement, subject to a £100 forfeit for any breach.

Enfield Autolette and Nimble Nine

The Enfield Autorette was short lived and quickly replaced by a more conventional four-wheeler with worm final drive, which was identified as the Enfield Autolette. It was originally fitted with an 8hp vee-twin engine, but the option for a 9hp four-cylinder engine was soon made available. It was in turn superseded in 1914 by the

The Enfield Autolette was available with either a twin-cylinder 8hp or a four-cylinder 9hp engine, three-speeds and worm drive. It was claimed to be absolutely reliable and to offer remarkably low running cost.

Alldays Midget light car, the Enfield version of which was known as the Enfield Nimble Nine.

Enfield-Allday

Although producing essentially the same car model, the two companies, Enfield Autocar and Alldays, had retained separate identities prior to production ceasing in 1916 as a consequence of the war. With the resumption of manufacture after the war the distinction between the two companies was lost and a single range was offered under the name Enfield-Alldays Motors Ltd. Alldays & Onions had acquired land on the Waverley Estate in Small Heath, Birmingham in 1914 for a new factory and production of both cars and Allon motorcycles was concentrated at the company's Waverley Works.

Prior to the commencement of the war the company had intended to concentrate on a 10/12hp model and all the required jigs for volume manufacture had been produced. However, the car was not considered suitable for post-war austerity Britain, which required cheaper and more economical light weight vehicles. Plans for the 10/12hp model were dropped in favour of a revolutionary 10hp model designed by Arthur William Reeves, manager of the company's Matchless Works. This was the 10hp Bullet. It was fitted with a 5-cylinder air-cooled radial engine, which owed much to wartime aero engine development. The original capacity was 1,207cc (62 x 80mm) although quoted as 1,247, but this was soon increased to 1,452cc by increasing the bore from 62mm to 68mm. In the larger capacity form the output was claimed to be 26.6hp at 2,500rpm. The chassis was no less unusual than the engine. It was constructed using two triangulated pressed steel side-members and four tubular cross members used to mount the bodywork and to locate the long multi-leaf cantilever springs for the front and rear axles.

The promise of low running costs had stimulated considerable demand for the Bullet. When first announced it was expected to sell for £275, but the price had risen to £350 by the time a brochure had been produced. This was perhaps to be expected during the post-war period of high inflation and escalating costs. However, by August 1919 the company was forced to admit that the car was still undergoing trials and was not yet being supplied to customers. A year later, in August 1920, the price had risen to £550 and Alldays announced the project was to be abandoned due to production costs. Whilst escalating costs were undoubtedly a factor in the car's demise it is likely that reliability

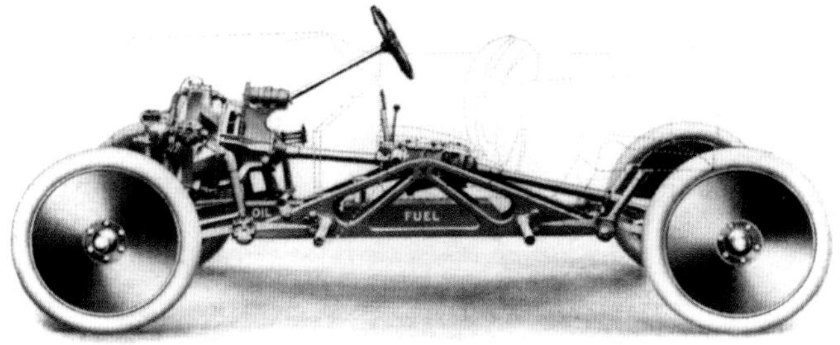

LEFT: Externally the four-seater Enfield-Allday Bullet appeared to be a conventional small car. RIGHT & ABOVE: However, the bodywork concealed very unconventional mechanics. The engine was a five-cylinder air-cooled radial The chassis consisted of a pressed steel backbone frame with cantilever suspension.

problems with the extremely advanced engine were of greater concern.

In 1921 two Italian brothers, A.C. (Gus) and Harry Bertelli, who were employed by Alldays, were given responsibility for the design and production of a replacement for the abandoned Reeves radial car. Gus was responsible for the mechanism of the car and Harry for the bodywork. Designated the 10/20hp, it was powered by a lively four-cylinder engine. Sporting versions of this model achieved modest sporting successes, but commercial success was initially marred by reliability problems. For 1922 and 1923 the company offered the choice of either the 10/20hp or a 12hp model with various body options. The parent company, which was by now in financial difficulties brought about mainly from the £120,000 spent on racing and the innovative Bullet car, was placed in the hands of receivers. Although the company found buyers, being re-registered on 8 July 1925 as New Alldays & Onions Ltd., the manufacture of motor cars was discontinued.

Enfield Trade Name

As part of his advice over the manufacture of the Autorette by the Enfield Autocar Company Ltd, Mr Fulton had expressed concern over the transfer of the mark or trade name 'Enfield' from the 'Enfield Cycle Co.' to the 'Enfield Autocar Co.' He noted, 'the Enfield Cycle Co. had a reputation for skilled manufacture which was a personal asset. The public depended upon the name 'Enfield' as a guarantee that the goods they bought were made by that Company in whose skill and integrity they had confidence. The 'Enfield Autocar Co.' was neither the old Company nor the old managers or workmen. Therefore, to put the name 'Enfield' upon goods not made by the old Company was something very like perpetuating a fraud upon the public'.

With the manufacture of Enfield cars coming to an end in 1925,the Enfield Cycle Company Ltd was able to regain the Enfield trade name. The directors' report of the Enfield Cycle Company Ltd for the year ending 31 August 1924 recorded: 'The Directors have purchased the Release of Right to use Enfield name and trade mark from the Liquidator of the Enfield Allday Motors (1923) Limited, and recommend that the amount should be written off'.

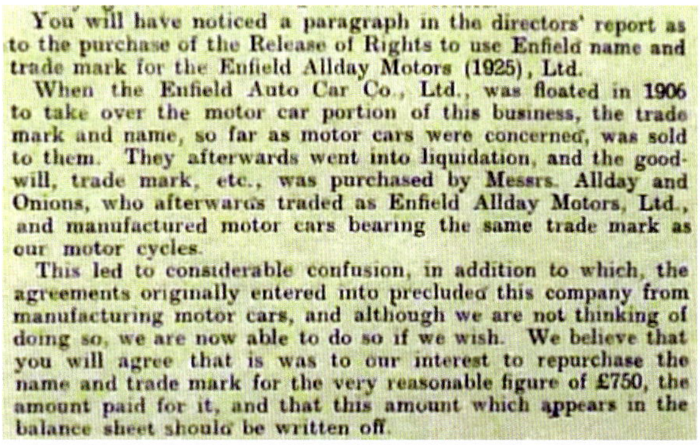

You will have noticed a paragraph in the directors' report as to the purchase of the Release of Rights to use Enfield name and trade mark for the Enfield Allday Motors (1925), Ltd.

When the Enfield Auto Car Co., Ltd., was floated in 1906 to take over the motor car portion of this business, the trade mark and name, so far as motor cars were concerned, was sold to them. They afterwards went into liquidation, and the goodwill, trade mark, etc., was purchased by Messrs. Allday and Onions, who afterwards traded as Enfield Allday Motors, Ltd., and manufactured motor cars bearing the same trade mark as our motor cycles.

This led to considerable confusion, in addition to which, the agreements originally entered into precluded this company from manufacturing motor cars, and although we are not thinking of doing so, we are now able to do so if we wish. We believe that you will agree that is was to our interest to repurchase the name and trade mark for the very reasonable figure of £750, the amount paid for it, and that this amount which appears in the balance sheet should be written off.

LEFT: An extract from the report of the chairman of the Enfield Cycle Company Ltd at the 29th ordinary general meeting of the Enfield Cycle Company Ltd held on 27 October 1924.

Advertisement dating to November 1921.

Enfield Autocar Production Car Models 1906 - 1916

Model	Production Years	Capacity cc	Type	Bore mm	Stroke mm	Wheelbase ft.in	Track ft.in	Chassis cwt
16/20	1906	4084	4 cyl	100	130			
24/30	1906-08	5881	4 cyl	120	130			
15	1907	3261	4 cyl	95	115	8’6”	4’2”	
25	1907	6107	4 cyl	120	135	9’10”	4’7½”	
18/22	1908-10	3261	4 cyl	95	115	9’3”	4’4”	14¾ (ch)
10	1909-12	1630	2 cyl	95	115	7’6”/8’0”	4’0”	12½ (ch)
30/35	1909-10	6107	4 cyl	120	135	10’0”	4’1”	
16	1910-11	2510	4 cyl	86	108	8’6”	4’2”	14 (ch)
12	1911	1944	4 cyl	75	110	8’6”	4’2”	13 (ch)
20	1911-12	3613	4 cyl	100	115	9’3”	4’2”	15½ (ch)
12	1912-14	2174	4 cyl	76	120	9’0”	4’3”	12 (ch)
14/16	1912	2510	4 cyl	86	108	9’0”	4’6”	13 (ch)
16/20	1912-14	3016	4 cyl	86	130	9’6”	4’6”	14 (ch)
14.3	1913-16	2177	4 cyl	76	120	9’3”	4’3”	
18.4	1913-16	3012	4 cyl	86	130	10’0”	4’6”	
24.9	1913-16	4082	4 cyl	100	130	10’6”	4’6”	20 (ch)
8	1914	1069	2 cyl	86	92			
9	1914	547	2 cyl	59	100	7’9”	4’0”	12½ (ch)
10	1914-15	1094	4 cyl	59	100	7’9”/8’0”	4’0”	
10	1916	1169	4 cyl	61	100	8’0”	4’0”	11½ (ch)

Alldays Production Car Models 1906 - 1916

Model	Production Years	Capacity cc	Type	Bore mm	Stroke mm	Wheelbase ft.in	Track ft.in	Chassis cwt
4	1899-1904	498	1 cyl	84	90			
7	1904-08	926	1 cyl	102	114	6’ 0”	10	
8	1906-07	1189	1 cyl	114	114	6’ 0”/6’ 6’’	3’ 8’	
10	1905	1418	2 cyl	95	101	6’ 0”/6’ 6”	4’0”	11½
16	1906	3402	4 cyl	95	120	8’ 9”	14	
20	1906-07	3259	4 cyl	95	114	9’ 9”	4’ 6”	
10/12	1906-11	1630	2 cyl	95	114	8’6”	4’ 3”	11½ (ch)
20/25	1906-11	3261	4 cyl	95	115	9’ 6”	4’ 6”	16 (ch)
12/14	1907	2438	4 cyl	84	110	8’ 0”	4’ 1¼”	12
14/16	1907	3053	4 cyl	90	120	8’ 4’’	4’ 1¼’’	
14/18	1908-12	2495	4 cyl	86	108	8’ 6’’/9’ 3”	4’6”	13 (ch)
30/35	1911-14	4890	6 cyl	95	115	10’ 6”	4’ 9”	18 (ch)
12/14	1912-16	2174	4 cyl	76	120	9’ 0’’/9’ 3½”	4’3”	12 (ch)
16/20	1912-16	3012	4 cyl	86	130	9’ 6’’/10’ 0”	4’6”	14
25/30	1912-16	4082	4 cyl	100	130	9’ 6’’/10’ 6”	4’6”	15½
8/10	1913	990	2-V	80	92	7’ 0’’	4’ 6’’	11 (ch)
10	1914	1069	2-V	86	92	7’ 0”	4’ 0”	9 (ch)
8/10	1915	1094	4 cyl	59	100	7’ 9”	4’ 0”	9
10	1916	1169	4 cyl	61	100	8’ 1½’’	4’ 0’’	11

CHAPTER FOUR

Early Motorcycles 1901 - 1909

The designers of the early motor bicycles fitted the engine in front of the steering head with direct belt-drive to the front wheel. This layout was first advocated in France in 1897 by the Werner brothers, Michel and Eugene, who called their machine the Motocyclette and had received wide acceptance, but these early machines were found to suffer instabilities referred to as 'side-slip'. The problem was a consequence of balance with the engine's location above the wheel, although this was not immediately appreciated. The French engineer Jules Lambert Joseph Gobiet thought he had a solution for the problem.

The bicycle had been enthusiastically taken up in Britain and by the later years of the Nineteenth century, English bicycles had become established as the best designed and engineered machines available; with the consequence that many of the early Continental manufacturers fitted English-made frames and components. Monsieur Gobiet had come to England to learn the trade and by 1900 was in charge of cycle building for Henri Guillaume's business at Old Rectory Mill, Alvechurch. He realised that pedal cycles, unlike motor bicycles, did not suffer from the dreaded side-slip and concluded that this was as a result of applying the power to the rear wheel of the pedal cycle. Thus if the drive from the motor of the Werner layout was instead taken to the rear wheel then side-slip would be eliminated. Aware of the interest being shown in powered machines by Enfield Cycle Co., Gobiet approached the company with his ideas. These proved sufficiently attractive to R.W.Smith for Gobiet to be taken on to the staff to develop the company's first powered two wheelers.

First Royal Enfield Motorcycles

Jules Gobiet designed the first Royal Enfield motorcycle in 1901. The layout followed the Werner arrangement with the engine, a 1½hp Minerva unit with automatic inlet valve, located in front of the steering head of a pedal cycle. The engine drove the rear wheel by a long twisted raw-hide belt, which crossed over at a position near the steering head to provide good belt tension. The belt was exposed and a simple shield was attached to the fuel tank to protect the rider's leg.

It is likely the prototype machine was constructed using an adapted heavy-weight bicycle frame using special thick gauge tubing with a considerably lengthened wheelbase. The front forks used 'D' section tube of great strength and were of the same design as used on Royal Enfield tricycles. It became immediately clear upon testing that the machine suffered from 'side-slip' to the same extent as the original 1897 Werner design. It was concluded that the likely cause of the poor handling problems was the high centre of gravity, inherent with the engine being mounted above the

The 1901 1½hp Minerva-powered Royal Enfield. Supporting the machine is Ernest Godbold, General Manager of the Enfield works and on his right is Walter Grew, manager of the Royal Enfield Motor Cycle Department. On his left are Daniel Doyle and the designer of the machine, Jules Gobiet.

front wheel, combined with the poor road surfaces of the day. To improve stability Gobiet and Smith investigated using a rear mounted engine before finally deciding that the answer lay in adopting a central, low down, engine position. Unfortunately the central engine featured on the recently announced new Werner machines and was protected by patent.

The original Gobiet design had serious handling problems, but these were no worse than most other machines available at the time and it was otherwise of sound design and construction. With no alternative design available and wishing to share in the growing two-wheeler market, Smith and Eadie decided to put the machine into production. The machine used a specially built frame and in several respects was quite advanced. It featured an automatic spray carburettor and band brakes for both wheels. It was announced at the Stanley Show in November 1901 and full details were published by Motor Cycling in its supplement of April 2nd 1902. The machine appeared to be well liked and the introduction to the article noted: 'The special features of this excellent machine are given prominent attention. They are the result of long and careful experiment.' Mindful of future advertising revenue, the article concluded: 'As showing the interest of the public, it may be recorded that whenever we stopped at a hotel or in the street, quite a number of people would come round with more or less intelligent questions. One of the most common was as to whether the steering was affected by the engine being in front. It is not; but you feel

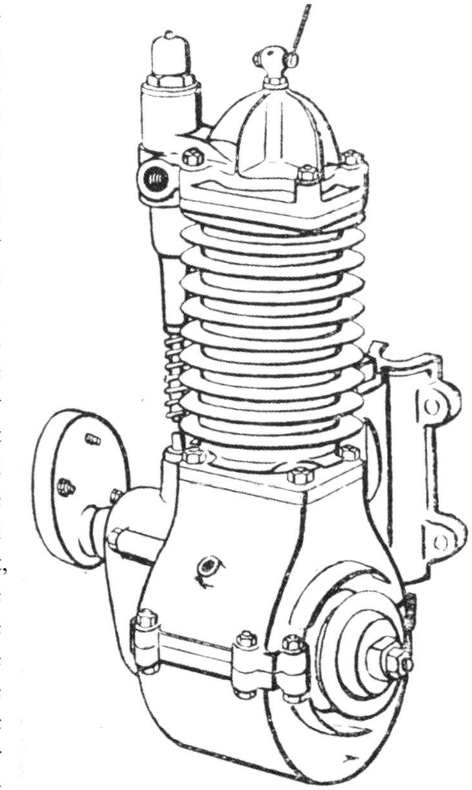

The 1½hp Minerva engine weighed 28lbs and had a claimed output at full speed of 1¾hp. The crankcase formed an integral clamp for mounting the engine to the headstock where, turning a disadvantage into a virtue, the makers claimed 'they obtain the maximum of air for cooling purposes'. The timing gear wheels were enclosed in a special outside case.

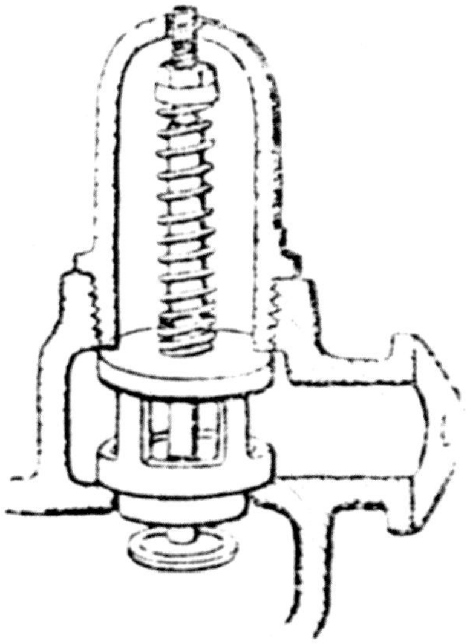

The automatic inlet valve was provided with a special casting with two seats to prevent air being drawn in through the top of the valve.

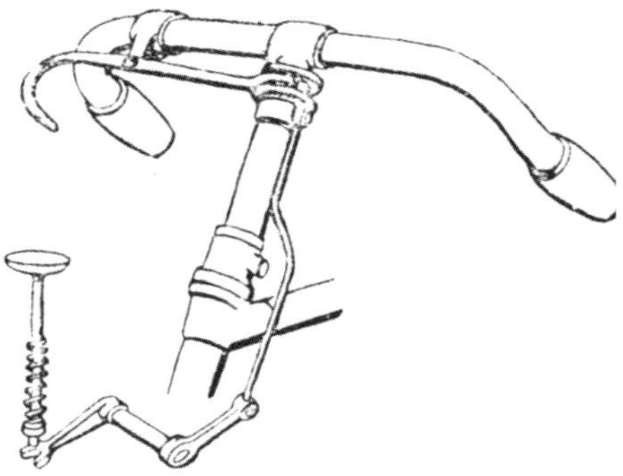

The exhaust valve lifter was operated by a hand operated lever on the right handlebar. The ignition switch was operated by the left hand.

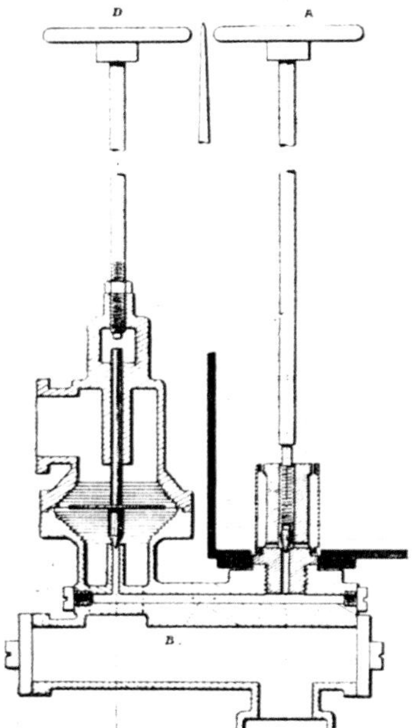

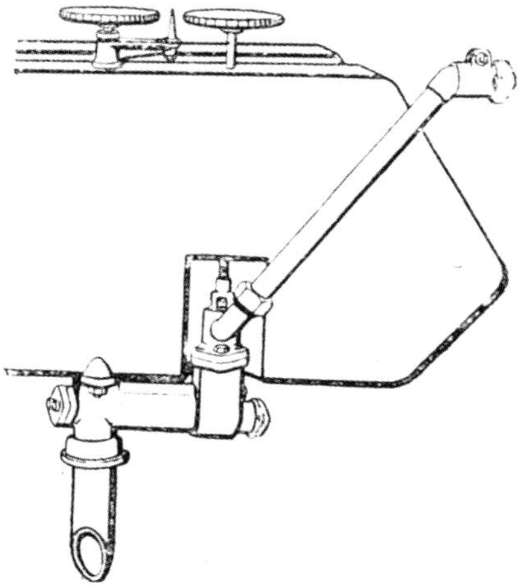

CARBURETTER FITTED TO TANK.

The carburettor operated automatically. Petrol flow into the carburettor was controlled by the adjustment knob A and associated needle jet, whereupon it flowed along passage B to the needle valve controlled by knob D, which was preset at the factory. The optimum position of jet A was obtained by trial and error; the best position for operation being marked with an O, and the shut-off position marked with an S. The tank had a capacity for one gallon of petrol and contained compartments for the induction coil and accumulator.

Advertisement from 1902 for the Gobiet-designed 1¾ hp Minerva powered Royal Enfield.

the weight of the engine naturally, and a little firmer grip on the handlebar is necessary.'

The sales brochure went further and justified the forward engine position in the following terms:

> 'The engine being in the vertical position there is no possibility of lubricating oil getting past the piston and into the combustion head.
>
> 'It is placed where it meets a full current of air which tends to keep the cylinder cool and so preserve the working conditions for the motor.
>
> 'Any splashing from the front wheel is caught by the over-hanging front guard.
>
> 'It is fitted to the main frame at a point where two lugs are closest and the resulting strain is thereby reduced to a minimum
>
> 'The bearings are wide and the life and durability of the engine are thereby greatly increased.'

The motorcycle designed by Gobiet in 1901 remained in production throughout 1902, pending the development of improved models. At this time the range of Royal Enfield-powered machines offered to the public comprised the 1901 Gobiet-designed motorcycle, a tricycle and 2¾hp or 3½hp versions of the quadricycle.

1903 Motorcycles

The Werner brothers had pioneered motorcycle development with their high engine, front wheel driven machine of 1897. Once the deficiencies of this design layout were identified they again pointed the way forward in 1901 with their new diamond framed machine, which achieved a low centre-of-gravity by mounting the engine centrally and using the engine crankcase to form the bottom frame member. Jules Gobiet and R.W.Smith, like many others in the industry, recognised that the new Werner layout provided a solution to the problem of side-slip. However, the Werner layout was protected by patents and motorcycle designers eagerly sought methods by which these could be circumvented. The James Cycle Company was the first to devise and market a solution in the form of the loop-frame, an arrangement in which the engine is supported by a frame member looping underneath the engine. Gobiet and Smith recognised that the loop-frame offered a practical design solution and by the time of the Stanley Cycle Show in November 1902 were able to display new models based upon this design.

In addition to the tricycle and the 2¾hp and 3½hp quadricycles the 1903 range of Royal Enfields, displayed at the Show, included three new motorcycle models. Two of the these were of the loop frame design and were fitted with Enfield-made engines. The first of these used an air-cooled 2hp engine of 239cc (65 x 72mm), with Longuemare spray type carburettor, and v-belt transmission. A single lever served the dual purpose as valve lifter and for ignition advance/retard. It was priced at 45 guineas.

The second of the loop-framed machines was fitted with a new water-cooled 2¼ hp, automatic-inlet-valve, engine of 277cc (70 x 72mm). It employed a novel form of chain drive, which was the subject of a patent taken out by R.W.Smith and Jules Gobiet in March 1902. The final drive was taken from the timing shaft and to provide the required gear reduction the driving shaft, instead of being at a two to one ratio, was reduced in speed by six to one. This required the necessary exhaust lifting and sparking cams being made in triplicate. For lubrication the mechanism was contained within an oil tight crankcase. The timing shaft extended outside the crank case and carried at its extremity a driving chain wheel, containing a spring clutch, from which the drive was conveyed to the back hub by a special chain. The drive chain encircled part of the circumference of the bottom bracket sprocket, which ran idle on the ball-bearing ratchet clutch when starting the engine. Thus the single chain could be used both for starting and to drive the rear wheel. An interesting feature was the small porthole in the combustion chamber which was fitted with thick glass through which the rider could observe the operation of the sparking plug whilst the engine was running. The cooling water was contained in a one and a half gallon compartment at the front of the fuel tank. It was circulated by thermo-syphon through a radiator, consisting of 18ft. of 3/8 inch aluminium piping coiled within the space between the tank and front down tube. The machine weighed 140lbs, complete with petrol and water. It was priced at 75 guineas (£78 15s). The riding speed was quoted as 30 miles per hour, however, this could be easily altered by changing the rear sprocket wheel.

The third new machine exhibited at the Stanley Show was fitted with a new 2hp Minerva side-valve engine, which was attached to the frame down tube in the usual Minerva position behind the front wheel. The inlet valve was mechanically operated and a Longuemare carburettor used. At 50 guineas, the machine appeared expensive compared to the 2hp Enfield-powered machine.

The 2¼hp water-cooled chain-drive Royal Enfield produced for the 1903 season.

The 1903 Royal Enfield 2¾hp employed chain final drive. The engine is air cooled and has an automatic inlet valve and a side exhaust valve.

The 1904 Range

Mention has already been made In Chapter 3 of the announcement in November 1903 of the new Hunt End factory, which had been constructed and equipped specifically for the manufacture of cars and motorcycles. On the surface this represented an endorsement of the car and motorcycle side of the business. The reality was that the returns on the investment were proving disappointing and the powered vehicles side of the business was being heavily subsidised by bicycle manufacture. The motorcar was still thought to represent a good business opportunity and the decision was made to concentrate resources on motorcars rather than motorcycles. Jules Gobiet,

Royal Enfield Chain Driven Motor Bicycle.

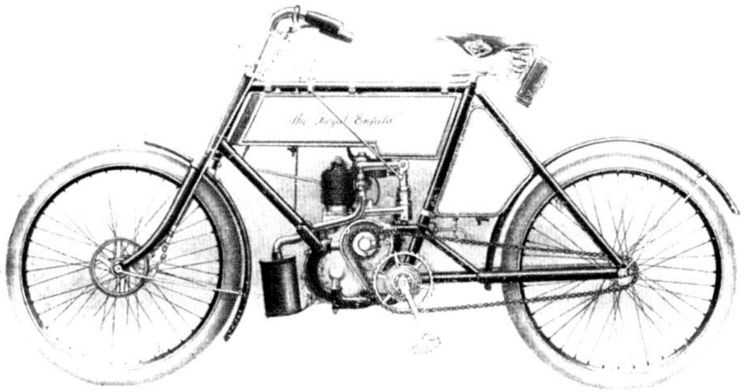

SPECIFICATION.

23-in. Frame, specially constructed throughout; 26-in. Wheels, fitted with 2-in. Tyres; two powerful Brakes; Royal Enfield Engine, 2½ horse-power, with patent Chain Drive; special Ignition by wipe contact and trembler on induction coil; Spray Carburettor; specially large Silencer; beautifully finished throughout.

For full specification and particulars see our Motor List.

Net Cash Price, 40 Guineas

Extra for metal gear case, 21/-

Royal Enfield Belt Driven Motor Bicycle.

SPECIFICATION.

23-in. Frame, specially constructed throughout; 28-in. Wheels, fitted with 2-in. Tyres; two powerful Brakes; 2¾ or 3½ horse-power Engine with Belt Drive; Spray Carburettor; specially large Silencer; mechanically operated inlet valve; beautifully finished throughout.

For full specification and particulars see our Motor List.

Net Cash Price 2¾-h.p., 40 Guineas.
" " " 3½ " 45 "

The 1904 model range. The water-cooled engine of the 1903 all-chain drive model was replaced by a less complicated air-cooled engine which allowed the machine to be offered at the significantly lower price of 40 guineas. The belt-drive machine was now available with either a 2¼ or 3½ horse-power Minerva engine.

who had been responsible for the first Enfield motorcycle, was perhaps disappointed by this change of emphasis and left the company at about this time to take up an appointment at the Premier Cycle Co Ltd. in Coventry.

The Royal Enfield range of cars and motorcycles for 1904 were displayed at the Stanley Cycle Show at the end of November 1903. The Minerva powered machine with forward mounted engine was discontinued but versions of the two loop frame designs remained available. The developments were minor and aimed at making the machines better value for money. The innovative chain-driven water-cooled model at 75 guineas had proved much too expensive to sell well and for 1904 was offered with an air-cooled 2½hp Royal Enfield engine at a much more modest 40 guineas. The belt drive machine was available with either a 2¾hp or a 3½hp engine Minerva engine. Both machines were fitted with spray carburettors. Front wheel braking relied on a Crabbe patent band brake, whilst the alternate options for the rear wheel were either an Enfield Coaster hub or a Bowden band brake with Hyde freewheel. Great attention had been devoted to the strength of the frame and front forks, and the bicycle was considered to present a very handsome appearance. Neither machine proved popular with the buying public.

First Sporting Success

One of the first racing successes by a Royal Enfield occurred at the New Brighton cycle-racing circuit on 1 August 1904. Riding his 2hp Royal Enfield-Minerva, L.Hadley achieved victory in a five-mile scratch race. In front of a crowd of 3-4000 spectators he averaged 44.5mph and crossed the line with a lead of almost half a lap.

1905 Models

By 1905 the demand for motorcycles was reducing and Enfield's limited their range to just two lightweight machines offering the choice of either

The 2¼hp Royal Enfield motorcycle for 1905 retained the Enfield patent two-speed gear.

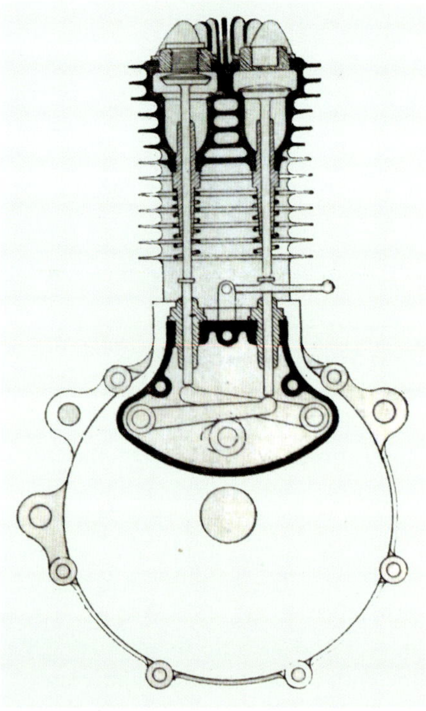

The 2¼hp (70 x 72mm) engine fitted to the 1905 Enfield motorcycles. The illustration shows the valve operation using rocking levers operated by cams.

belt or chain drive. Both used Enfield's, own vertical 277cc (70 x 72mm) single-cylinder engine in a diamond frame. The engine, which developed 2¼hp at 1,800rpm, was unusual in respect of the method of actuating the mechanically operated valves by rocking levers working on cams. The chain-drive model boasted the Enfield patented two-speed gear and clutch and weighed a modest 111lb. The belt-drive machine weighed only 90lb. Despite the need to reduce production costs and control prices, the machines were finished to Modèle Riche standard with attractive crimson lining and plated wheel rims.

The Motorcycle Doldrum Years

The over production of machines in previous years, as manufacturers, sought to enter or consolidate their position in the motorcycle market, combined with a fall in demand due to rising unemployment, had had a profound effect on the motorcycle industry. Manufacturers struggled to survive and many without more diverse interests to fall back on went out of business. The Enfield Cycle Company was fortunate as its business was founded on pedal cycle manufacture and the demand for pedal cycles remained firm and the pedal cycle business highly profitable. For 1906 the company decided to leave the unsettled motorcycle business for a period to concentrate on cars and bicycles. It would be 1910 before Royal Enfield returned to the motorcycle market.

Departure of Albert Eadie

At the annual meeting of the Enfield Cycle Company on 7th October 1907, Albert Eadie announced his departure from the Company. Albert had been the driving force behind the Eadie and Enfield companies, but the pressure of work had taken its toll on his health. He retained his shareholdings in the Enfield Cycle Company Ltd, but disposed of all other assets. Mr J.W.Davis was appointed as the new general manager.

CHAPTER FIVE

Model Year 1910

By 1908 the British motorcycle market was showing signs of recovery and the time was opportune for the Enfield Cycle Co. to consider restarting motorcycle manufacture. The disposal of the Enfield Autocar Company to Alldays & Onions in April 1908 had left the company's designer, the Swiss engineer Werner Guillon, seeking alternative employment. He was taken on by the Enfield Cycle Co. to work under R.W.Smith with the brief to design and develop a new Royal Enfield motorcycle. The Swiss company Motosacoche, through its enterprising London-based salesman, Osborne de Lissa, was at this time actively promoting its products including new v-twin engines. It is hardly surprising that these proved attractive to the Swiss engineer in drawing up his new motorcycle design. The early history of Motosacoche and the collaboration with the Enfield Cycle Company is summarised at the end of this chapter.

2¼hp Lightweights

The new Royal Enfield, designed by W.Guillon and R.W.Smith, was revealed to the public at the Stanley Show in November 1909. It was fitted with a Motosacoche 60-degree v-twin side-valve engine of 297cc (52mm x 70mm), rated at a nominal 2¼hp. Ignition was by high tension magneto and a spray type carburettor was fitted to a T-shaped inlet manifold located between the cylinders. The inlet manifold and exhaust pipes were held in position by a single bolt. The mechanically operated valves were located on the inside of the vee. It was

The 297cc v-twin Motosacoche engine fitted to the Royal Enfield Lightweight Motor Bicycles, Models 140 and 145, provided an ability to drive at between 4 and 35 mph and could easily obtain an average speed of 25mph,

The Royal Enfield 2¼-H.P. Lightweight Motor Bicycle.

MODEL No. 140.

SPECIFICATION.

FRAME.	19½in., of exceptionally strong design, fitted with foot rests and spring forks.
WHEELS.	26in. × 26in.
TYRES.	Continental Motor Cycle, 26in. × 2in.
MUDGUARDS.	Specially designed, extra strong and wide, with front extension and side shields.
CARRIER.	Exceptionally strong and light, fitted with large square tool bag, containing full kit of tools in leather roll-up.
STANDS.	Back and Front – the former fitted to fork ends, and when not in use secured to mudguard by efficient spring attachment.
NUMBER PLATES.	Back and Front of good size.
BRAKES.	Powerful front rim actuated from handle-bar—Rear pedal brake operating from lug on frame and acting on Belt rim.
SADDLE.	Lycett's L77 Motor Cycle Saddle.
FINISH.	Frame parts cozlettized and enamelled stone color or black to order, tank panelled and lined — front wheel plated, with enamelled centre, back wheel enamelled all over.
ENGINE.	Royal Enfield Twin Cylinder, "V" pattern, bore and stroke 52 × 70m/m producing 2¼ H.P. — mechanically operated valves (interchangeable).
IGNITION.	Magneto, shaft driven, most accessible, and protected from wet and mud by aluminium case.
CARBURETTOR.	Specially constructed and semi-automatic, very economical and easy to manipulate.
PETROL CONSUMPTION.	100 miles to gallon.
TANK.	Very strong—3 compartments (petrol capacity, 8½ pints—oil, 1½ pints—paraffin, ½ pint), fitted with oil and petrol gauges, inclined pumps and large fillers. All outlets for petrol, oil, and paraffin, are covered with fine mesh wire gauze.
TRANSMISSION.	⅝in. "V" Belt on deep section pulley—belt rim securely laced to wheel, ensuring perfect truth.

Weight—Completely equipped as above, about 100 lbs.

ROYAL ENFIELD LIGHTWEIGHT MOTOR BICYCLE.

Model No. 140.

Lightweight Motor Bicycle Model 145

The specification of Model 145 differed from that of Model 140 by the fitment of Dunlop studded tyres and a Brooks 130/3 saddle. It was priced at £42 compared with £39 18s for the Model 140. The standard colour of the machines was khaki.

The editor 'BICYCLING NEWS' wrote:- 'The Enfield Motor Bicycle is a gem of neat design - as dainty and complete little mount as one could wish for.'

Mr F.Sanders, Worcester, wrote:- 'I am delighted with my Enfield Motor Bicycle, it is thoroughly efficient and has never given me the slightest trouble.'

a tidy, well engineered design and built to a high standard.

The engine was fitted centrally, low down, within a loop-frame, with the magneto located on a bracket just behind the rear down tube. The frame top tube was attached to a lug at the bottom of the steering head and a strut was fitted between this tube and the top of the steering head to stiffen the steering head. The front fork was of the Druid design with side springs. The tank, containing compartments for petrol, oil and paraffin, had sight tubes set into the side to record levels. Engine lubrication was by hand pump. It was recommended after a day's ride to open a drain tap at the bottom of the crankcase to discharge the dirty oil and swill-out using paraffin. The paraffin was injected using an oil can and it was recommended that extra shots should be injected into the cylinders overnight, especially in cold weather, to assist starting.

The machines were single speed with direct drive to the rear wheel by a v-belt; the final drive ratio was 6.5:1. Pedals, with chain drive to the rear wheel, were provided for starting and to provide manual assistance on uphill stretches of road. Footrests were attached to a strut extending forward

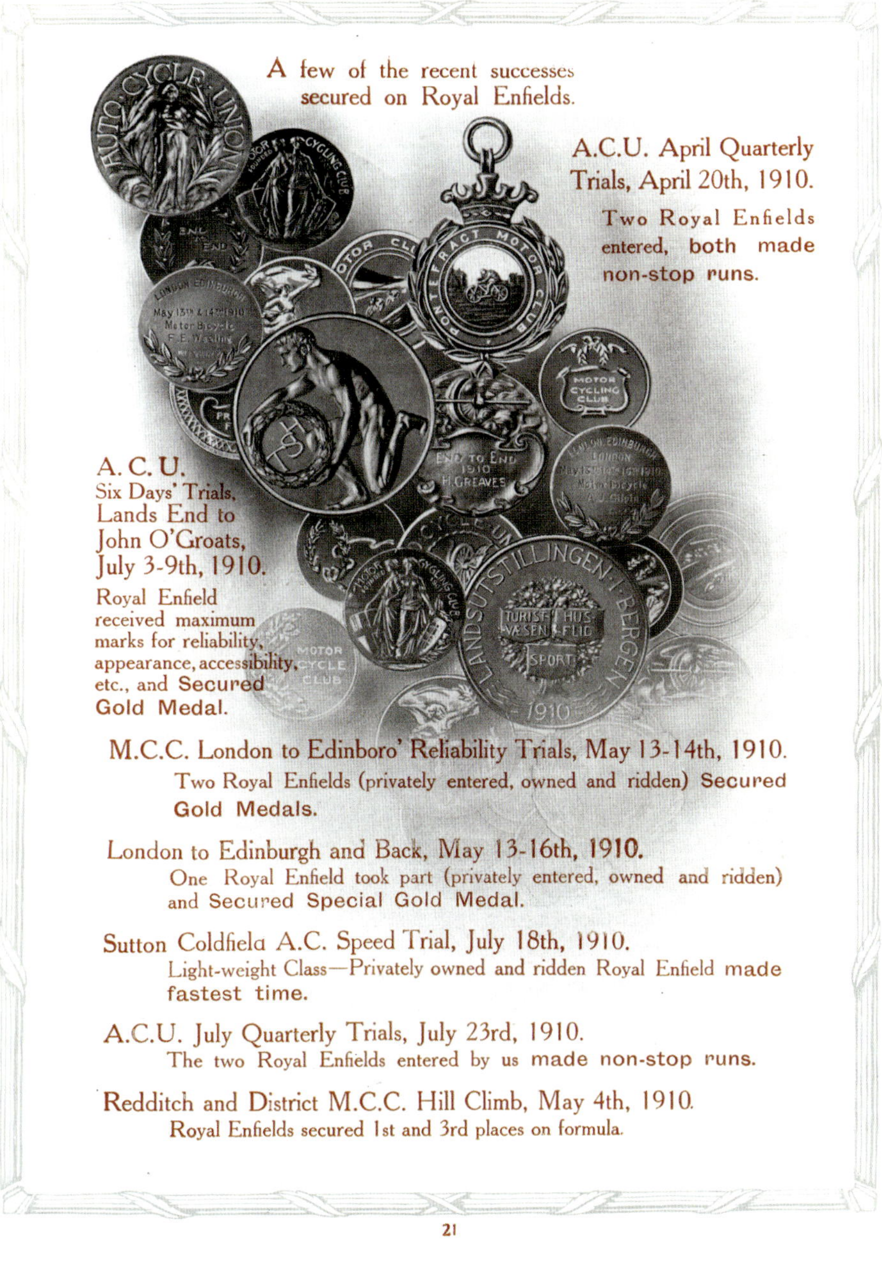

A few of the recent successes secured on Royal Enfields.

A.C.U. April Quarterly Trials, April 20th, 1910.
Two Royal Enfields entered, both made non-stop runs.

A. C. U. Six Days' Trials, Lands End to John O'Groats, July 3-9th, 1910.
Royal Enfield received maximum marks for reliability, appearance, accessibility, etc., and Secured Gold Medal.

M.C.C. London to Edinboro' Reliability Trials, May 13-14th, 1910.
Two Royal Enfields (privately entered, owned and ridden) Secured Gold Medals.

London to Edinburgh and Back, May 13-16th, 1910.
One Royal Enfield took part (privately entered, owned and ridden) and Secured Special Gold Medal.

Sutton Coldfield A.C. Speed Trial, July 18th, 1910.
Light-weight Class—Privately owned and ridden Royal Enfield made fastest time.

A.C.U. July Quarterly Trials, July 23rd, 1910.
The two Royal Enfields entered by us made non-stop runs.

Redditch and District M.C.C. Hill Climb, May 4th, 1910.
Royal Enfields secured 1st and 3rd places on formula.

21

RIGHT: The success of the Models 140 and 145 in the numerous reliability trials held throughout the year was recorded in the sales brochure for 1911.

from the front down tube and provided attachment for a front wheel stand. The 2¼hp machines were available as Model 140 or 145 according to the choice of tyres and saddle.

The Enfield engineers already appreciated that the future lay with chain, rather than belt, drive, and the 2¼hp 297cc v-twin would be the last belt drive Royal Enfield.

2¾hp Machines

The 2¼hp machines proved popular and obtained a number of awards in the popular trials of the day. However, greater power and variable gearing were required if sales were to be maintained. During the year W.Guillon and R.W.Smith began the development of a Royal Enfield with a larger 2¾hp Motosacoche v-twin engine with the option of a two-speed gear. The new 2¾hp was displayed alongside the 2¼hp machine at the Olympia Cycle and Motor Cycle Exhibition in November 1910. Both machines would be offered for sale to the public during 1911 with the smaller machine being phased out during the year.

Financial Statement

The financial results for the year ending 31st August 1910 showed a net profit of £6,728, compared with £12,458 for 1908-09 and for the first time since 1903 money was not transferred to the reserve account, which was currently standing at £33,000. The dividend on the ordinary shares remained at 5 per cent. Since registration in 1896 the ordinary shares had received a 10% dividend for 3 years, followed by 5% for 5 years and 6% for one year before rising to 10% for 1905-6 and 1906-7. Profits fell again during 1907-8 resulting in a 5% dividend before recovering in 1908-9 when a dividend of 6% was paid. Financial year 1909-10 was the company's worst since its registration, however, from this point the company would enjoy a period of sustained growth and profitability which can be attributed to the manufacture of motorcycles. A dividend of 7% had been maintained on the preference shares throughout.

Frank Smith

The eldest of R.W.Smith's four sons, Frank Smith was born on 7 August 1886. He was educated at Quinton College and Birmingham University before serving an apprenticeship with R.W.Ward and BSA. He joined Enfields in 1909 and became joint managing director of the Enfield Cycle Company alongside his father in 1914.

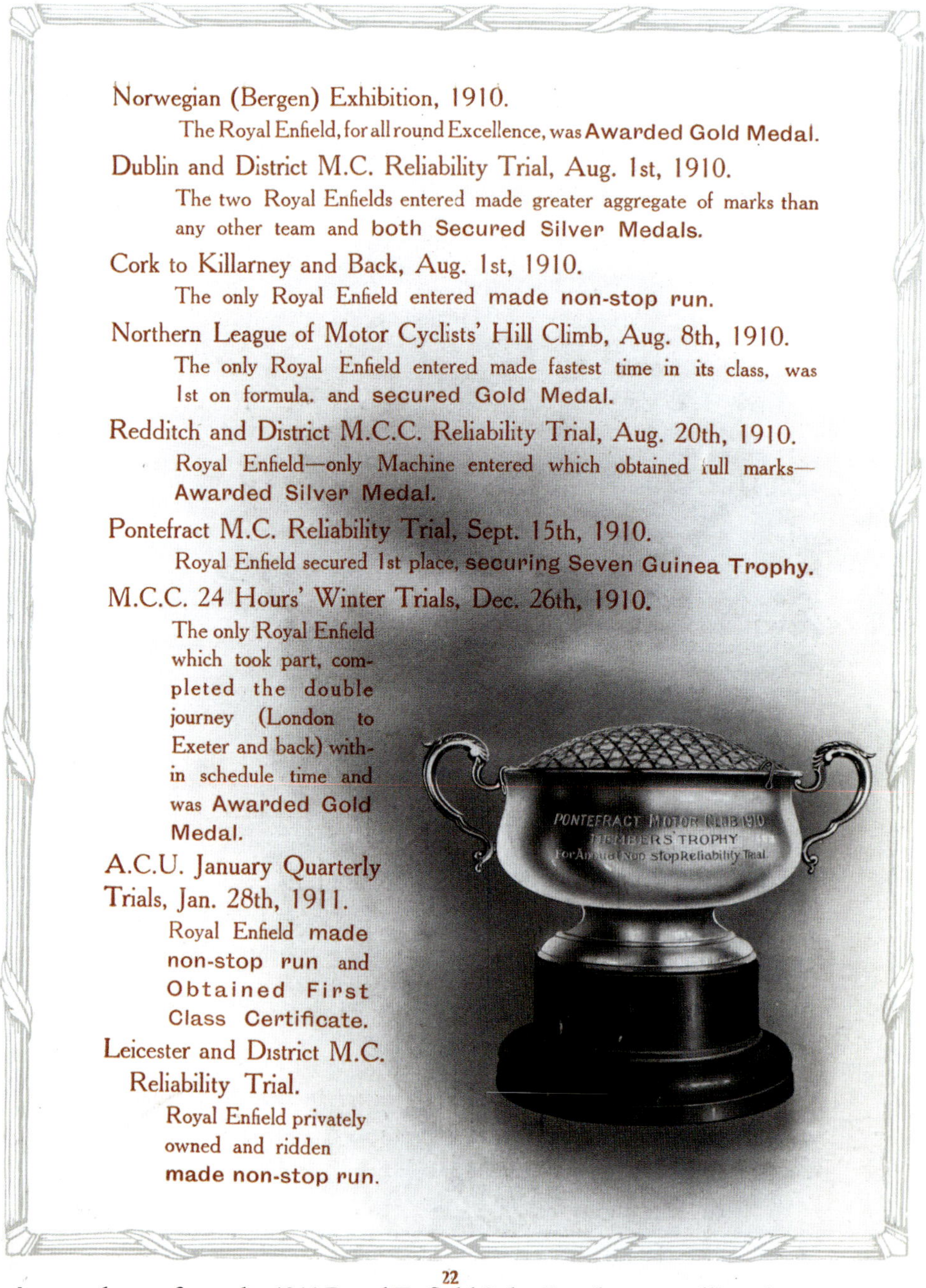

Norwegian (Bergen) Exhibition, 1910.
The Royal Enfield, for all round Excellence, was Awarded Gold Medal.

Dublin and District M.C. Reliability Trial, Aug. 1st, 1910.
The two Royal Enfields entered made greater aggregate of marks than any other team and both Secured Silver Medals.

Cork to Killarney and Back, Aug. 1st, 1910.
The only Royal Enfield entered made non-stop run.

Northern League of Motor Cyclists' Hill Climb, Aug. 8th, 1910.
The only Royal Enfield entered made fastest time in its class, was 1st on formula, and secured Gold Medal.

Redditch and District M.C.C. Reliability Trial, Aug. 20th, 1910.
Royal Enfield—only Machine entered which obtained full marks—Awarded Silver Medal.

Pontefract M.C. Reliability Trial, Sept. 15th, 1910.
Royal Enfield secured 1st place, securing Seven Guinea Trophy.

M.C.C. 24 Hours' Winter Trials, Dec. 26th, 1910.
The only Royal Enfield which took part, completed the double journey (London to Exeter and back) within schedule time and was Awarded Gold Medal.

A.C.U. January Quarterly Trials, Jan. 28th, 1911.
Royal Enfield made non-stop run and Obtained First Class Certificate.

Leicester and District M.C. Reliability Trial.
Royal Enfield privately owned and ridden made non-stop run.

22

A second page from the 1911 Royal Enfield Sales Brochure extolling the successes achieved by the Royal Enfield 2¼hp Lightweight Motor Bicycles during 1910.

Motosacoche

In 1899 the brothers Armand and Henry Dufaux designed a small four-stroke engine complete, with its own fuel system and exhaust attachment. They mounted the assembly within a triangular sub-frame that could be fitted directly within the frame of an ordinary pedal cycle. It was identified as 'une moto sa coche' or 'a motor in a bag,' and was sold under the name 'Motosacoche'. The capacity was 241cc, although this was subsequently increased to 290cc. The automatic inlet valve followed conventional design practice. The mechanical exhaust valve, however, was more unconventional being operated by a cranked lever actuated by eccentric grooves machined into a face of the flywheel.

A company, Motosacoche, was established in Rue Acacias, Geneva, Switzerland and the engines were sold under the name MAG, an abbreviation for Motosacoche, Acacias, Geneva. The engine package was a well engineered solution to motorising the pedal bicycle and quickly gained popularity on the Continent. In 1903 the firm H & A Dufaux (England) Ltd was established in London to market the Motosacoche in Great Britain. Two years later the Dufaux brothers departed from the company in order to concentrate on building airplanes.

The Motosacoche created some considerable interest in Great Britain, but few sales. The situation was not improved until 1908 and the arrival of the Australian Osborne Louis de Lissa. He had been employed as a draughtsman at the Rover Company and joined H & A Dufaux (England) Ltd as a travelling salesman. Such were de Lissa's marketing and sales skills that he quickly expanded the business. He lost little time in arranging for the manufacture of complete Motosacoche machines to be sold through the British base.

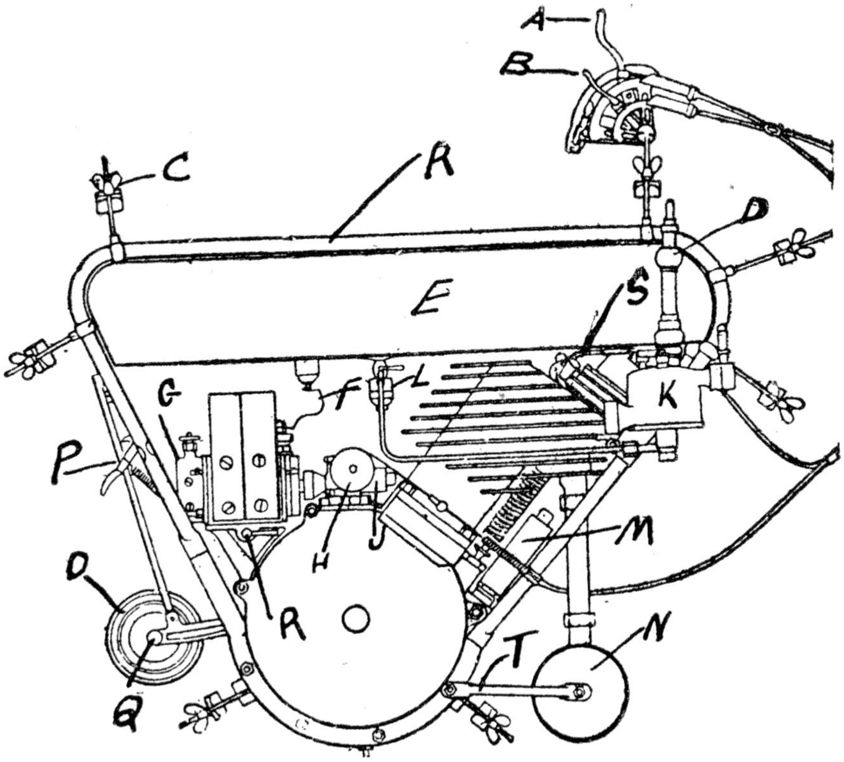

The general arrangement of the Motosacoche unit designed for fitment into a bicycle frame.

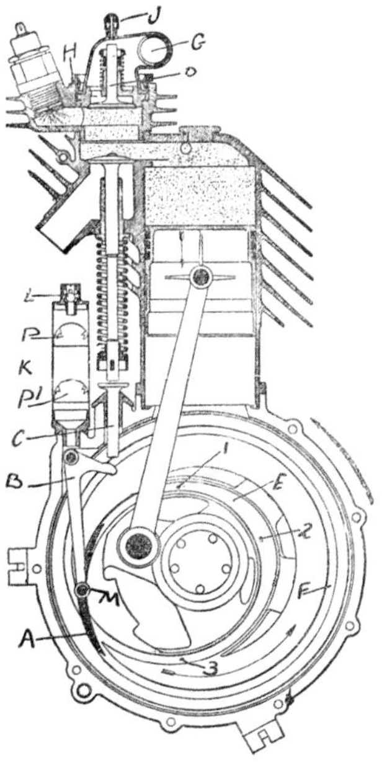

A section through the Motosacoche 2½hp engine of 1908. The engine dispensed with the usual half-speed shaft and gear wheels and operated the exhaust valve through two grooves cut into the flywheel. The inner groove, E, was made slightly eccentric and the outer groove, F, circular. The black crescent-shaped piece follows the grooves and operates the cam.

Although marketed as a Motosacoche, the first machines were built by the Rover bicycle company of Coventry thereby ensuring that the 'engine in a bag' was incorporated into a high quality purpose built bicycle. De Lissa was also able to persuade the Swiss company to diversify its range of engines and to commence the manufacture of complete machines. In 1912 H & A Dufaux was renamed Motosacoche (GB) Ltd with de Lissa in charge.

Gradually the engines became bigger and more powerful, twin cylinders were produced and the initial idea of a motorized pedal-bike was given up. The factory built complete motorbikes under the "Motosacoche" label, but kept on selling M.A.G. engines to many well known manufacturers in France, England, Germany, Austria and Italy. Factories were also established in France and Italy for the production of Motosacoche motorcycles under licence.

An illustration of the 1913 3½hp Motosacoche, reproduced from the sales brochure. There is a marked similarity with the Royal Enfield Models 150 and 160, which are illustrated on page 100. The Motosacoche is fitted with a larger 496cc (64 x 77 mm) engine than the 344cc Royal Enfield Models 150 and 160, but in other respects the machines appear identical.

Osborne de Silva would certainly have visited the Enfield Cycle Company to promote his company's wares. He claims to have influenced the company to recommence motorcycle production although it is likely that such a move was already planned. He would certainly have been influential in persuading the Enfield Cycle Company to fit MAG engines to Royal Enfield machines as the company's Swiss engineer, Werner Guillon, would have been very receptive to the use of a Swiss product. Very soon a collaboration was established with the Enfield Cycle Company for the supply of components to assemble a complete machine. A 423cc (62 x 70mm) v-twin machine exhibited on the Motosacoche stand at the 1908 Stanley Show is thought to have been constructed at the company's London base using Enfield cycle fittings. The Enfield Cycle Co. first exhibited their 2¼hp Lightweight machine with MAG engine at the 1909 Stanley Show and an identical machine, in Motosacoche livery, was exhibited at the show by the Dufaux company.

Guillon and Smith were credited with the design of the Enfield 2¼hp but it is not known to what extent de Lissa influenced their thinking. This early cooperation between the Enfield and Motosacoche appears to have extended through to 1918. There is a striking similarity between the Royal Enfield and Motosacoche v-twin motorcycles during this period, as witnessed by the two Motosacoche machines illustrated. It is thought that Royal Enfield supplied the Swiss firm with complete machines less engine, but including the Enfield patent two-speed gear. Confirmation for this conclusion is contained in the Enfield Cycle Company works' register for the 6hp twin, now held by the Vintage Motor Cycle Club The register suggests that machines were being dispatched to Motosacoche AG as late as April 1918. The register does not record the engine number, as would be normal practice, and it may be concluded that the JAP engine fitted by Enfields at this time was excluded from the transaction.

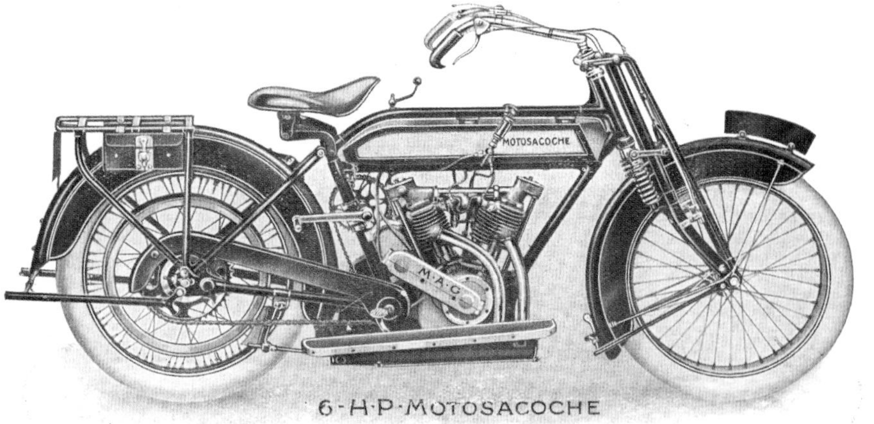

An illustration from the Motosacoche sales catalogue for 1915 of the 6hp Model.

CHAPTER SIX

Model Year 1911

Building on the success of the 1910 Lightweight 2¼hp models Monsieur Guillon developed an improved version with a larger Motosacoche engine and chain drive, which was unveiled at the Motor Cycle Show in November 1910.

Although the new Motosacoche engine had a similar specification to the earlier engine, it was a totally new design. Still a 60 degree side-valve v-twin with cooling fins cast to run parallel to the ground, the engine was now of 344cc capacity (54mm x 75mm) with a nominal rated output of 2¾hp. The side-valves were no longer located between the cylinders but in the front of the front cylinder and behind the rear cylinder. Fuel was supplied by a lightweight Brown and Barlow carburettor attached to a T-manifold between the cylinders. The high tension Bosch magneto, which was bevel driven from the timing gear, was located in an accessible position behind the engine. The exhaust pipes passed forward to a single silencer located forward of the engine. A foot operated exhaust cutout was provided to gain improved performance whilst riding through unpopulated areas.

1911 Model Range

Model 140	2¼hp	297cc	Belt-drive	£39 18s
Model 145	2¼hp	297cc	Belt-drive	£42
Model 150	2¾hp	344cc	Chain-drive single-speed	£43 1s.
Model 155	2¾hp	344cc	Chain-drive single-speed	£45 3s
Model 160	2¾hp	344cc	Chain-drive two-speed	£52 10s

LEFT: The 344cc v-twin Motosacoche engine fitted to Royal Enfield Lightweight Models. The bevel gear, shown on the right, is driven off the timing gear for the rear cylinder and connects, via a universal joint, to the high tension magneto.

Single-Speed 2¾hp Model 150

The frame was of loop design with two top tubes. The uppermost tube was curved upwards to attach directly to the top of the steering head, without use of a strut as with the Models 140 and 145. The design resulted in a very low build allowing the rider, when in the saddle, to place both feet on the ground. The front fork was a new Royal Enfield design, covered by patent No. 25522/10. It had a single central spring acting in tension when receiving a road shock and in compression upon recoil. It was claimed that the machine was at all times floating on the spring and that all road shocks were effectively absorbed. The Royal Enfield Spring Fork had been tested and shown to be both strong and light.

Drive was by chain rather than by v-belt, which was generally preferred by most manufacturers at this period. Chains removed all problems associated with belt stretching and slipping, but chains lacked the belt's

RIGHT: This catalogue is believed to date from late 1910 and illustrates both Models 140 and 150 together with the range of twenty-five different pedal cycles.

ROYAL ENFIELD
MADE LIKE A GUN

$2\frac{3}{4}$ H.P. MOTOR BICYCLE.

Model No. 150.

FRAME $19\frac{1}{2}$in., fitted with footrests, strong carrier, front and rear stands, and spring forks.

WHEELS 26 in. × 26 in.

ENGINE Royal Enfield Twin-cylinder, "V" pattern, $2\frac{3}{4}$ h.p.

BORE AND STROKE 54 × 75 m/m.

VALVES Mechanically operated.

IGNITION Magneto, shaft driven.

CARBURETTOR Brown & Barlow.

PETROL CONSUMPTION Approximately 100 miles to the gallon.

TYRES 26 in. × 2 in. Clincher Motor Cycle.

SADDLE Motor cycle, specially designed, with large seat and strong springs.

BRAKES Powerful front rim brake and rear foot brake actuating on belt rim.

TRANSMISSION Hans-Renold Chains.

PRICE - Net Cash - 41 Guineas.

Model No. 155.

As above, but equipped with Dunlop studded tyres and Brooks' 130/3 Saddle.

Price - Net Cash - 43 Guineas.

ability to absorb the engine's power fluctuations. Enfield had exhaustively studied the problem over the previous twelve months and had devised a slipping clutch mechanism to smooth the power output; the first development of a cush drive. The primary chain drove a countershaft located at the bottom bracket, which was mounted in an eccentric for chain adjustment. The secondary drive chain was adjusted by moving the rear wheel within slotted fork ends. Pedaling gear was provided for starting and a Royal Enfield Free Hub fitted to the back hub.

The fuel tank was similarly styled as the previous year's models with sight glasses for fuel and oil levels. The innovation was to attach it by two lugs brazed to the middle tube of the frame. The tank could now be constructed with the metal seam at the top and the possibility of leakage reduced.

In the absence of a belt rim, the foot brake operated on a special rim fitted to the right hand side of the back wheel. The front stirrup brake was hand operated. The hand controls, together with the handlebars clearly demonstrated Enfield's attention to detail. The two arms of the handlebars were attached to a lug at the steering stem and swept back to be in easy reach of the rider. Inverted hand levers were used for both the front brake and exhaust valve-lifter, with the control cables passing through the handlebars as far as the steering head. A particularly neat fitting on the right handlebar was used to mount the two carburettor levers and the cables again routed through the handlebar. The ignition advance/retard lever was placed on the right hand side of the tank.

TOP LEFT: Specification of the Models 150 and 155 reproduced from the early 1911 season's sales catalogue.

LEFT: The Models 150 and 155 illustrated in the early catalogue . On later models the inlet manifolds and exhaust arrangements would be altered and the enclosed secondary chaincase replaced by an open design.

Single Speed Chain Drive 2¾ hp Royal Enfield Lightweight Model 150. The early illustrations show a full rear chain guard.

The late 1911 season sales catalogue listed only the 2¾hp Models 150 and

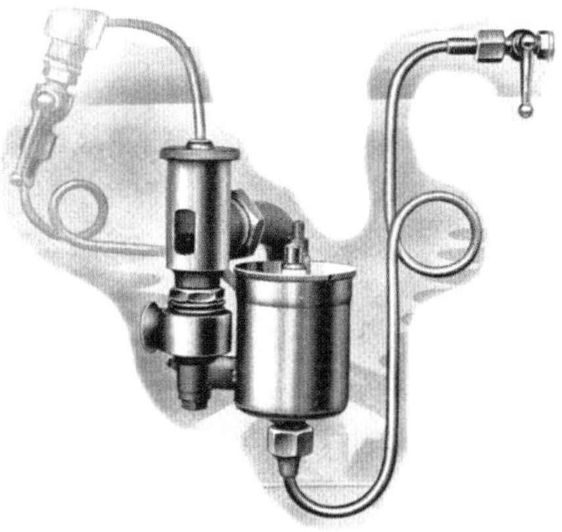

The Brown and Barlow Lightweight carburettor. The hot air inlet at the bottom of the mixing block is turned to receive air from the front cylinder. In hot weather it can be rotated to receive cooler air.

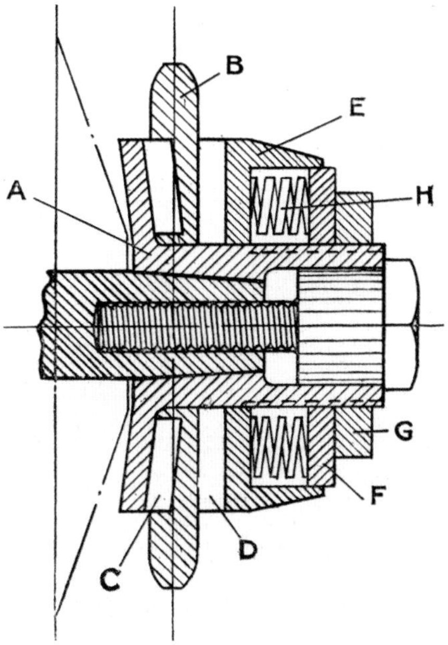

Chain Sprocket and Slipping Clutch.

A Bush mounted on engine shaft.
B Engine sprocket -free to rotate.
C & D Friction washers.
E Spring box containing 12 springs.
F Washer.
G Nut

The spring box E and washer F are keyed to slide on bush A but not rotate. The spring compression is adjusted such that slippage between the sprocket B and the friction washers is sufficient to absorb engine shocks.

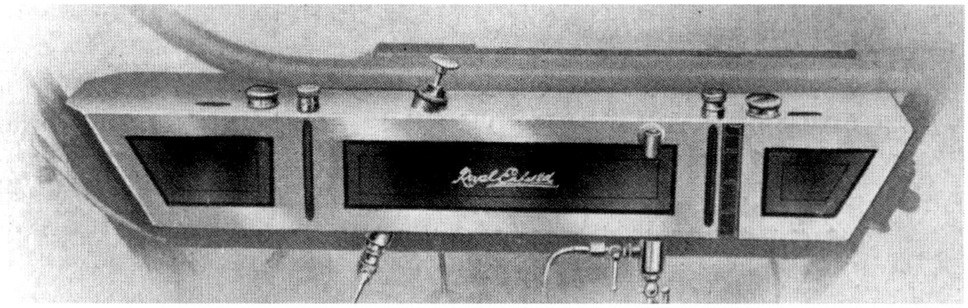

The fuel tank introduced for 1911 models

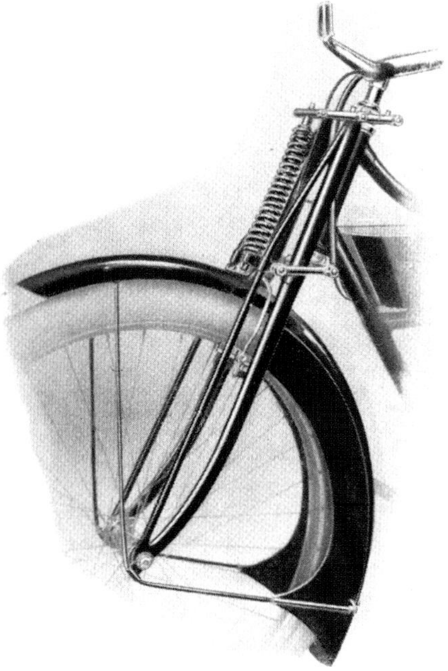

The Royal Enfield patented spring fork. Elongated fork link plates locate the top anchorage yoke for the spring forward of the fork pivot. Thus the fork spring acts in tension when responding to road shocks. Both front and rear mudguards were well valanced.

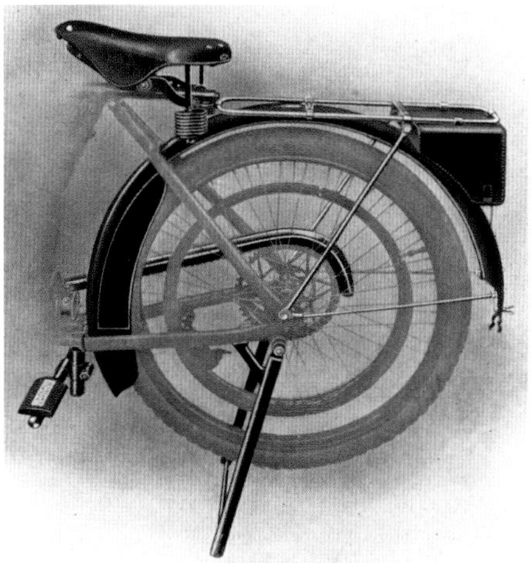

The rear carrier and toolbox. The rear mudguard was well valanced for rider protection. The illustration shows the position of the foot rests for the Model 160.

The machine weighed approximately 120 pounds. It was painted using khaki enamel and the bright parts nickel plated.

Single-Speed 2¾hp Model 155

The Models 150 and 155 were identical in all respects other than the tyres, the Model 150 was fitted with Dunlop tyres whereas the Model 155 used Clincher tyres, and benefited from a Brooks 130/3 saddle. At £45 3s, the Model 155 was guineas more expensive than the Model 150.

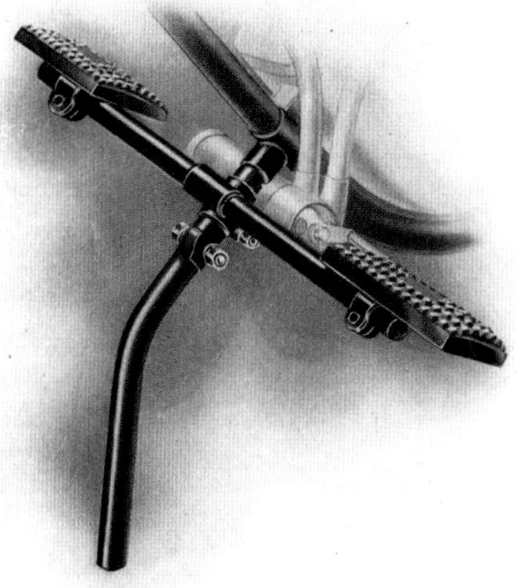

The arrangement of the foot pedals. Also shown is the front stand in the lowered position.

Two-Speed 2¾hp Model 160

Also introduced in November 1910, at the same time as the single speed Models 150 and 155, was a two-speed and free engine-version, the Model 160. The advantages of the new two-speed gear model were spelt out in the brochure:

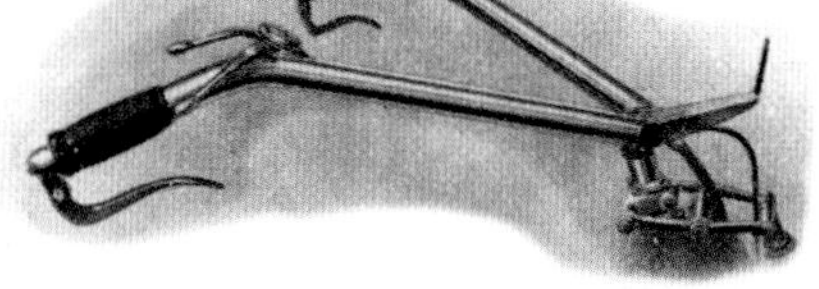

The stylish handlebar arrangement, which would remain a feature of most Royal Enfield models produced through to the mid 1920s.

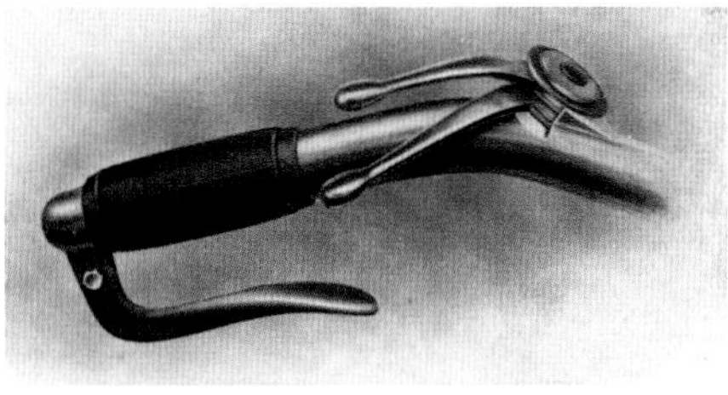

The hand brake and carburettor controls. A tidy appearance was created by routing the Bowden cables operated by these levers, plus that of the left-hand decompressor lever, through the handlebars to emerge at the steering head.

Engine and two-speed gear arrangement of the Model 160. Neat chain cases, split horizontally, concealed the chains and protected them from dirt and wet.

The advantages of a reliable two-speed and free engine device are obvious to those with only the smallest experience of motor cycling.
To be able to start the engine with a single turn of a handle as with car practice, sit astride one's stationary mount, slip in the clutch and glide away at 4 miles an hour, picking up to forty almost at once, if desired, is a boon appreciated to the full by all who have experienced it.
We claim that our two-speed and free-engine device is absolutely reliable, simple, efficient and silent, and adds considerable to the hill climbing powers and flexibility of the machine.

The general specification of the Model 160 was as the same as the models 150 and 155. The addition of the two-speed gear increased the weight of the machine by approximately 5 pounds and increased its price by 7 guineas (£7 7s). The machine was enamelled navy grey.

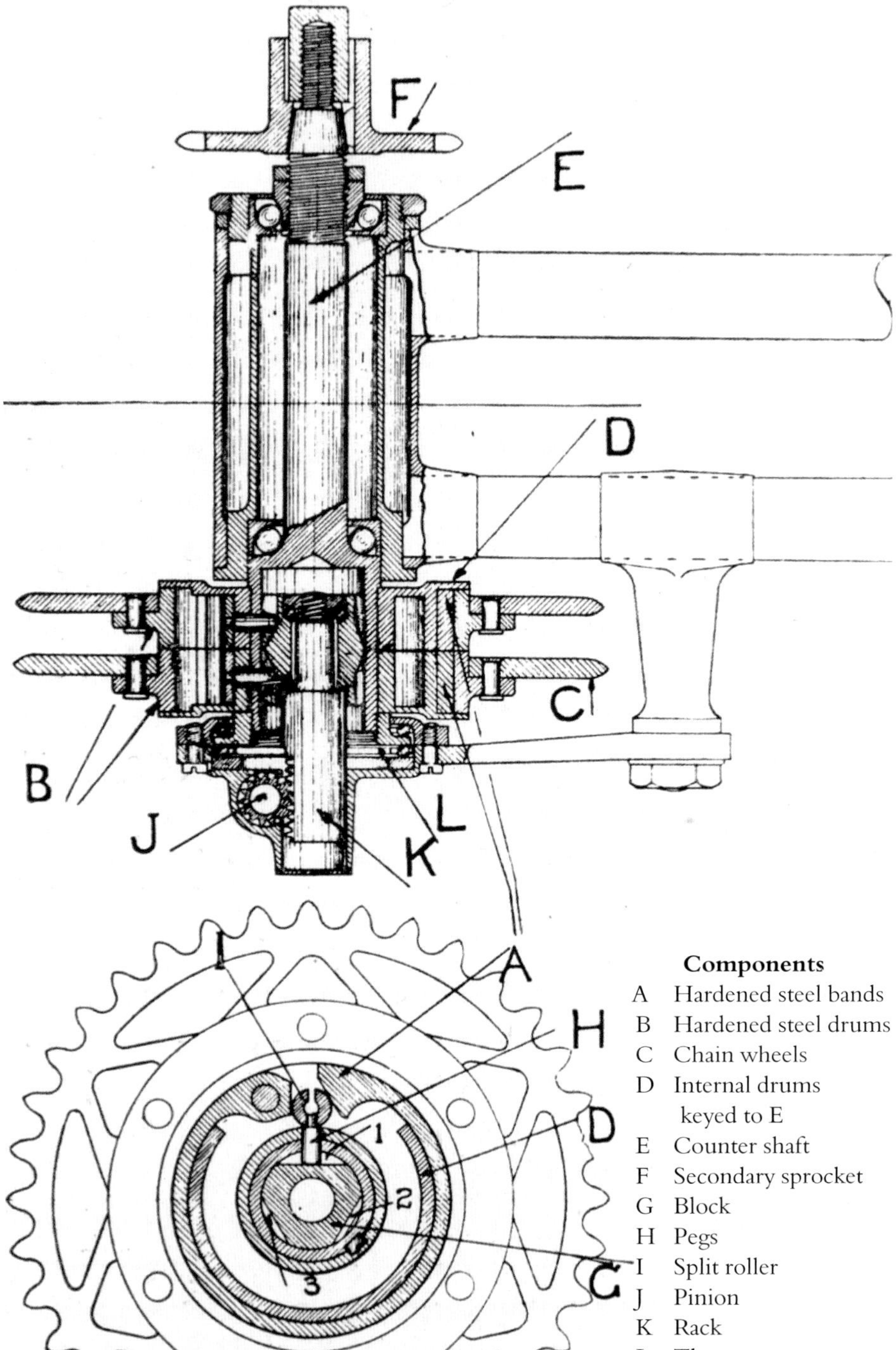

Components
A Hardened steel bands
B Hardened steel drums
C Chain wheels
D Internal drums keyed to E
E Counter shaft
F Secondary sprocket
G Block
H Pegs
I Split roller
J Pinion
K Rack
L Thrust race

The Royal Enfield Patent Two-speed Gear and Free-engine.

The Two-Speed Gear Mechanism

The two-speed gear employed two primary chains, one for each gear, running on drive sprockets keyed to the engine shaft which transmit power via unequal sized chain wheels running on a counter shaft in the bottom bracket. The secondary drive to the rear wheel was by chain on the opposite side of the machine.

The free engine position is obtained when both of the driven chain wheels rotate freely on the counter shaft. Either one or the other of the two gears is obtained according to which of the driven chain wheels is locked to the counter shaft whilst the other remains free to rotate.

The mechanism of the gear selection may be understood from the diagram. The hand lever is used to turn pinion J, which in turn moves the rack J and forces the block G containing a cam in either direction according to the gear required. The movement of the cam is to force one of the pegs H against a split roller I. This forces open the band A so that it engages with the hardened steel drum B, which is rotated by the engine, to obtain drive. The roller I being slit allows a smooth clutch action as the band A picks up the drive.

Three separate pairs of cams were cut into the block G. These were of different heights and selected to take up any wear that occurred.

2¼hp Models 140 and 145

The 297cc single-speed belt-drive models were offered alongside the new 2¾hp chain-drive machines at the start of the year. The specification appears to have been unchanged except for use

Two Speed and Free Engine 2¾ h.p., Royal Enfield Lightweight.

Specification—Model No. 160.

ENGINE - - - -	Royal Enfield Twin Cylinder "V" Pattern, bore and stroke 54 × 75 m/m, producing 2¾-h.p.—mechanically operated inlet valves, adjustable tappets, large silencer with cut-out. Lodge plugs (British made.)
TWO SPEED & FREE ENGINE MECHANISM - -	Royal Enfield exclusive design (patents applied for); internal clutch type, handle starting.
TRANSMISSION -	Hans Renold Chain.
IGNITION - - -	Bosch Magneto, shaft driven, waterproof terminals.
CARBURETTOR -	Brown & Barlow Lightweight.
TANK - - - - -	Two compartments—capacity: petrol 1 gallon; oil, 1 quart.
PETROL CONSUMPTION -	About 100 miles to gallon.
FRAME - - - -	19½in., long wheel base, exceptionally strong, fitted with two pairs of foot rests and spring forks (registered design).
TYRES - - - -	Dunlop Studded 26in. × 2in.
MUDGUARDS -	Strong and wide, plain section, with front extension and side shields.
CARRIER - - -	Very light, made of specially strong steel tube.
STANDS - - - -	For Front and Back Wheels.
HANDLEBAR - -	Shaped to give comfortable riding position—carburettor control and lamp bracket integral with bar; concealed cables.
BRAKES - - - -	Powerful front rim actuated from handlebar, and rear pedal brake.
SADDLE - - - -	Brook's 130/3.
TOOL BAG - - -	Strong, conveniently placed beneath carrier, full set of tools in roll-up.
FINISH - - - -	Enamelled Naval Grey; usual parts heavily plated.
WEIGHT - - - -	Approximately 125 lbs.

Price, net cash, 50 Guineas.

2¼hp 297cc V-twin Models 140 and 145

Engine: 60° Side-valve Twin
Capacity: 297cc
Bore x stroke: 52mm x 70mm
Carburettor: AMAC multiple jet
Ignition: Bosch magneto
Frame: Single tube loop, Enfield forks
Transmission: Single speed, V-belt
Tank: 8½ pints petrol, 1½ pints oil, ½ pint paraffin
Suspension: RE spring front fork
Tyres: 26 x 2in
Model 140, Continental
Model 145, Dunlop
Brakes: Front stirrup,
Rear dummy rim
Saddle: Model 140, Lycett L77
Model 145, Brooks 130/3
Weight: 100lbs
Finish: Khaki, panelled tank
Price: Model 140, £39 18s
Model 145. £42

of a revised tool box as fitted to the 2¾hp models. The prices were unchanged.

With the introduction of their chain-drive models, Enfield had rendered obsolete overnight the belt drive machines, with their inherent problems of belt stretch and slipping. The reasons to retain the older models are not clear, but it is likely that the company was being cautious. Motorcyclists at the time would not buy into new technical innovations until proven in the various trials and speed events of the day. Aware of this reluctance, Enfield Cycle Company offered both chain- and belt- drive machines whilst chain drive received acceptance. As soon as the merits of the chain-drive models had been proven and demonstrated worthy of the higher price, the company dropped production of the belt-drive machines.

Equipment and Accessories

Royal Enfield machines were supplied with a tool kit, suitable for maintenance and servicing, contained in a tool roll. Horn and lighting equipment were not fitted as standard, but available as optional extras. The handsome curled horn was available in two sizes at 7s 6d and 10s 6d. It was declared to provide a deep and penetrating tone and to be an effective road clearer. The Royal Enfield acetylene lighting set was priced at 30s inclusive of the generator. Other accessories included a front mudguard mud flap at 10s 6d, a storage wallet to fit between the curve of the top frame tube and the tank for 5s and Royal Enfield Cylinder Oil at 1s 6d per quart tin.

Leather wallet to carry small items in the space between the curve of the frame top tube and the tank.

The Enfield Cycle Company employed several very talented and innovative design engineers able to solve a wide range of problems, including as here a cricket ball bowling machine.

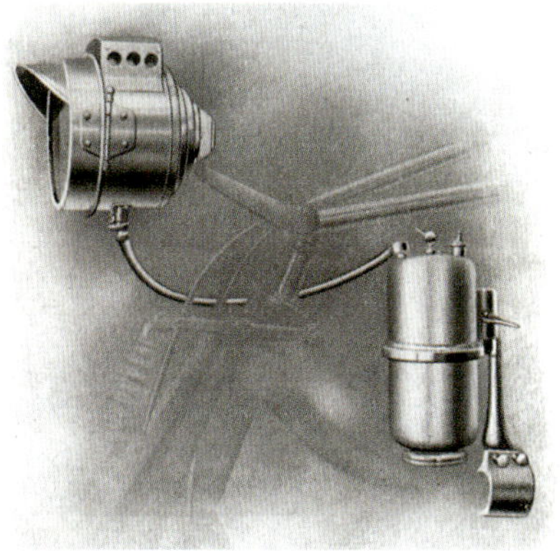

The Royal Enfield headlamp and acetylene generator set.

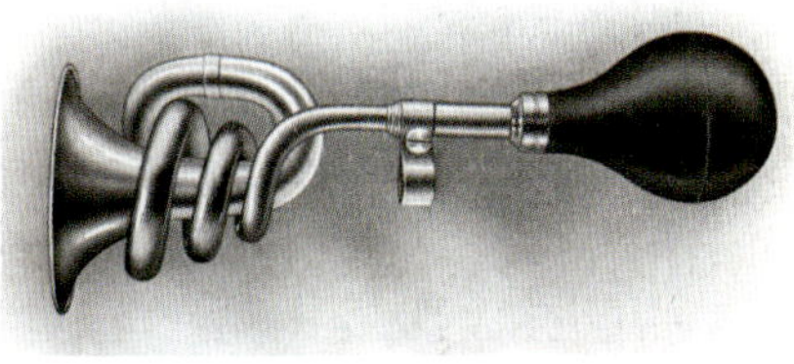

The Royal Enfield motor horn.

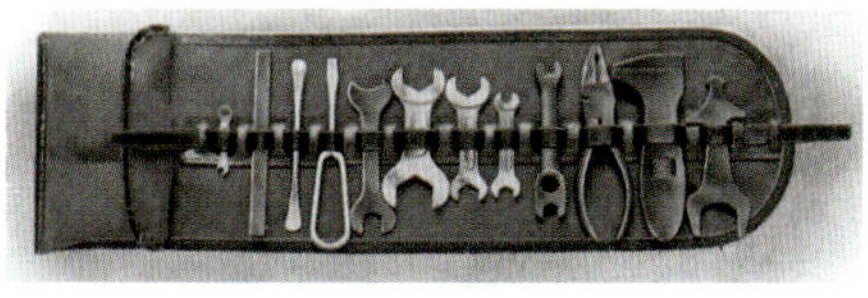

The standard tool roll. A chain rivet extractor was available for an additional 3 s 6d.

Bert Colver and Racing Successes

H.V. (Bert) Colver had joined Harry and Charlie Collier of Matchless in 1904 to form part of the Matchless 'works' racing team. The three men provided a complementary mix of talents with Harry carrying out the design, Charlie having the best riding skills and Bert having a talent for tuning. As a consequence the team proved highly successful gaining numerous awards in TT and track events. From 1910, with the rebirth of the Royal Enfield motorcycle brand, the Enfield Cycle Co would have been looking for a suitable competition rider to promote their range of Motosacoche powered machines. Osborne de Lissa, manager of the London branch of Motosacoche, had become acquainted with Colver at Brooklands and, aware of the Enfield requirement, persuaded Colver to race the new 2¾hp Royal Enfield during 1911.

Bert Colver's first race appearance on a 2¾hp Twin was at a British Motor Cycle Racing Club's race meeting on the 7th June 1911 in the Junior Brooklands One-Hour TT. This was a scratch race for machines eligible for the new Junior category in the Isle of Man TT, which restricted capacities for singles to 300cc although twins up to 340cc were allowed owing to their perceived greater unreliability. Bert made a good start and was in third place at the end of the first lap, but his near standard production machine was soon outclassed by the more highly tuned race machines. The race was won by Oliver Godfrey at a new record distance for the hour of 54 miles 726 yards. Bert was forced to stop on his eighth lap with a sooted up sparking plug. He was further delayed by a puncture on the eleventh, but nevertheless covered a creditable distance of 44 miles within the hour.

In 1911 the Isle of Man TT was held for the first time over the demanding 37½ mile Snaefell or Mountain Circuit; the race being over four laps of the circuit, a total distance of 150 miles. The Enfield Cycle Company entered both Bert Colver and H Greaves on 2¾ hp twins. Colver was forced to retire, but Greaves, who had previously found success in long distance trials, including the ACU's Land's End to John o'Groat's in July 1910, did well to achieve fifth place.

Colver's 2¾hp Twin was a very different machine when it reappeared at Brooklands on the 26 August 1911. No longer a standard production machine, it had benefitted from Bert's tuning abilities and had been modified for track use. He achieved the first ever Brooklands victory for Royal Enfield covering 54m 122yds in the hour, just 604yds short of the 350cc hour record. He returned to Brooklands in September for a Junior One-Hour event, but despite a promising start had to settle for fourth place.

Bert Colver's 2¾hp machine as it appeared on its return to the Brooklands track for the BMCRC race meeting on 26 August 1911. The magneto is positioned above the engine in the gap created between the tanks. Rigid forks were were popular for track racing, but would have resulted in an uncomfortable ride.

Lieut. Harry T.Bennett, R.N., of HMS Triton wrote to Motor Cycling: 'Having ridden an Enfield Motor Cycle since June and having completed nearly 3,000 miles without any mechanical failure, I feel I am justified in making a few remarks as to my experience with this machine. I have climbed many a hill without pedalling on the Enfield that has baulked me on a 3½hp machine. For the novice there is nothing to think about except to oil the machine every six or seven miles and prevent exceeding the speed limit. On several occasions I have reached a speed of 42½ miles by Cowey Speedometer on the level, and have been able to ride so slowly that a person could walk alongside the machine.'

Bicycling News reported: 'This lightweight Motor Bicycle (the Royal Enfield) is becoming more and more popular the longer and better it is known, and one of our staff, who has had much riding on it and a lot of experience of other makes during the last seven years, declares it to be one of the best and handiest machines upon the market at the present time . . . . there certainly is no machine that can be more deftly managed at times of emergency than the Royal Enfield. The fame it has gained upon the road has been considerably enhanced by its success in all sorts of tests and trials in competition.'

CHAPTER SEVEN

Model Year 1912

The Royal Enfield range for 1912, announced at the Olympia Show held in November 1911, consisted of three basic models. The 2¾hp lightweight solos, Models 150 and 160, introduced twelve months previously were retained. They were joined by an open framed model powered by a 2½hp single-cylinder engine, and a 6hp v-twin sidecar combination.

All models, with the exception of the single-speed model 150 were fitted with the Royal Enfield two-speed gears, clutches and sprung footboards instead of pedals. The new open frame 2½hp machine was intended for use by ladies and professional gentlemen.

1912 Model Range

Model 150	2¾hp	344cc	Chain-drive single-speed	£43 1s.
Model 160	2¾hp	344cc	Chain-drive two-speed	£52 10s
Model 170	2½hp	241cc	Open-frame two-speed	£52 10s
Model 180	6hp	771cc	Sidecar two-speed	£84

The 1912 Motor Cycles Brochure was a lavish production of twenty eight pages. In addition to detailed specifications of the various models, it gave illustrations and explained the operation of the various components. Three pages were devoted to listing recent sporting successes on Royal Enfield machines, with further pages devoted to press reports and testimonials from satisfied owners.

Royal Enfield Works, Redditch. Illustration from the 1912 Sales Brochure.

Royal Enfield Cush Drive

The new models were fitted with the Royal Enfield Patent Cush Drive in the rear hub, which it was claimed ensured 'perfectly smooth and sweet running, longer engine life, and reduces tyre wear to a minimum'. The driving sprocket was made with a cup on its inside into which were set three metal vanes. A cap, fastened to the wheel, also incorporated three metal vanes. This fitted within the sprocket wheel so that all six vanes were on a common plane and radiating out from a common axis. The spaces between the vanes were filled with rubber blocks or buffers to serve as a cushioning device. The power was transmitted through three of the buffers and the shock of any recoil was taken by the other three buffers.

The cush drive proved highly effective and did much to help popularise chain drive. Enfield-made cush drives and wheel hubs would be used by numerous motorcycle manufacturers.

The Royal Enfield Patent Cush Drive. The sprocket is shown on the left and the cup, which was fastened to the wheel hub, and rubber buffers is on the right.

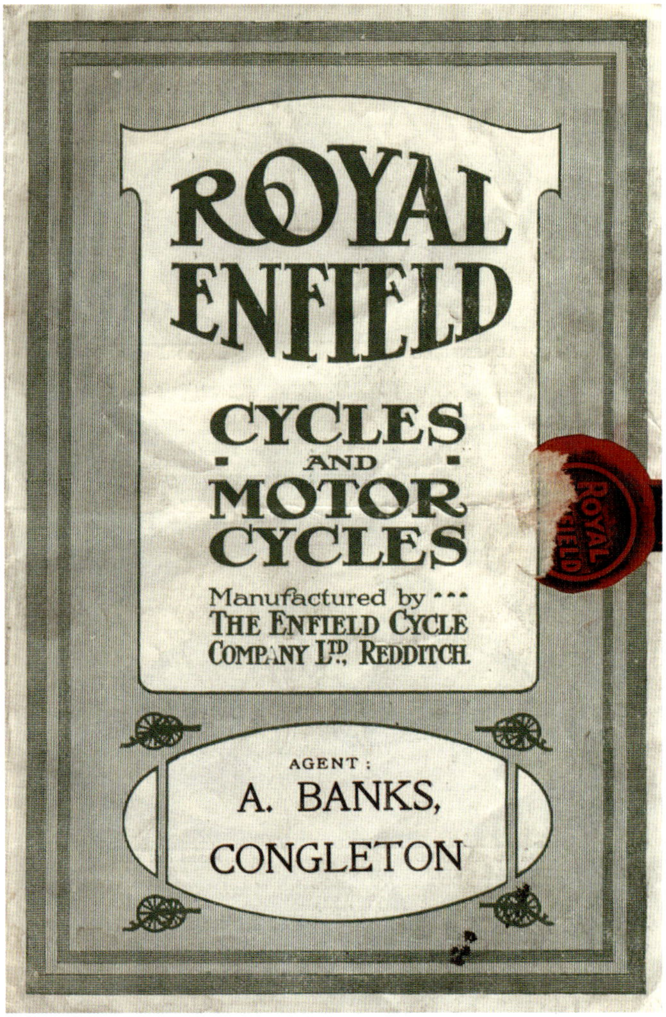

During 1912, the Enfield Cycle Company also produced a sales leaflet to promote the combined range of Royal Enfield motorcycles and bicycles. Nineteen different bicycles and four motorcycle models were described.

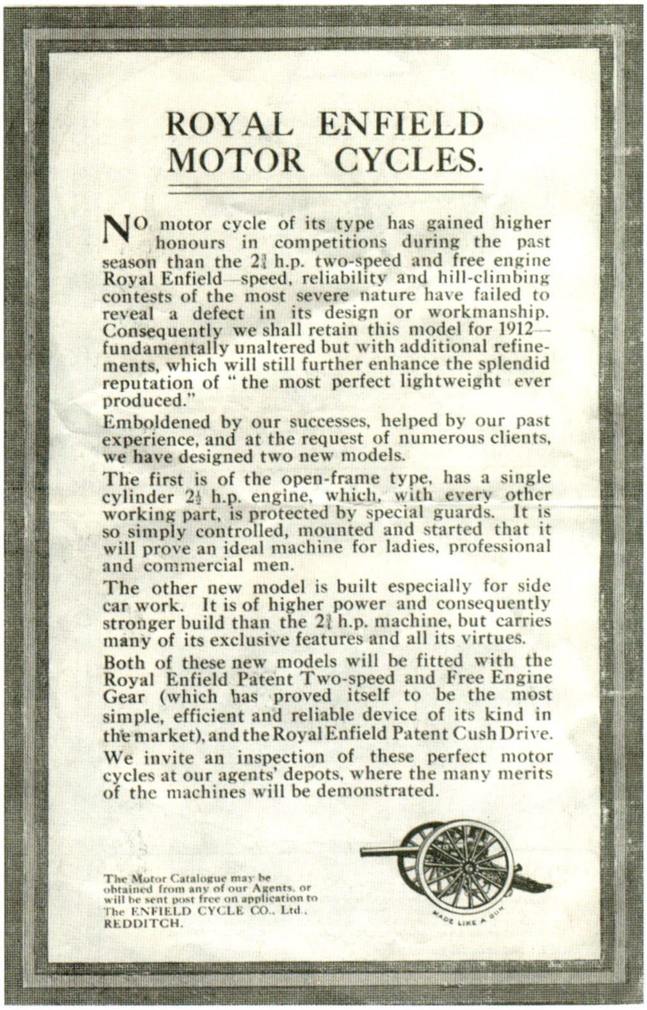

ROYAL ENFIELD
MOTOR CYCLES.

NO motor cycle of its type has gained higher honours in competitions during the past season than the 2¾ h.p. two-speed and free engine Royal Enfield—speed, reliability and hill-climbing contests of the most severe nature have failed to reveal a defect in its design or workmanship. Consequently we shall retain this model for 1912—fundamentally unaltered but with additional refinements, which will still further enhance the splendid reputation of "the most perfect lightweight ever produced."

Emboldened by our successes, helped by our past experience, and at the request of numerous clients, we have designed two new models.

The first is of the open-frame type, has a single cylinder 2¼ h.p. engine, which, with every other working part, is protected by special guards. It is so simply controlled, mounted and started that it will prove an ideal machine for ladies, professional and commercial men.

The other new model is built especially for side car work. It is of higher power and consequently stronger build than the 2¾ h.p. machine, but carries many of its exclusive features and all its virtues.

Both of these new models will be fitted with the Royal Enfield Patent Two-speed and Free Engine Gear (which has proved itself to be the most simple, efficient and reliable device of its kind in the market), and the Royal Enfield Patent Cush Drive.

We invite an inspection of these perfect motor cycles at our agents' depots, where the many merits of the machines will be demonstrated.

The Motor Catalogue may be obtained from any of our Agents, or will be sent post free on application to The ENFIELD CYCLE CO., Ltd., REDDITCH.

MADE LIKE A GUN

The virtues of the Royal Enfield motor cycles and bicycles were proclaimed in the combined sales leaflet. The page summarising the motorcycles makes clear, perhaps with justification, that the 2¾ hp two-speed and free-engine model was the 'most perfect lightweight ever produced'.

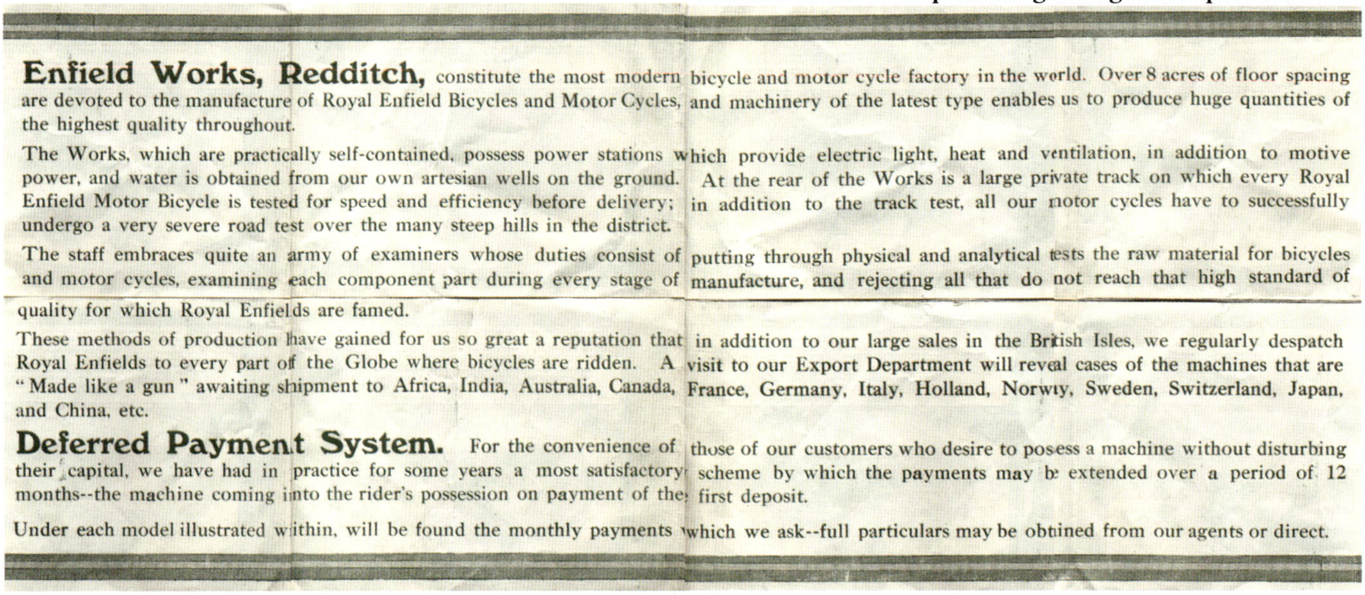

Enfield Works, Redditch, constitute the most modern bicycle and motor cycle factory in the world. Over 8 acres of floor spacing are devoted to the manufacture of Royal Enfield Bicycles and Motor Cycles, and machinery of the latest type enables us to produce huge quantities of the highest quality throughout.

The Works, which are practically self-contained, possess power stations which provide electric light, heat and ventilation, in addition to motive power, and water is obtained from our own artesian wells on the ground. At the rear of the Works is a large private track on which every Royal Enfield Motor Bicycle is tested for speed and efficiency before delivery; in addition to the track test, all our motor cycles have to successfully undergo a very severe road test over the many steep hills in the district.

The staff embraces quite an army of examiners whose duties consist of putting through physical and analytical tests the raw material for bicycles and motor cycles, examining each component part during every stage of manufacture, and rejecting all that do not reach that high standard of quality for which Royal Enfields are famed.

These methods of production have gained for us so great a reputation that in addition to our large sales in the British Isles, we regularly despatch Royal Enfields to every part of the Globe where bicycles are ridden. A visit to our Export Department will reveal cases of the machines that are "Made like a gun" awaiting shipment to Africa, India, Australia, Canada, France, Germany, Italy, Holland, Norway, Sweden, Switzerland, Japan, and China, etc.

Deferred Payment System. For the convenience of those of our customers who desire to posess a machine without disturbing their capital, we have had in practice for some years a most satisfactory scheme by which the payments may be extended over a period of 12 months--the machine coming into the rider's possession on payment of the first deposit.

Under each model illustrated within, will be found the monthly payments which we ask--full particulars may be obtained from our agents or direct.

Several pages of the sales leaflet were devoted to praising the Enfield Works, which at the time was as advanced as any in the world, and the quality of the Enfield product.

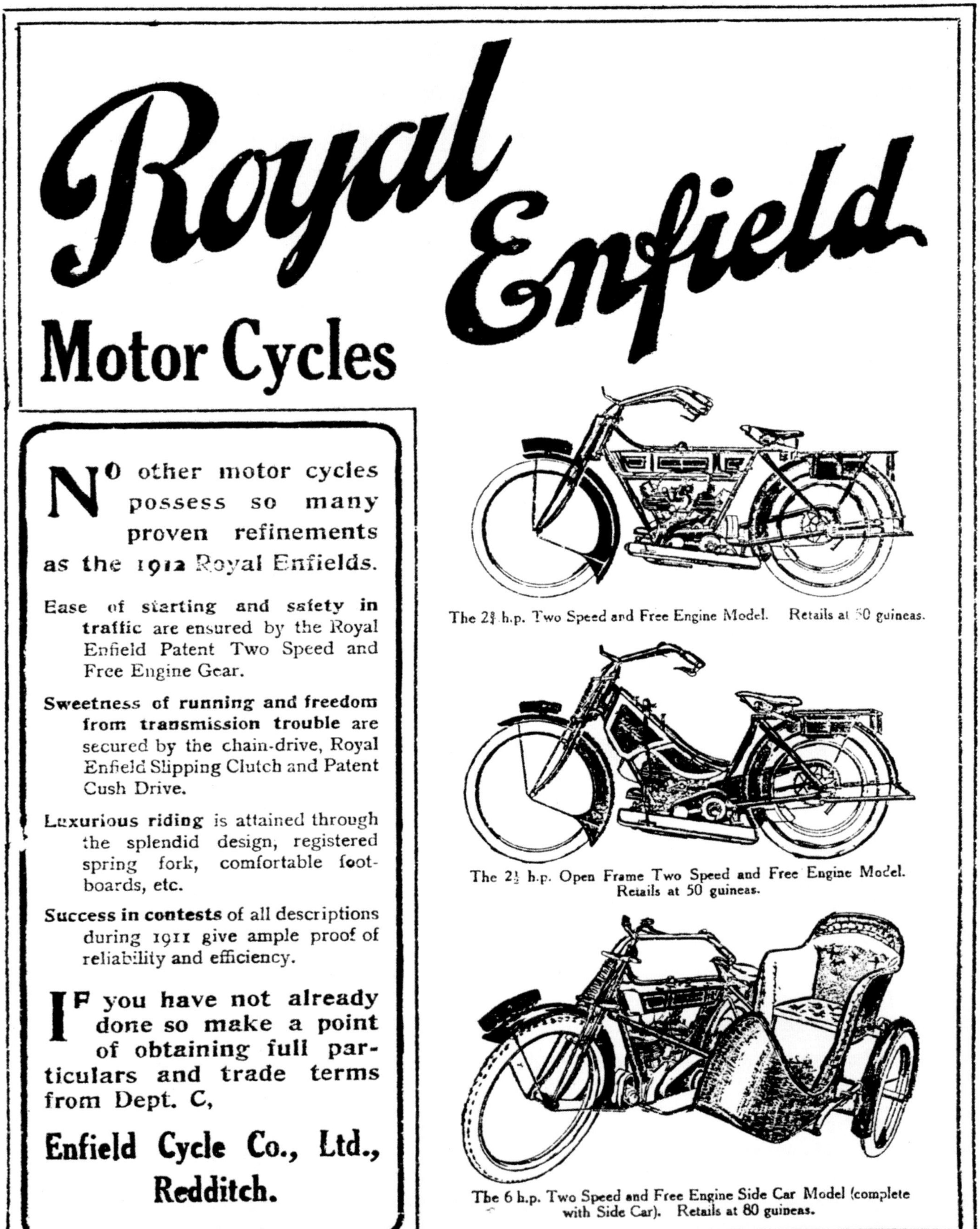

Full page advertisement published in The Cycle and Motor Cycle Trader in December 1911 for the new 1912 range of Royal Enfield motorcycles.

2 ¾ hp Lightweights

General Specification
Engine: 344cc side-valve
Bore and stroke: 54 x 75mm
Lubrication: automatic drip
Ignition: Bosch magneto
Frame: Royal Enfield loop design
Front fork: Royal Enfield
Handlebar: Royal Enfield patent
Tank capacities: petrol 1 gallon
oil 1 quart
Brakes: Front rim, hand operated
Rear dummy rim, pedal operated
Tool bags: Two leather pannier bags containing set of tools

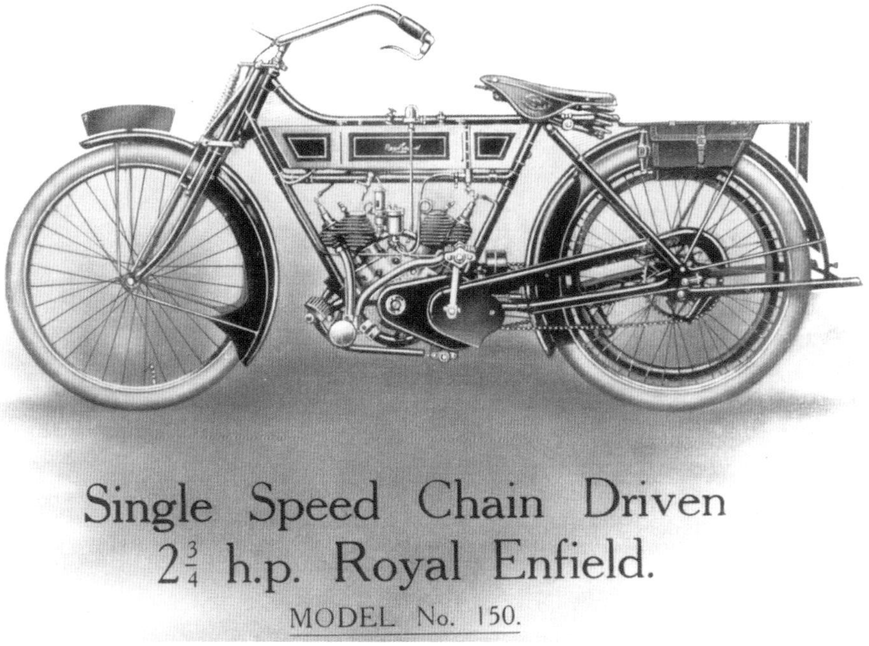

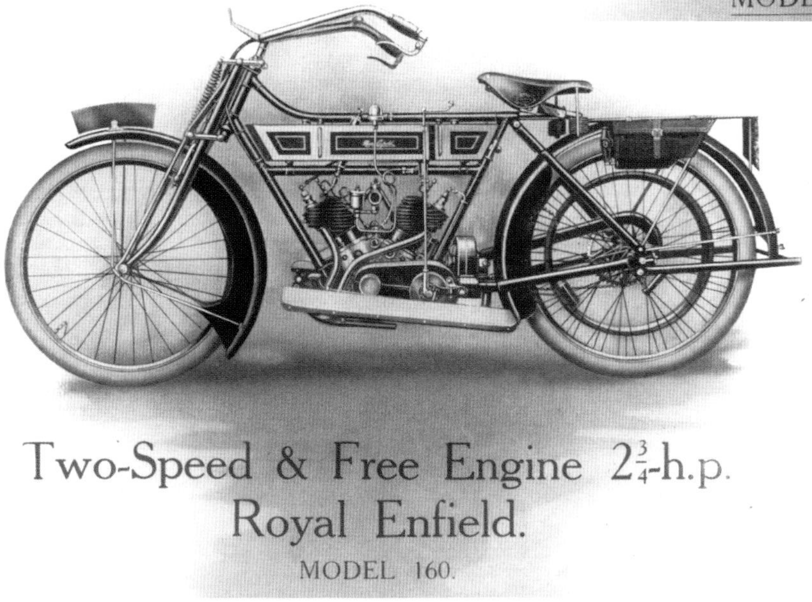

Model Specific Specifications

Model 150
Transmission: Single speed, chain drive, slipping clutch
Carburettor: Brown & Barlow
Tyres: Clincher 26 x 2
Fuel consumption: 110mpg approx.
Weight: 120lbs approx.

Model 160
Transmission: Two-speed & free engine; handle start
Carburettor: AMAC multi-jet
Tyres: Dunlop
Front 26 x 2, rear 26 x 2¼
Fuel consumption:120mpg approx.
Weight: 130lbs approx.

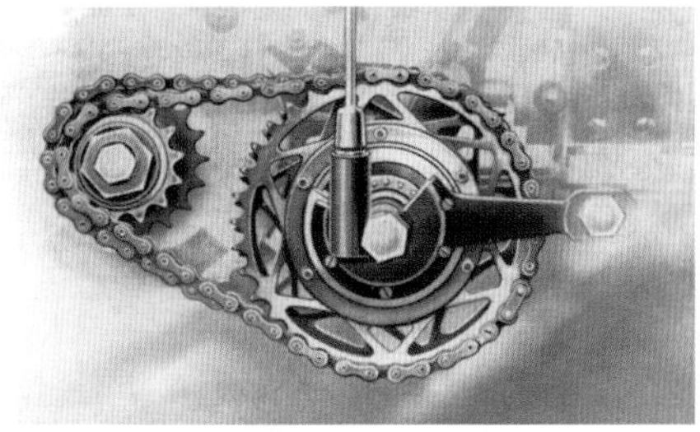

The Royal Enfield Patent Two-speed and Free Engine Gear, see page 91, was fitted to all 1912 machines except the Model 150.

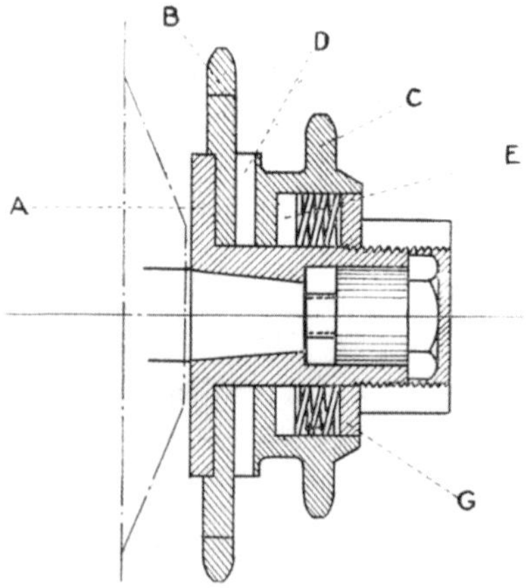

The Royal Enfield slipping cluth mechanism fitted to the 2¾hp Lightweight Model 150. See page 91

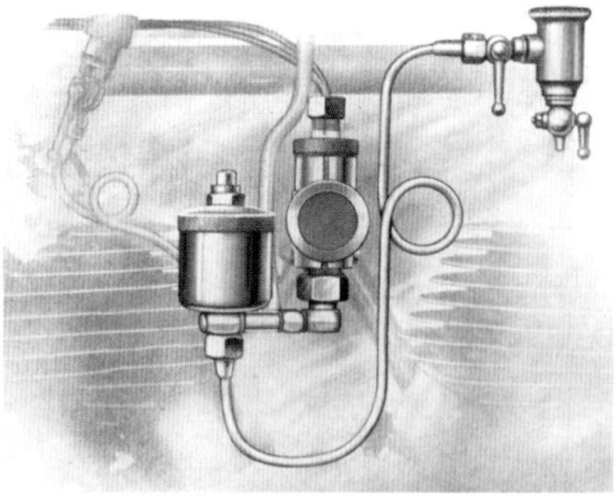

The 1912 models were fitted with an AMAC Multiple jet carburettor.

ABOVE: The Motosacoche 2¾hp 60-degree v-twin side-valve engine, with bore and stroke of 54 x 75 mm, fitted to the Models 150 and 160.

An artist's interpretation of testing a 2¾ hp model at 50 miles an hour on the private test track at the rear of the Enfield Works. The illustration was included in the 1912 Brochure.

RIGHT: The fuel tank on both the 2¾ and 6 hp models included separate compartments for oil and petrol. Sight glasses were set into the sides of the tank to indicate levels. Oil flow to the engine was by hand pump and adjusted by a valve and sight glass. A very efficient petrol filter formed part of the fuel tap.

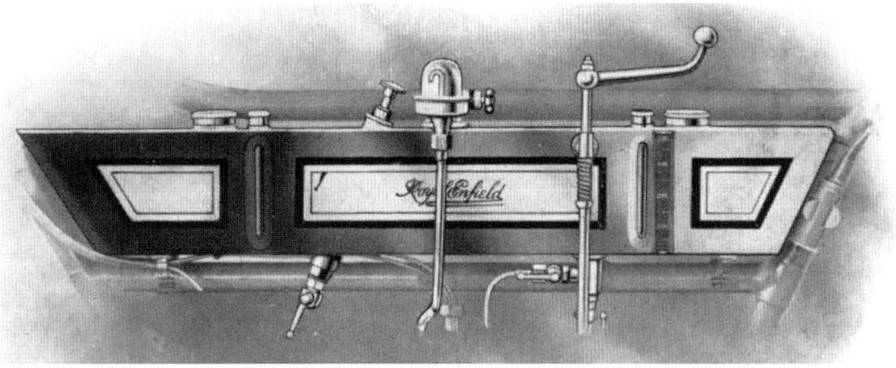

Mr C. P. Finn's splendid performance on his 2¾hp Model secured him a Gold Medal in the Harrogate and District Team Trial, in addition to many more prizes in other contests, and his inclusion in the Royal Enfield sales catalogue for 1912.

A feature of the Sales catalogues of the period was the inclusion of testimony from Royal Enfield Riders praising the virtues of the machines. The following are extracted from the various comments made:

'No Devonshire hills that the Royal Enfield will not take.'

'I climbed Porlock using the top gear only.'

'My 2¾ hp has now done 3,000 miles and has called for no attention.'

'I am delighted with the ease of riding and control.'

'The gears are a marvel of neatness and efficiency.'

Wm. Bennetts of Cornwall commented on his 2¾ hp model : 'I am certain that I shall not wish to be taken over the roads faster than my Royal Enfield can take me.'

The Press were equally enthusiastic praising the ease of control owing to the efficient type of change-speed gear and free engine. Motor Cycle noted that all semblance of jerkiness is removed on account of the friction clutch on engine shaft. They were similarly surprised at the speed that could be obtained and praised the hill climbing qualities. The Irish Cyclist went so far as to claim that 'With its 2¾ hp twin-cylinder engine, two-speed gear and free engine, the Royal Enfield is as near the ideal machine as any I know of.'

Open Frame 2½hp Model 170

The convenience of the open or drop frame for ladies wearing skirts had been appreciated from the advent of the safety bicycle and most bicycle manufacturers offered specific ladies versions of their popular machines. The open frame motorcycle offered similar advantages to the lady motorcyclist, but it also presented practical design difficulties if the accepted central position of the engine was to be retained. With relatively few lady riders, few manufacturers were prepared to invest in the design of a ladies motorcycle; the Enfield Cycle Company were amongst the first.

The perceived merits of the new open frame Model 170 were identified in the sales brochure:

> The elegant design of this mount and the thorough manner in which all working parts are guarded so that the rider's clothes cannot possibly become entangled and always remain perfectly clean, make an immediate appeal to motor cyclists: it would be difficult to imagine a machine more suitable for the use of ladies, medical and commercial men.
> It is mounted and dismounted with the greatest ease, the engine develops surprisingly high power, and this, in combination with our Patent Two-speed and Free Engine Gear, makes an ideal lightweight machine, extremely flexible, easy to control and comfortable as an armchair. Engine shocks are reduced to a minimum by the Royal Enfield Patent Cush Drive, so ensuring sweet and easy running and long tyre life.

The rear mudguard was well valanced and the rear section could be swung back to facilitate tyre repair.

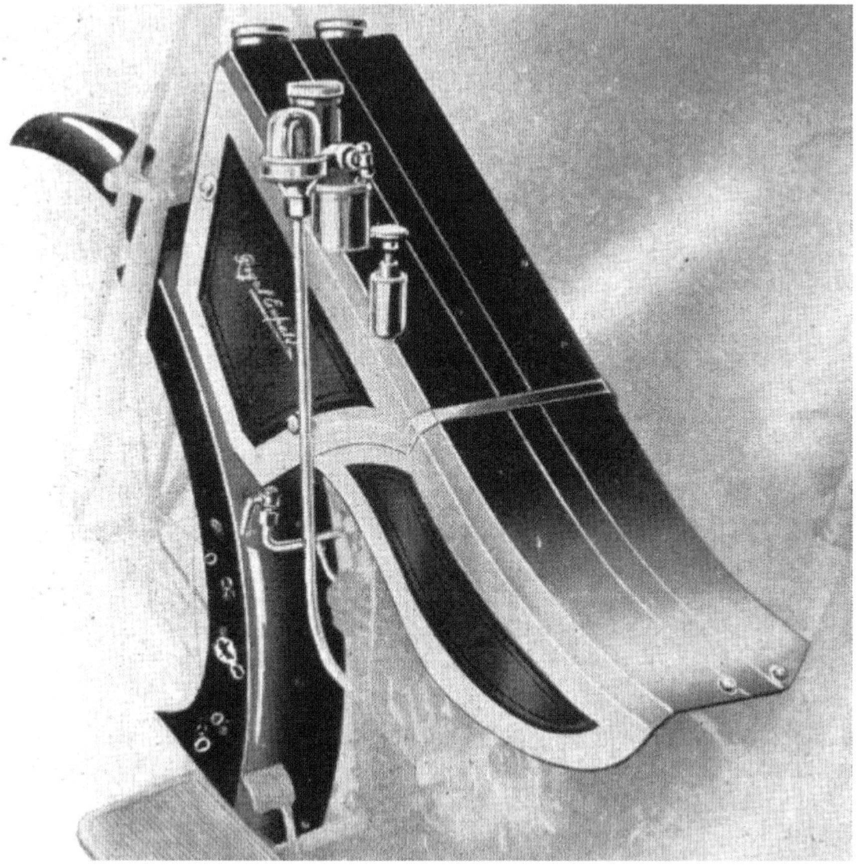

The rider was well protected by the well valanced front mudguard, tank and engine guards.

The Model 170 was fitted with a 241cc (64mm x 75mm) single cylinder Motosacoche engine. The cylinder sloped forward at 30 degrees from the vertical and the cooling fins were similarly inclined to be parallel to the ground. The engine and working parts were all covered by detachable covers. The tank was made in two parts and offered a combined fuel capacity of 1½ gallons and 1 quart of oil.

The machine was priced at £52 10s, the same as the two-speed 2¾hp Model 150. It would be remain in the Royal Enfield catalogue for only one year; it would be another eleven years before Royal Enfield offered the public another open frame machine, the 1924 two-stroke Model 201A

Two-Speed and Free Engine $2\frac{1}{2}$ h.p. Open Frame Royal Enfield.

MODEL No. 170.

Engine	Royal Enfield Single Cylinder; bore and stroke 64 × 75 m/m; mechanically operated inlet valve; large silencer.
Two-Speed & Free Engine	Royal Enfield Patent; internal clutch type; foot pedal control; handle starting and declutching hand lever.
Transmission	Hans Renold Chain, in conjunction with Royal Enfield Slipping Clutch and Patent Cush Drive.
Ignition	Bosch Magneto; gear driven.
Carburettor	AMAC Multiple Jet.
Lubrication	Automatic drip feed and auxiliary hand pump.
Tank	Made in two parts; petrol capacity $1\frac{1}{2}$ gallons; oil capacity 1 quart
Petrol Consumption	Approximately 120 miles to gallon.
Frame	Royal Enfield open frame design; with efficient detachable guardings over engine and other working parts; fitted with footboards and registered spring forks.
Tyres	Dunlop studded, 26 × 2.
Mudguards	Strong and wide; front extension; side shields to front and rear guards.
Carrier	Made of specially strong steel tube.
Stands	For front and rear wheels.
Handlebar	Royal Enfield registered design; carburettor control and lamp bracket integral with bar; concealed cables.
Brakes	Powerful front rim actuated from handlebar, and rear pedal brake.
Saddle	Large seat, with excellent springs.
Tool Bags	Two leather pannier bags, containing set of tools in roll up.
Finish	Enamelled Naval grey; usual parts heavily plated.
Weight	About 120 lbs.

6hp Side Car Model 180

The Model 180 was the first of the large capacity Royal Enfield V-twin sidecar machines. In coming years Enfield's sidecar machines would establish an enviable reputation for their rugged reliability, particularly for arduous commercial applications. This first machine, however, was designed specifically to carry a passenger.

The sidecar frame was of Enfield design. It was manufactured using weldless steel tube. The lugs for attaching the sidecar formed an integral part of the motorcycle frame, thus solving problems of sidecar alignment, and allowing the sidecar to be attached or detached in a few minutes. The sidecar body, which was constructed with a wickerwork front and side door, was mounted to the chassis by coil springs at the front and 'C' springs at the rear. A 650 x 65 mm motor-car tyre was fitted to the rear wheel to reduce tyre wear and the cost of tyre maintenance.

The engine was a 771cc (76mm x 85mm) v-twin JAP. It was started by a hand crank. Lubrication relied upon a drip sight feed with an auxiliary hand pump.

The fuel tank on the 6hp Model, as with the 2¾hp Model, was attached by two lugs brazed to the frame lower top tube. Supporting the weight from the bottom allowed the tank joint to be made at the top, thus reducing the possibility of leakage

The machine was fitted with the Royal Enfield two-speed and free engine gear. The transmission benefited from the use of the Royal Enfield Slipping Clutch and patent Cush Drive.

RIGHT: The illustration in the 1913 Sales Brochure showing the method of attaching the sidecar frame to the motorcycle and the springs for mounting the body. The illustration was produced from the above photograph by 'ghosting-out' the motorcycle.

The 6hp v-twin JAP engine of the Model 180. The engine was started by a hand crank clamped to the frame down tube and connected to the engine by chain and sprocket.

Works photograph of the 6hp Side Car Model 180 showing the sidecar frame and the three-point attachment to lugs on the motorcycle frame, which allowed easy removal and re-attachment without realignment problems. The photograph shows a rare near-side view of the 6hp machine.

6-h.p. Side Car Royal Enfield,

Fitted with Royal Enfield Patent Two-speed and Free Engine Gear.

MODEL No. 180.

Engine	Royal Enfield Twin Cylinder; bore and stroke 76 × 85 m/m; mechanically operated inlet valves; exceptionally large silencer situated under footboard.
Two-Speed & Free Engine	Royal Enfield Patent; internal clutch type; handle starting.
Transmission	Hans Renold Chain in conjunction with Royal Enfield Slipping Clutch and patent Cush Drive.
Ignition	Bosch Magneto; waterproof; chain-driven; controlled from handlebar; waterproof terminals.
Carburettor	Amac Multiple Jet.
Lubrication	Automatic drip feed and auxiliary hand pump.
Tank	Two compartments; petrol capacity 1½ Galls. Oil capacity 1½ qts.
Petrol Consumption	Approximately 80 mile to a gallon.
Frame	Royal Enfield design; plenty of ground clearance; side car attachments are built into frame, which is fitted with footboards and our registered spring forks.
Tyres	Dunlop studded 26 × 2½ on front and side car wheels; 650 × 65 m/m grooved voiturette on rear wheel.
Mudguards	Strong and wide, front extension and side shields. Back guard is divided so that a portion will swing down to facilitate tyre repair
Carrier	Made of specially strong steel tubing.
Stands	For front and rear wheels and Side car.
Handlebar	Royal Enfield registered design; carburettor and magneto controls, and lamp bracket integral with bar; concealed cables.
Brakes	Powerful front rim actuated from handlebar, and rear pedal brake.
Saddle	Large and comfortable pan seat with excellent springs.
Tool Bags	Two pannier bags with complete set of tools in roll up.
Finish	Best black enamel, usual parts heavily plated.
Side Car	Chassis, Royal Enfeld design; very strong. Body work, cane, torpedo front and side door, beautifully upholstered, complete with apron and mat, excellently sprung

PRICE—Complete with Side Car - - Net Cash

ABOVE & RIGHT: Two photographs from the factory photograph albums showing 1912 6 hp Side Cars being used as intended. Both machines have been equipped with acetylene lighting sets with separate units on the motorcycle and sidecar. The machine on the left has a separate acetylene generator, whilst the machine on the right uses an integral lamp/generator unit.. The passenger on the left-hand machine has been given control of the bulb horn.

Military Tests

An investigation of the military role of motorcycles for communications and other duties in the event of future hostilities was carried out in early 1912. Royal Enfield and all other leading manufacturers were invited by the Transport Committee of the War Office to take their latest production models for appraisal in speed trials at Brooklands on Monday, 29 January. Bert Colver took along a 2¾hp Twin and achieved the second fastest lap for a 350cc machine of 38.7mph.

Sporting Activities

Prior to 1912 the twin cylinder motorcycles had been allowed a capacity advantage in the IOM Tourist Trophy Races owing to their perceived inferior reliability compared with single cylinder machines. That this was no longer appropriate was demonstrated convincingly in 1911 by the success of P.J.Evans on his Humber 340 twin in the Junior race and by Oliver Godfrey on his Indian v-twin in the Senior race. Consequently the regulations were revised for the 1912 event and capacity classes restricted to 350cc and 500cc for the Junior and Senior events, respectively. Whether it was because of this change in regulation, or possibly because of the more arduous nature of the Mountain Circuit as opposed to the shorter and less demanding St John's Circuit, several manufacturers, including Royal Enfield, decided to boycott the 1912 event. Thus the first appearance of Bert Colver's new racing machine was at Brooklands in July 1912. The new 2¾hp Royal Enfield machine was being prepared for an attempt on the 350cc one-hour record. The 344cc side-valve engine of the standard machine was replaced by an experimental v-twin engine, designed by W.Guillon, with overhead inlet valves, side exhaust valves and mechanical lubrication. The capacity was 424cc (60 x 75mm). It had an outside flywheel and the magneto, which was chain driven, was mounted above the engine in the space between the oil and fuel compartments of the tank.

The race debut of the new machine was to be the Brooklands Junior 150-Mile Race to be held on 24 August, but heavy rain caused a postponement. When it was held three weeks later, the race attracted fifty starters. The pace was set by Les Bailey (Douglas) and Sam Wright (Humber) who were both riding special racing machines and able to lap the circuit at around 55mph. Bert with his less highly tuned machine was able to lap steadily at 50mph before having to stop to clean a plug on lap 6. He completed a further six laps before being forced to retire with a broken valve. He was more successful four weeks later in the Junior Hour Race for 350cc machines at the BMCRC Open Championship Meeting where he obtained a creditable second place having covered a distance of 53 miles and 10 yards.

Prototype ioe v-twin Engine

The experimental engine fitted to Colver's race machine was the prototype for a new 3hp Royal Enfield, which would be announced at the time of the Olympia Show in November

1912 and would form part of the Royal Enfield range for 1913. The bore of the production engines was increased from 54mm to 60mm, but retained the same 75mm stroke, to produce a swept volume of 424cc.

Hunt End Works

To keep abreast with the demand for Royal Enfield motorcycles, in early 1912 the Enfield Cycle Company reacquired the Hunt End works, which had stood idle since the purchase of the Enfield Autocar Company by Alldays & Onions in 1908. The works became a subsidiary of the Redditch works and were reequipped with new machinery to ensure the company was able to maintain the excellent quality of its machines. The works would subsequently be sold in 1918 to Alcad, manufacturer of nickel cadmium batteries.

Bert Colver used a 6 hp Side car machine with a simple plank body to transport his racing machine. The machine being transported is his Brooklands racer. The machine was based upon the standard 344cc side-valve Model 160 and fitted with an experimental 2¾hp ioe engine designed by W.Guillon .

This works photograph is of the prototype 3hp Twin Royal Enfield, which would form part of the 1913 model range. The engine is installed in a modified 2¾hp frame with a straight top tube. The production versions announced in November would have a fuel tank sloping down at its end and a revised top tube designed to provide a lower saddle height. An unusual feature of the design was the replacement of the conventional oil tank, at this period usually incorporated as part of the fuel tank, by a glass container located behind the seat tube.

Mrs R. Lord (née Miss Muriel Hind,), was a well-known lady motor cyclist and a regular contributor to the leading journals on matters affecting motor cycling for ladies. She frequently featured in Royal Enfield promotion material and was depicted with her open frame Model 170 in both the 1912 and 1913 sales brochures.

ROYAL ENFIELD

MADE LIKE A GUN

13,000 miles— and always satisfactory!

A splendid instance of remarkable sidecar service under severe conditions.

The above rider is Mr. H. March, of Hampton Court. The machine is a standard 6 h.p. Royal Enfield sidecar combination.

"From 1st August, 1912, to 6th December, 1913, I covered 13,000 miles on my 6 h.p. Royal Enfield sidecar combination. The machine was on the road EVERY day in ALL weathers, and I always carried goods or passengers. During the whole time I had only one break-down, and this was due to an old chain (the chain had been used for over 10,000 miles). For the whole 13,000 miles my running expenses have been less than 1¼d. per mile."

Just another instance of Royal Enfield reliability. Service such as this—13,000 miles, at a cost of 1¼d. per mile, mostly with three passengers and driver—is not uncommon for Royal Enfields. All Royal Enfield models are fitted with the Enfield patent Two-speed and Free-engine Gear, Chain Transmission, and Cush Drive Hub.

6 h.p. Sidecar Combination - net cash, 80 Guineas.

Write for the Royal Enfield Motor Catalogue.
Please mention "Motor Cycling."

The ENFIELD CYCLE Co., Ltd., REDDITCH,
and 48, Holborn Viaduct, E.C. Agents everywhere.

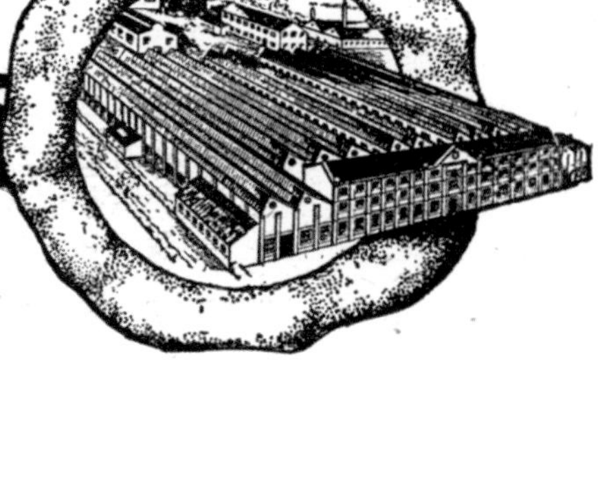

CHAPTER EIGHT

Model Year 1913

A three model range for 1913 was announced at the Olympia Show held in November 1912. The 2¾hp Model 160 solo machine and the 6hp Model 180 sidecar combination had both proved to be extremely popular during 1912 and were retained. They were joined by a new 3hp machine, fitted with a 424cc v-twin engine with overhead inlet valves, which had been developed from the 2¾hp Guillon-designed engine raced at Brooklands by Bert Colver.

All the machines were fitted with all-chain drive and the Royal Enfield patent two-speed and free engine countershaft gear. In addition they were all fitted with a kick-starter coupled to the engine countershaft on the nearside of the machine and a new drop frame to give a lower riding position. An improved exhaust system was also introduced across the range.

1913 Model Range

Model 140	3hp	424cc	Solo	£52 10s.
Model 160	2¾hp	344cc	Solo	£50
Model 180	6hp	771cc	Side-car	£84
Model 180	with coach-built body			£86 2s.

The 1913 Season brochure was a lavish publication of thirty-two pages, which provided full specifications and technical details of the three Royal Enfield models. Also included were 'Riding Hints', a summary of 1912 sporting successes and owners' testimonials.

The sales catalogue for the 1913 season carried illustrations of the company's Redditch works and head offices and the recently reacquired and reequipped Hunt End works.

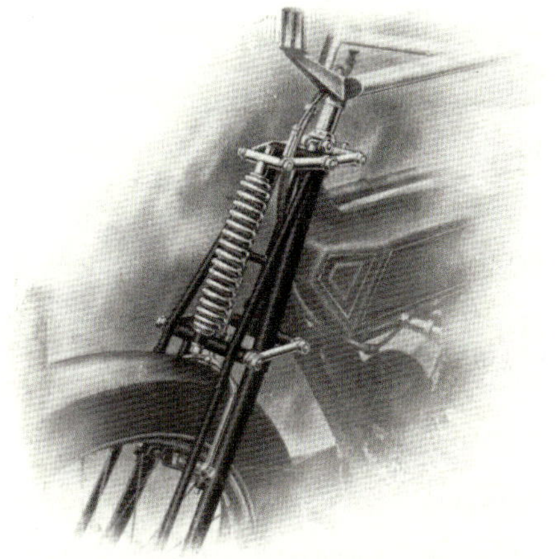

The Royal Enfield front fork fitted to the 3hp Model 140.

The Guillon-designed 424cc (60mm x 75mm) ioe v-twin fitted to the Model 140.

3hp Model 140

The new machine represented Guillon's ongoing development of his earlier designs. The engine was still a 60 degree v-twin with 75mm stroke, but was now bored out to 60mm to give an increased capacity of 424cc. It was fitted with overhead inlet valves housed in a dome directly above the side exhaust valves. The timing gear consisted of a single cam operating each inlet valve and a second cam each exhaust valve. The method of attaching the cylinder barrels to the crankcase was unusual; it consisted of a double clamp within the vee between the cylinders and a single clamp on the opposite side of each cylinder.

The other interesting innovation was the introduction of an automatic dry-sump lubrication system; one of the very earliest known examples of its use. Oil was fed under pressure from a geared oil pump to the big-end bearings via a hollow crankpin and thus distributed to each cylinder. The oil accumulating in the crankcase sump was then pumped out and returned to a glass tank fitted behind the seat tube of the frame.

The engine was fitted with a kick-starter. The pedal was fitted on the right hand side of the machine just behind the magneto. Sharp downward movement of the lever caused it to engage with one of a train of magneto gears by means of a cone clutch.

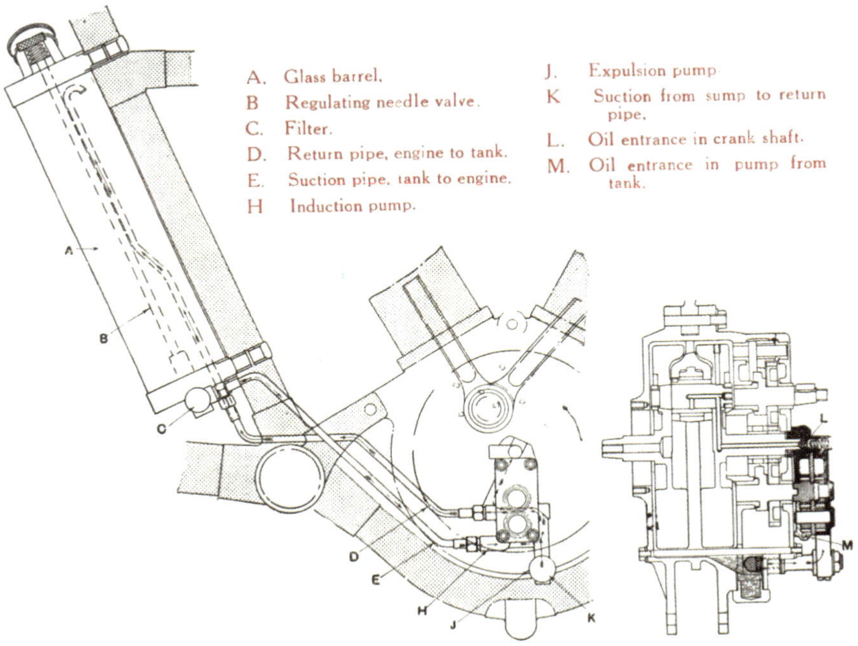

The dry-sump lubrication system of the Guillon-designed 424cc ioe v-twin.

Royal Enfield
3 h.p. Two-speed and Free Engine Model.

Model 140.

ENGINE . . . Royal Enfield twin-cylinder, 60 × 75 m.m., M.O. overhead inlet valves, with working parts completely enclosed and dust-proof.

TWO-SPEED AND FREE ENGINE GEAR. Royal Enfield Patent design; expanding clutch type, operated by the hand lever shown in illustration.

KICK-STARTER An entirely new type of "kick-starter" is fitted. The machine starts easily whilst the rider is seated in the saddle.

TRANSMISSION Roller chains, ½ in. pitch, in conjunction with the Royal Enfield cush drive hub.

CUSH DRIVE HUB. The Royal Enfield patent cush drive hub is fitted to the rear wheel, ensuring, in conjunction with the chain transmission, a perfectly smooth and supple drive.

IGNITION . . . New Bosch waterproof magneto, gear driven. Controlled from handlebar.

CARBURETTOR AMAC multiple jet.

LUBRICATION . Royal Enfield patent mechanical forced feed lubrication system. This marks an important improvement in the lubrication of motor cycles. Fully described on page 19.

TANK . . . . Fitted with large glass-topped filler, with gauze strainer and needle valve to regulate flow of petrol. Lubricating oil is carried in separate tank. Petrol capacity 1¼ gallons.

FRAME . . . Royal Enfield design, dropped at back to give low saddle position; fitted with spring footboards and registered spring forks.

MUDGUARDS . Strong and wide, with front extension; side shields on both guards.

TYRES . . . . Dunlop studded: front 26 in. × 2¼ in., back 26 in. × 2¼ in.

HANDLEBAR . Royal Enfield; concealed cables; carburettor control and lamp bracket integral with bar.

BRAKES . . . Powerful front rim brake actuated from handlebar, and rear brake operated by foot pedal.

SADDLE . . . Specially large and well-sprung pan seat.

TOOL BAGS . Two leather pannier bags, containing complete set of tools.

CARRIER . . . Made from weldless tubular steel and strongly attached.

STANDS . . . Two stands fitted (one for each wheel). Both stands fasten securely with spring clips.

FINISH . . . . Enamelled in best black enamel; tank artistically panelled in two shades of green. Usual parts heavily plated.

Price, net cash - 50 Guineas.

xx. MOTOR CYCLING. 7th October, 1913.

The 3 h.p. ROYAL ENFIELD–positively the best solo motorcycle ever evolved, as performances at Brooklands prove–and some of its features

1. THE MECHANICAL LUBRICATION. There are no hit or miss methods in Enfield lubrication. The oil is carried in a tank at the back of the seat tube and pumped from thence mechanically to the engine. When the engine is running there is a continuous circulation of oil thoroughly lubricating each cylinder.

2. THE ENFIELD 3 h.p. SILENCER. Not every motorcyclist realises the importance of an efficient silencer. Unless the exhaust is cleanly carried away overheating and carbonisation quickly tell their tale. The silencer on the 3 h.p. Enfield has two expansion boxes of ample size, effectually and efficiently silencing the exhaust.

3. THE 3 h.p. ENFIELD ENGINE & POWER UNIT. The standard engine has a bore and stroke of 60 mm. and 75 mm. respectively, giving a capacity of 425 c.c. On this capacity enormous power is developed. The overhead valves are housed in a dome directly over the exhaust valves, and the rocker arm is completely cased in and protected. There is no more reliable power unit made than this.

4. The ENFIELD TWO-SPEED & FREE-ENGINE COUNTERSHAFT GEAR. Unquestionably the most reliable gear fitted to any motorcycle. No pedals to push—both gears operated by a simple hand lever. All parts "get-at-able." Chain drive is fitted throughout. No belt troubles and worries on the ENFIELD. Many leading makers now fit the ENFIELD Two-speed Gear to their models because they can get none better.

5. THE ENFIELD CUSH-DRIVE HUB. This shock-absorbing device is fitted in the rear wheel. The illustration on the right is the driving end of the hub: the driving sprocket is shown on the left. The drive from the countershaft to the rear wheel is transmitted through a series of rubber buffers, and is remarkably smooth and supple.

Write now for a copy of the complete Royal Enfield Motor Catalogue, which describes all the Royal Enfield Models.

Address: Dept. "H"—

THE ENFIELD CYCLE CO., LTD., REDDITCH.

And 48, HOLBORN VIADUCT, LONDON, E.C. Agents Everywhere.

ROYAL ENFIELD

·MADE LIKE A GUN·

2¾hp Two-speed & Free-engine Model 160

New on the Model 160 for 1913 was the fitment of a kick-starter mechanism. Located at the front of the engine, it was operated by the right foot and connected to the crankshaft by sprocket and chain.

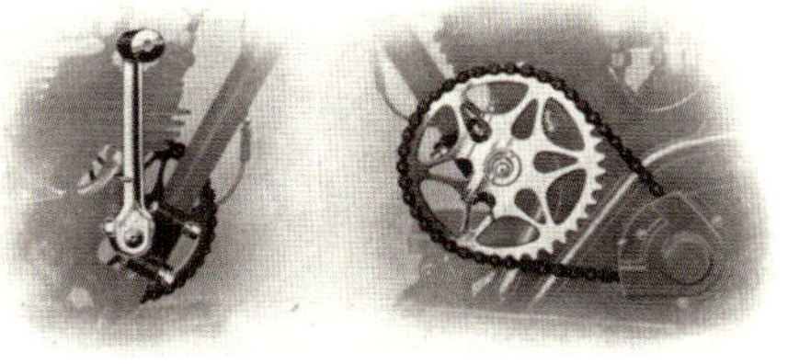

The kick-starter mechanism of the Model 160.

Specification of Model 160

Engine type: 60° Side-valve v-twin
Bore and stroke: 54mm x 75mm
Capacity: 344cc
Carburettor: AMAC multiple jet
Ignition: Bosch magneto
Gearbox: Two-speed, RE patent
Transmission: Chain, RE slipping clutch & cush drive
Tank: 1 gallon petrol, 1 quart oil
Suspension: RE spring front fork
Tyres: Dunlop studded
Front 26 x 2in, rear 26 x 2¼in
Brakes: Front stirrup, rear dummy rim
Saddle: Well sprung leather seat
Finish: Naval grey, panelled tank
Price: £84

2¾hp Two-speed and Free Engine Model 160.

Controls and Starting Procedure for the 2¾hp Twin

The recommended starting procedure for the 2¾hp, Model 160, much of which will be familiar to riders of early machines, is briefly summarised below.:

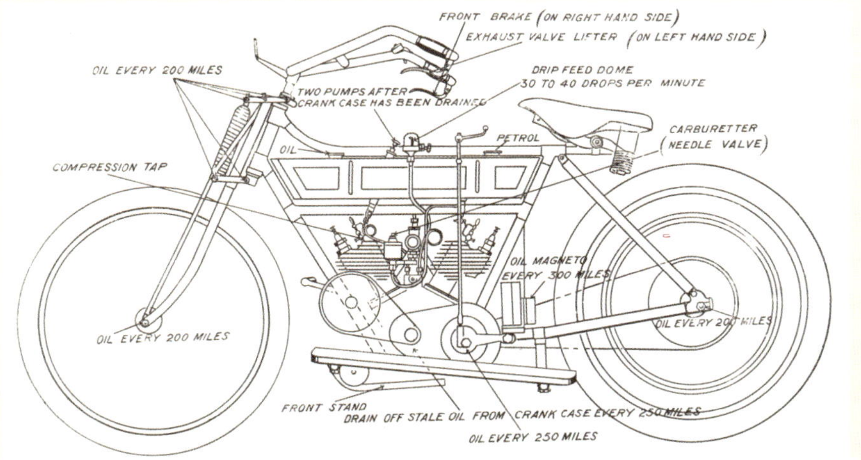

* Pump with the hand pump two full charges of oil to the engine.
* Turn on the petrol tap underneath the tank and press down the plunger on top of the carburettor until the carburettor floods.
* Place the two-speed gear in free engine position.
Slightly retard magneto and partly open throttle and extra air levers.
* A sharp kick down on the kick-starter should cause the engine to fire immediately.
* With the engine running, the rider may take his seat in the saddle, noticing the position of the back brake pedal, in case an application of it is necessary.
* The two-speed gear lever is then gradually moved towards the low gear position when the machine will pick up speed.
* At this point the gear lever should be brought back fully into the low gear position and the extra air lever gradually opened; experience will dictate the best position of the lever.

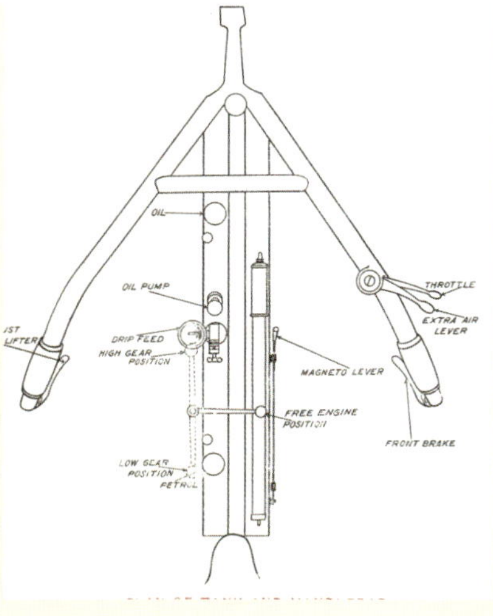

6hp Side-car Combination Model 180

Specification

Engine type: 60° Side-valve v-twin
Bore and stroke: 75mm x 85mm
Capacity: 771cc
Carburettor: AMAC multiple jet
Ignition: Bosch magneto
Gearbox: Two-speed, RE patent
Transmission: Chain, RE slipping clutch & cush drive
Tank: 2 gallons petrol, 3 pint oil
Suspension: RE twin spring front fork
Tyres: Palmer Cord 650 x 65mm
Brakes: Front stirrup, rear dummy rim
Saddle: XL'ALL
Finish: Black, green panelled tank
Price: £84

The 6hp Model 180 side-car combination was offered as standard with a wicker-work side-car body, but was also available with a coach-built body made in the Enfield works. The sales information for this bodywork noted: 'The outline and dimensions allow of the fullest comfort for the passenger. It is roomy and fitted with a side-opening door flush with the contours of the side-car body. There are no crevices, as with the wicker bodies, for the cold air to penetrate, and the general appearance is most pleasing, whether viewed from the front or side.' The upholstery was in green 'Pegamoid' and a waterproof apron was provided to protect the passenger's knees. The coach built side-car combination cost £86 2s; 2 guineas more than the standard machine with wicker side-car body.

The Model 180 was 'essentially a combination for touring and travelling'. For carrying luggage on these occasions a suitable grid was available for an additional 2 guineas (£2 2s). It was manufactured of weldless steel tube and attached at the back of the sidecar frame at four separate points.

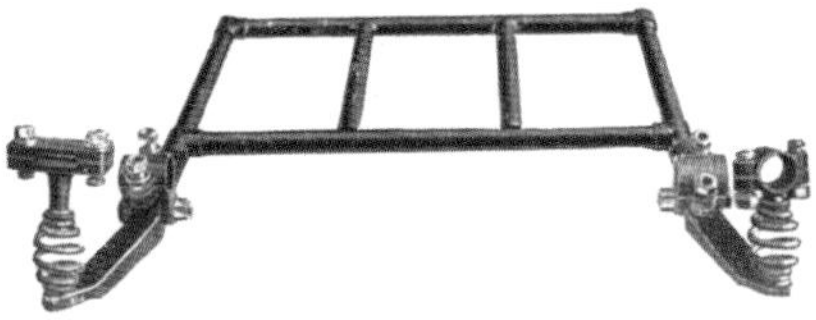

The wicker-work side-car body was of a more pointed design than when first introduced for the 1912 Model Year.

The coach-built side-car body was introduced for the 1913 Model Year.

ABOVE & LEFT: The substantial luggage grid available for attachment to the rear of the side-car frame.

Works photographs of the coach-built and wicker side-car bodes with optional hood and windscreen fitted

6hp Model 180 Trade

In response to the increasing use of motor vehicles for trade delivery purposes, Enfield offered on request a special carrier body for fitting to a slightly modified 6hp side-car chassis. The body was made in-house by the company's coach builders.

The 1913 Competition Season

The successes achieved by Enfield works riders persuaded the directors to agree to return to the Isle of Man for the 1913 Junior Tourist Trophy Race. Bert Colver remained with Enfields to develop a reduced capacity version of the new 3hp IOE model.

Five of the reduced-stroke machines were prepared for the TT. Bert Colver, H.Greaves and E.Keyte formed an official Enfield works team whilst D.Alexander and C.M.Down entered as amateurs. The event consisted of two separate races; a two-lap race on the Wednesday and a four-lap race on the Friday for riders surviving the earlier race. The event

The Enfield managing director, R.W.Smith, was at Brooklands to witness the Enfield works team's success in winning the Manufacturers Team Prize in the Six-Hour Multi-Class Race, and was photographed with the team. The riders are E.Keyte (64), H.Greaves (63) and H.Colver. R.W.Smith is standing between Keyte and Greaves, whilst the Enfield company secretary, T.C.Patchett, is between Greaves and Colver.

A Performance which proves the Superiority of the 3 h.p. Royal Enfield for Sustained Riding at Speed

Messrs. Keyte, Greaves, and Colver, all on 3 h.p. Enfield Models, who finished second, third, and fourth in the Six Hours' Race—remarkable consistency.

Unprecedented 3 h.p. Enfield Team Performance.

OPEN SIX HOURS' RACE (350 c.c. Class), B.M.C.R. Club Meeting, July 16th, 1913.

THREE 3 h.p. ROYAL ENFIELDS competed.
THREE 3 h.p. ROYAL ENFIELDS finished.
THREE 3 h.p. ROYAL ENFIELDS Broke Previous Six Hours' Record.

Distances Covered.

2nd in Race, ...	H. V. Colver, 3 h.p. Enfield	302 miles,	1,521 yards.
3rd ,, ...	E. H. Keyte, 3 h.p. Enfield	281 ,,	893 ,,
4th , ...	H. Greaves, 3 h.p. Enfield	279 ,,	648 ,,

Fully recognising this marvellous performance, the Brooklands Motor Cycle Racing Club awarded Three Special Gold Medals to the successful Enfield riders.

The remarks at the foot of the following page, *re* engine capacity, apply also to the 3 h.p. Royal Enfields ridden in the above race.

ABOVE & FACING PAGE: The success of the Enfield team at Brooklands was applauded in the 1914 sales brochures.

The most brilliant Victory of the Year. The 3 h.p. Royal Enfield Team vanquishes "all comers."

Messrs. Colver, Greaves, and Iron, whose 3 h.p. Enfields covered 155 miles between them, thus winning the One Hour Team Race against all comers—a remarkable performance.

What the Royal Enfield Team accomplished.

ONE HOUR TEAM RACE, B.M.C.R.C. Meeting, September 13th, 1913.

	Total Distance covered.			
1st—Enfield 3 h.p. Team	**155**	**miles**	**684**	**yards.**
2nd—*——Team (one 7 h.p., two $3\frac{1}{2}$'s) -	129	,,	1,730	,,
3rd—*——Team (three $3\frac{1}{2}$'s) - -	120	,,	523	,,
4th—*——Team (three $3\frac{1}{2}$'s) - -	106	,,	199	,,

In two teams only two of the riders finished, in one team only one rider finished, but in the winning Royal Enfield team all three riders finished without a hitch or a semblance of trouble.

The engines fitted to the 3-h.p. Enfields for this race were of 350 c.c., actually 75 c.c. smaller than our Standard 3-h.p. engine. These engines were made smaller for this race because it was originally intended to divide it into classes—for engines under 350 c.c., under 500 c.c., and any size over. This makes the Royal Enfield performance all the more remarkable, as Standard size engines, being of larger capacity, would naturally have improved upon these distances.

The works rider, H Greaves, with 3hp Twin Model 140.

proved a disappointment with only two of the five Enfield riders, Down and Greaves, finishing in sixteenth and seventeenth positions.

The official works team of Colver, Greaves and Keyte gained greater success in the Six-Hour Multi-Class Scratch Race organised by the BMCRC at Brooklands on 16 July 1913. The team was entered on their reduced stroke (350cc) versions of the 3hp ioe model in order to compete in Class B for 350cc solo machines. With ten different categories of machine, including cyclecars, large capacity V-twin solos and sidecar combinations, it was to prove an exciting and spectacular race. George Stanley on his Singer took an early lead in Class B and at the end of the first hour was ahead of Hugh Mason (NUT), Colver and Greaves. However, his lead had been wrested away by Mason at the end of the second hour and he would retire shortly afterwards. Mason maintained his lead and at the end of the fourth hour had completed 82 laps, whilst Colver in second place and Greaves in third had both covered 74 laps; Keyte was in fourth place on 70 laps. At the end of the race Mason had a lead of 23 miles. Colver took second place at an average of 50.26mph, with Keyte and Greaves in third and fourth place, respectively. The Enfield team received the Manufacturers Team Prize for the first works team to finish the race intact.

The Enfield riders returned to Brooklands to compete in the remaining BMCRC meetings of 1913. D.Iron and Bert Colver were unsuccessful in the 350cc Ten-Mile Record Race in August. However, this was compensated by a major success at the September meeting. The Enfield team of H.Colver, H.Greaves and D.Iron, on their modified 3hp twins, covered an aggregate distance of 155 miles 684 yards in one hour to win easily against machines of almost twice their capacity. The fastest rider, Greaves, covered 55 miles 1,579 yards. Colver and Greaves returned to the Brooklands track for the BMCRC meeting on 8 October. In the One-Hour Passenger Race, Colver covered a distance of 40 miles 98 yards on his modified 3hp machine, with sidecar attached, to set a new world record.

With the race won, he quickly detached the sidecar to join the start line for the following 350cc Solo One-Hour Scratch Race. He finished in fourth place having ridden 52 miles 885 yards. H. Greaves achieved a third place with 54 miles 1,042 yards.

Management Changes

By 1913 R.W.Smith had been appointed managing director in place of J.W.Davis. Lord Ernest Seymour remained the company chairman and G.W.Cartland and T. Evans were the remaining board members.

Increase in the Capital of the Company

The New Enfield Cycle Co Ltd, subsequently the Enfield Cycle Co Ltd, had been set up with a nominal share capital of £125,000. However, the considerable increase in the Company's business made it desireable that the capital of the Company should be increased. On 20th October 1913 an Extraordinary General Meeting of the Enfield Cycle Company was held in the Grand Hotel, Birmingham, to agree an increase of the Capital of the Company to £150,000 by the creation of 25,000 7 per cent 'B' Preference shares of £1.

A subsequent Extraordinary Meeting of shareholders was called two years later on 14th October 1915 to increase the capital of the Company by a further £25,000 by the issue of 25,000 additional Ordinary Shares of £1 each.

The comfortable pan saddle fitted to the 6hp Twin.

The works riders Bert Colver and H Greaves were also expected to promote other Enfield products. They are photographed here in the firm's experimental cyclecar of 1913.

A 6hp Twin serving the requirements of the fire service.

The two photographs above are from the factory albums and show the 6hp Side-car being used as intended for family transport. The upper photograph is of a machine fitted with the 1913 season wicker-work side-car. The other machine also has a wicker-work body, but this appears to have been woven to meet the specific family requirements.

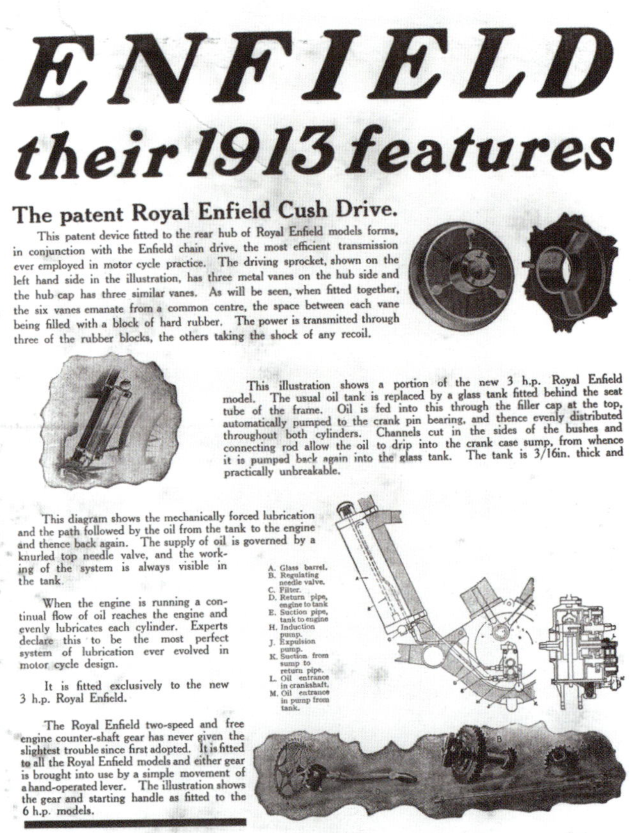

Three promotional adverts produced by the Enfield Cycle Company during 1913.

ABOVE In addition to the lavish publication illustrated on page 109, the Enfield Cycle Company also published a cheaper publication for wider circulation which contained specifications of the three Royal Enfield models. It was noted that 'Full particulars of past successes, complete technical descriptions of all Royal Enfield models and other interesting matter is contained in the Royal Enfield Motor Book, sent free on request.'

CHAPTER NINE

Model Year 1914

1914 Model Range				
Model 140	3hp	424cc	Solo	£52 10s.
Model 150	3hp	424cc	T.T.	£52 10s
Model 160	2¾hp	344cc	Solo	£45
Model 180	6hp	771cc	Sidecar	£84
Model 180 with coach-built body				£86 2s.

The 1914 range of Royal Enfield motorcycles first shown to the public at the Olympia Show in November 1913 contained few surprises. The three basic models had sold well during 1913, a consequence both of the quality of the product and the excellent publicity achieved through the sporting successes of Royal Enfield riders, both at home and abroad. The three models were retained with only detailed modifications. The Open Frame 2½hp Model 170 had proved less popular and was discontinued. The racing achievements of Royal Enfield riders at Brooklands and at the Isle of Man TT had been particularly noteworthy and to capitalise on these a new sports version of the 424cc ioe Model 140 was made available. Designated the 3hp T.T. Roadster it was aimed at the fast road rider, but was easily adapted for competition use.

The absence of an economy lightweight machine represented a notable gap in the Royal Enfield range at this period. In future years the Enfield Cycle Company would be recognised for its lightweight two-strokes and the first of these was developed during the year.

The Enfield Cycle Company's promotional material for 1914 maintained the high standard set by the previous years' publications. Of thirty two pages in length it gave full specifications and described the technical operation of the Royal Enfield models. Also included were summaries of sporting achievements and owner testimonials. A separate leaflet (RIGHT) was produced listing in greater detail the previous year's sporting successes.

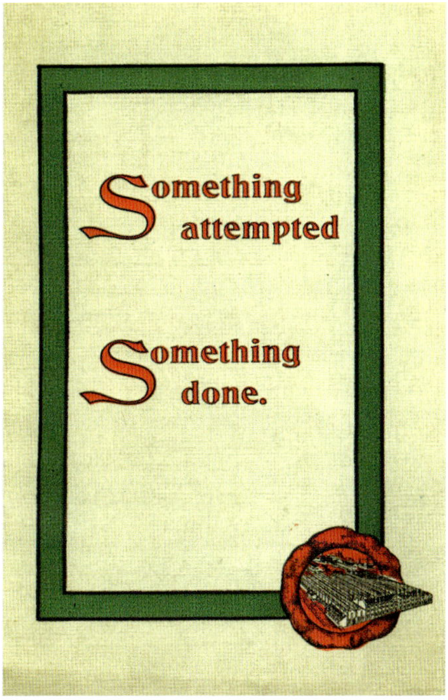

The very considerable sporting successes achieved by Royal Enfield riders, both at home and abroad, during 1913 were recognised by a special publication.

Aluminium Primary Chaincase

The Enfield Cycle Company was amongst the first manufacturers to appreciate the benefits of protecting the primary drive chain from water and road grit and was amongst the first manufacturers to fit sheet metal primary gear cases. For the 1914 season the 6hp model was fitted with an attractive cast aluminium chain case. It was split horizontally and the top half was easily removed for adjustment or replacement of the primary chain.

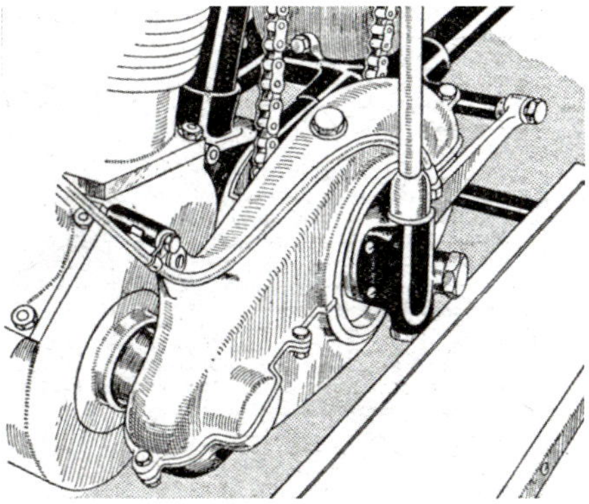

The cast aluminium chaincase fitted to the 1914 6hp models.

2¾hp Side-valve Twin, Model 160

2¾hp Model 160 Specification
Engine: 60° Side-valve v-twin
Capacity: 344cc
Bore x stroke: 54mm x 75mm
Carburettor: AMAC multiple jet
Ignition: Bosch magneto
Kick-starter: Chain to crankshaft
Gearbox: Two-speed, RE patent
Transmission: Chain, RE cush drive
Tank: 1¼ gallons petrol, 1 quart oil
Suspension: RE spring front fork
Saddle: Well sprung leather
Tyres: Dunlop studded
front 26 x 2in, rear 26 x 2¼in
Brakes: Front stirrup,
rear dummy rim
Finish: Naval grey, panelled tank
Price: £47 5s

Revised Kick-starter for 3hp Model 140

The kick-starter mechanism originally fitted to the 3hp Model 140 was of an ingenious design and located on the right side of the machine. The action of pressing the pedal caused engagement with one of the magneto gear wheels by means of a cone clutch. For the 1914 season it was replaced by an improved mechanism, similar to that of the 2¾hp model. The lever was connected by chain directly to a clutch on the drive shaft. The lever automatically disengaged to return to the original position.

The prototype design of the revised kick-starter mechanism for use with the 3hp Twin.

3hp Twin Model 140

3hp Model 140 Specification
Engine: 60° ioe Twin
Capacity: 424cc
Bore and stroke: 60mm x 75mm
Carburettor: AMAC multiple jet
Ignition: Bosch magneto
Gearbox: Two-speed, RE patent
Transmission: Chain, RE cush drive
Tank: 1½ gallons petrol
Suspension: RE spring front fork
Saddle: Brooks B170
Tyres: Palmer Cord 26 x 2¼in
Brakes: Front stirrup,
rear dummy rim
Saddle: Large leather
Finish: Black enamel, green
panelled tank
Price: £52 10s

3hp T.T. Roadster, Model 150

The T.T. Roadster was based very closely on the design of the Standard 3hp model and built with sporting styling but little extra tuning. It was claimed to be the ideal machine for competition riding. It used the standard 3hp engine, with its automatic lubrication system, but the AMAC carburettor was replaced by a Senspray unit, which was claimed to give better results at speed. The Enfield patent two-speed and free engine countershaft gear and cush drive rear hub was fitted, but with the kick-starter omitted.

3 h.p. Royal Enfield T.T. Roadster Model.

Model 150.

SPECIFICATION.

Engine – – – – Royal Enfield twin cylinder, 60×75 m/m; M.O. overhead inlet valves, with working parts completely enclosed and dust-proof.

Two-Speed and Free Engine Gear. Royal Enfield patent design; expanding clutch type.

Transmission – – Roller chains, Hans Renolds extra heavy, ½in. pitch, in conjunction with the Royal Enfield cush drive rear hub.

Cush Drive Hub – The Royal Enfield patent Cush Drive Hub is fitted. (See page 20.)

Ignition – – – New Bosch waterproof magneto, chain driven. Controlled from handlebar.

Carburetter – – Latest pattern Senspray.

Lubrication – – Royal Enfield patent mechanical forced-feed lubrication system. Fully described on page 25.

Tank – – – – Fitted with large glass-topped filler, and gauze strainer. Needle valve to regulate flow of petrol. Lubricating oil carried in a separate tank. Petrol capacity, 1½ gallons.

Frame – – – – Royal Enfield loop frame, dropped at back to give low saddle position; fitted with spring footboards and registered design spring forks.

Mudguards – – Strong and wide, with front extension and side shields on rear guard.

Tyres – – – – Palmer Cord, 26in. × 2¼in.

Handle-bar – – Royal Enfield new design T.T. handlebar.

Brakes – – – – Powerful inverted lever front rim brake; rear brake operated by foot pedal.

Saddle – – – – Brooks' B170, specially large and well-sprung.

Tool Bags – – Two leather pannier bags, containing complete set of tools.

Carrier – – – Made from weldless tubular steel and strongly attached.

Stands – – – – Two stands fitted (one for each wheel). Both fasten securely with spring clips.

Finish – – – – Enamelled in best black enamel; tank artistically panelled in two shades of green. Bright parts heavily plated.

A lightweight unvalanced front mudguard was fitted and, contrary to the brochure specification given on the previous page, the design did not include a front stand. The foot-boards were retained but reduced in length. The most obvious change was the new type of semi 'T.T.' handlebar, which had been specially designed for the model.

6hp Side-Car Combination, Model 180, with coach built body The body was built in-house at the Enfield Cycle Company's coach-building factory.

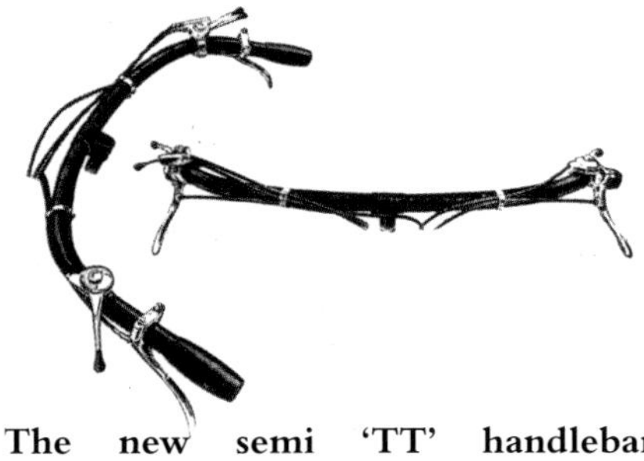

The new semi 'TT' handlebar designed to provide a more sporting appearance for the Model 150, T.T Roadster.

An offside view of the 346cc engine of the 1914 Works TT machine.

6hp Side-car Combination Model 180

The 6hp model represented good value for money and had achieved success in hill climbs and reliability trials. It was selling well, but had now been on the market for two years and minor improvements were appropriate. By this period wicker-work side-car bodywork was becoming dated and although it remained available and would do so for a further year, emphasis was placed upon the promotion of the coach-built body, which was constructed in-house at the company's own coach-building factory.

The most striking innovation on the 6hp model was the cast aluminium alloy primary drive casing, which has already been described and is illustrated on the previous page. Rider comfort was improved through the fitting of footboards with springs at the front end, Palmer-Cord cycle-car type tyres and re-designed girder-type forks, which were now fitted with twin side-by-side springs.

The sidecar body was now suspended on leaf springs rather than coil springs as previously. The rear luggage carrier was also redesigned and was now a standard fitting rather than a two guineas optional extra. A special Colonial version of the sidecar chassis was also available to provide extra ground clearance for poor road conditions.

6hp Model 180 Specification

Engine: JAP Side-valve v-twin
Capacity: 771cc
Bore x stroke: 76mm x 85mm
Carburettor: AMAC multiple jet
Ignition: Bosch magneto
Gearbox: Two-speed, RE patent
Transmission: Chain, RE cush drive
Tank: 2 gallons petrol, 3 pint oil
Suspension: RE twin-spring front fork
Stands: Front, rear & sidecar wheel
Saddle: XL'ALL Leather pan-saddle
Tyres: Palmer Cord 'Cycle-car' 650 by 65mm on all wheels
Brakes: Front stirrup, rear dummy rim
Toolbags: Two, complete with tools
Finish: Black, green panelled tank
Price: With either wicker-work or coach built body £86 2s

Off-side view of the 6hp Model 180 showing the valve side of the engine and the handle starting arrangement.

Works photographs of the 771cc JAP side-valve twin engine fitted to the Royal Enfield 6hp Twin. The left hand photograph shows the handsome aluminium chaincase concealing the primary chains, which was a Royal Enfield innovation.

The frame of the 6hp Twin incorporated three integral lugs for the sidecar attachment.

The 1914 design of sidecar chassis. The body was mounted on 'C' springs at the front and rear to provide a more comfortable ride. The attachments to the motorcycle formed part of the frame itself and were very strong, and safe, whilst ensuring perfect alignment.

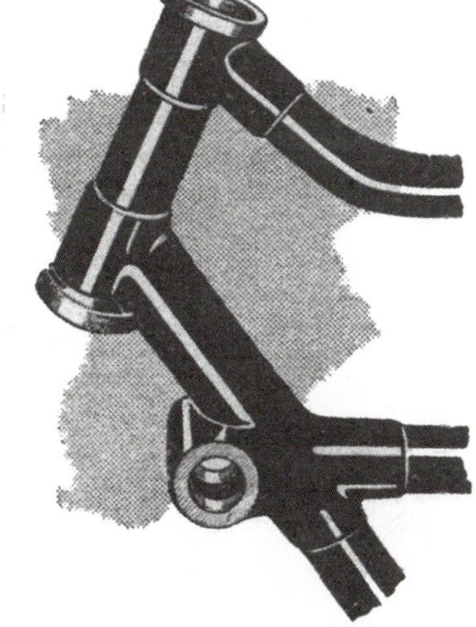

Detail of the new steering head of the 6hp Twin showing the one-piece sidecar and head lugs.

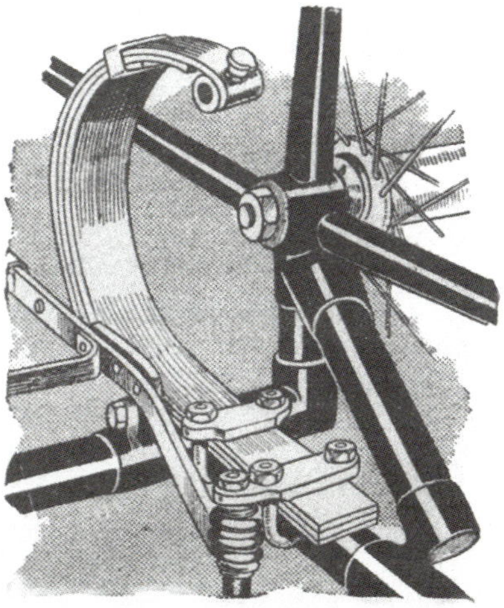

Detail of the sidecar chassis for 1914.

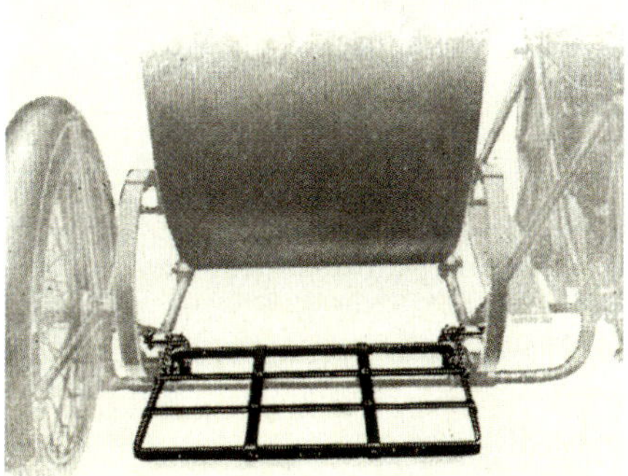

The luggage grid, which in previous years had been constructed using tubular steel, was re-designed using steel bar for use with the new sidecar chassis. Previously a 2 guineas extra, it was now included as standard with the passenger machines.

The brochure illustration of the 6hp Trade Delivery Side-car Model 180.

Model 180 6hp Trade Delivery Side-car

A special carrier body for fitting to a slightly modified side-car chassis, which was first made available in 1913, had proved popular. For the 1914 season the side-car chassis was specially adapted to take a roomy box body. The standard box body was made from well seasoned wood by the company's coach-builders and was considered most suitable for general delivery work. For lighter duties a special basket carrier body could be fitted. This was ideal for the delivery of bulky, but lightweight, parcels. It was suggested to be most suitable for laundry collection and delivery and for confectioners.

The 1914 sales brochure drew attention to the economic advantages of the trade sidecar:

> 'Running expenses vary with the locality and the class of goods carried, but making reasonable allowance for all possible items of upkeep, and also allowing a 10 per cent. Annual depreciation – it will be found that 1½d per mile more than covers the cost for an approximate annual mileage of 5,000 miles. There is no other light delivery vehicle which can compare with this for all-round cheapness and efficiency.'

The Royal Enfield Patent Spring Handlebar

The standard handlebar fitted to the Model 180 was of the Enfield registered swept back design with transverse strengthening bar. It was supplied without a lamp bracket as most owners preferred to use the bracket supplied with their choice of acetylene lamp.

The innovation for 1914 was a special spring handlebar, which was made available for an extra 21s. The handlebars were held in a two-piece lug, inside which was a strong spring. This allowed the handlebar to move vertically by several inches in order to absorb road shocks. It was claimed that when travelling at speed the handlebar remained at a level position although the head of the machine was in constant movement. It was only listed for the 1914 season.

The handlebars were finished in a special black finish claimed to be hard and permanent and a great improvement on the plated handlebar.

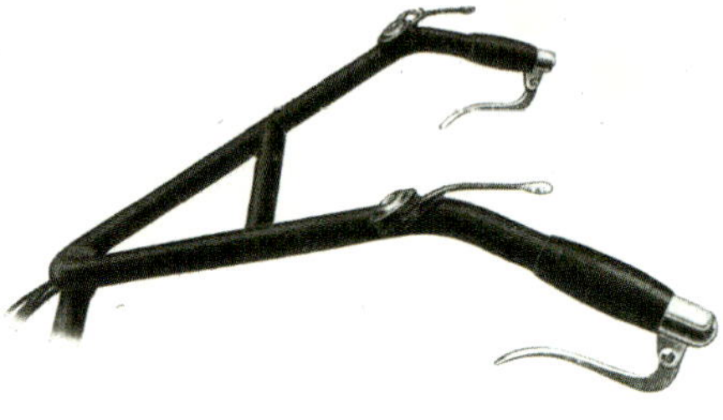

The standard handlebar had a very tidy appearance with cables contained within the tubing.

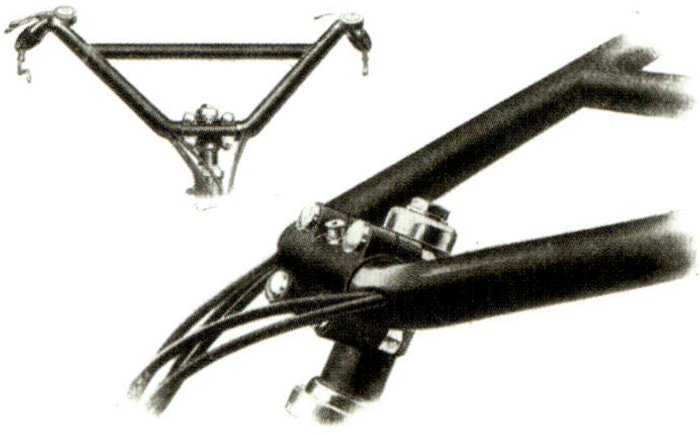

The Patent Spring Handlebar design.

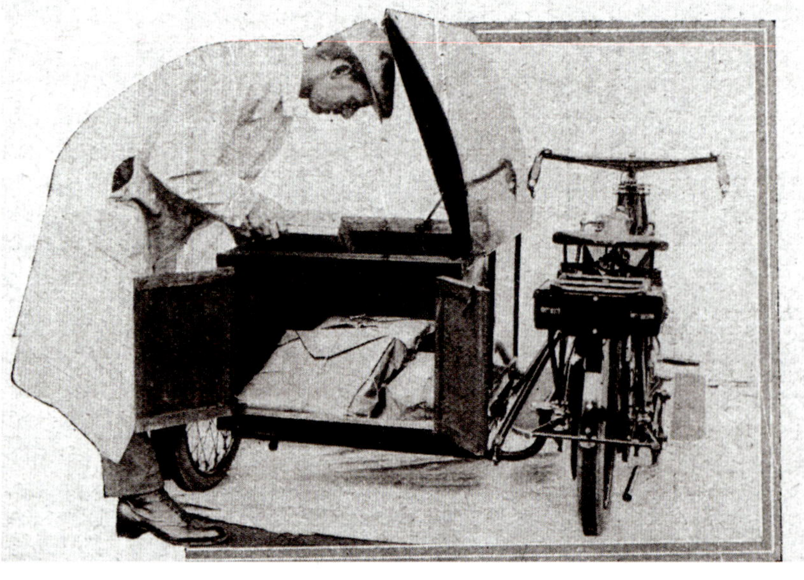

A factory illustration demonstrating the ease of loading a box side-car.

The JAP 6hp Twin Engine

John Prestwich produced his first motorcycle engine, a single cylinder engine of 293cc (70 x 76mm), in 1902. Very soon J. A. Prestwich of Tottenham was amongst the leading suppliers of proprietary engines to the motorcycle industry. The company's twin-cylinder engines quickly established an enviable reputation and by 1914 were being fitted by more than twenty manufacturers. It was no surprise when, in October 1913, the company announced that for 1914 it would specialise in twin-cylinder models. Four new twins were planned ranging from 5hp (70 x 85mm), 6hp (76 x 85mm), 8hp (85 x 85mm) and 10hp (85 x 95mm); the latter being intended for use in cyclecars. The improvements made to the 1914 engines comprised a simplified arrangement of valve timing, a new pressure-fed oiling system and a new design of exhaust valve lifter,

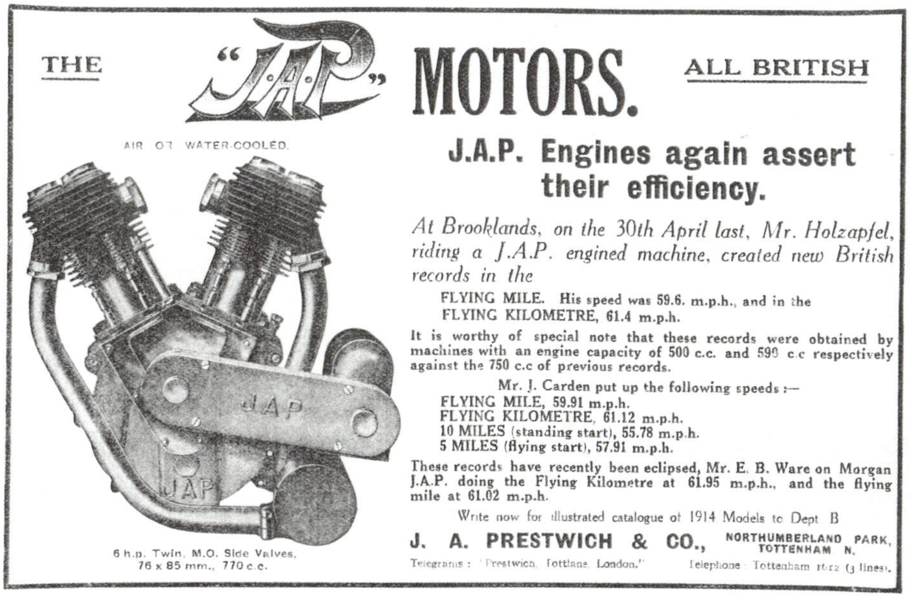

By 1914 JAP engines had demonstrated their worth in racing and record breaking. This advert dates from May 1914.

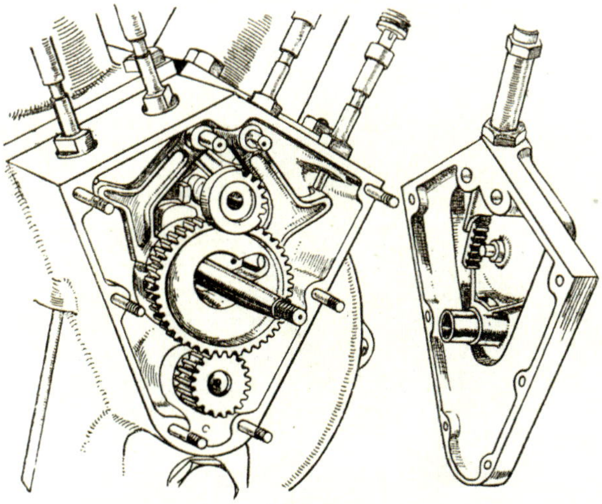

The valve gear of the 6hp JAP engine fitted to the Model 180 was a patented design with a single cam-shaft. By an ingenious arrangement of hardened steel levers linking the cams to the tappet rods, the four valves were operated by two cams, one operating the inlet valves and the other the exhaust valves.

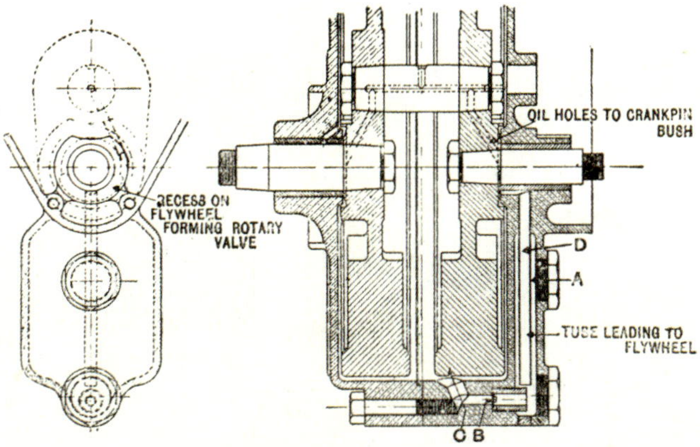

An additional method of lubrication was employed with the 6hp JAP engine for 1914. An oil box A, cast into the side of the crankcase, was connected to the crankcase by the passage C and non-return valve B. With the piston at the bottom of the down stroke, pressure in the crankcase forces oil through the valve B into the oil box A. When the piston is near the top of its stroke, the partial vacuum in the crankcase causes the oil to be sucked up the pipe D and pass through a recess under the bearing bush and into a recess in the flywheel. From here it is thrown by centrifugal force through passages in the flywheel to the main and big end bearings. The boss on the flywheel acts as a rotary valve to block the recess under the bearing bush when the piston is at the bottom of its stroke.

RIGHT: Hand cranking the 6hp Twin into life.

The Royal Enfield Redditch Works

The Enfield Cycle Company's original works at Hunt End had been frequently enlarged to meet the requirements of the expanding business. In 1907 new works were constructed in Hewell Road, Redditch and the Hunt End works sold. However with the expansion of the business, and despite extensions at Hewell Road, the company required larger premises. The Hunt End. works were reacquired in 1912 and by 1914 both factories were exceedingly busy. In May 1914 The Motor Cycle and Cycle Trader reported upon a visit to the Redditch works

At the time of the visit the area of the workshops and offices amounted to 1¼ hectare and a further 2 hectares of land were occupied by a test track, out buildings and land available for additional building. The Hunt End works were of comparable size. Despite the large factory space at its disposal and the adoption of up-to-date equipment and the latest methods, the volume of business was already proving taxing. The reporter was impressed that 'the Company could possibly deal with such a large output as the present with the rapidity and dispatch now achieved'.

The works used best practices for the period. The shops were substantially built and light and airy. Each block was ventilated separately with a continuous supply of ducted fresh air, which could be warmed in the cooler months. The company had its own power station. A Wilson pressure gas plant, consuming 1,300 tons of coal annually, supplied three large Westinghouse gas engines having a combined output of 570hp. These were adequate both to light the works and offices, 2,000 candle power arc lamps were used in the works, and to power the electric motors used to turn the overhead shafting used to drive the machinery.

The quantity of each of the Royal Enfield models to be produced over any given period was decided by the directors. This figure was used by the Drawing Office to draw up schedules specifying the amount of materials and the numbers of each particular part required for the selected models. Copies of the schedules were passed to all departments. The Buying Department arranged the supply of raw materials to the manufacturing departments, who were responsible to ensure parts were made available in the stores for the Assembly Department. The completed machines went to the testing and examination shop for stringent inspection and tests before being allowed to be passed to the Despatch Department to be packed for shipment.

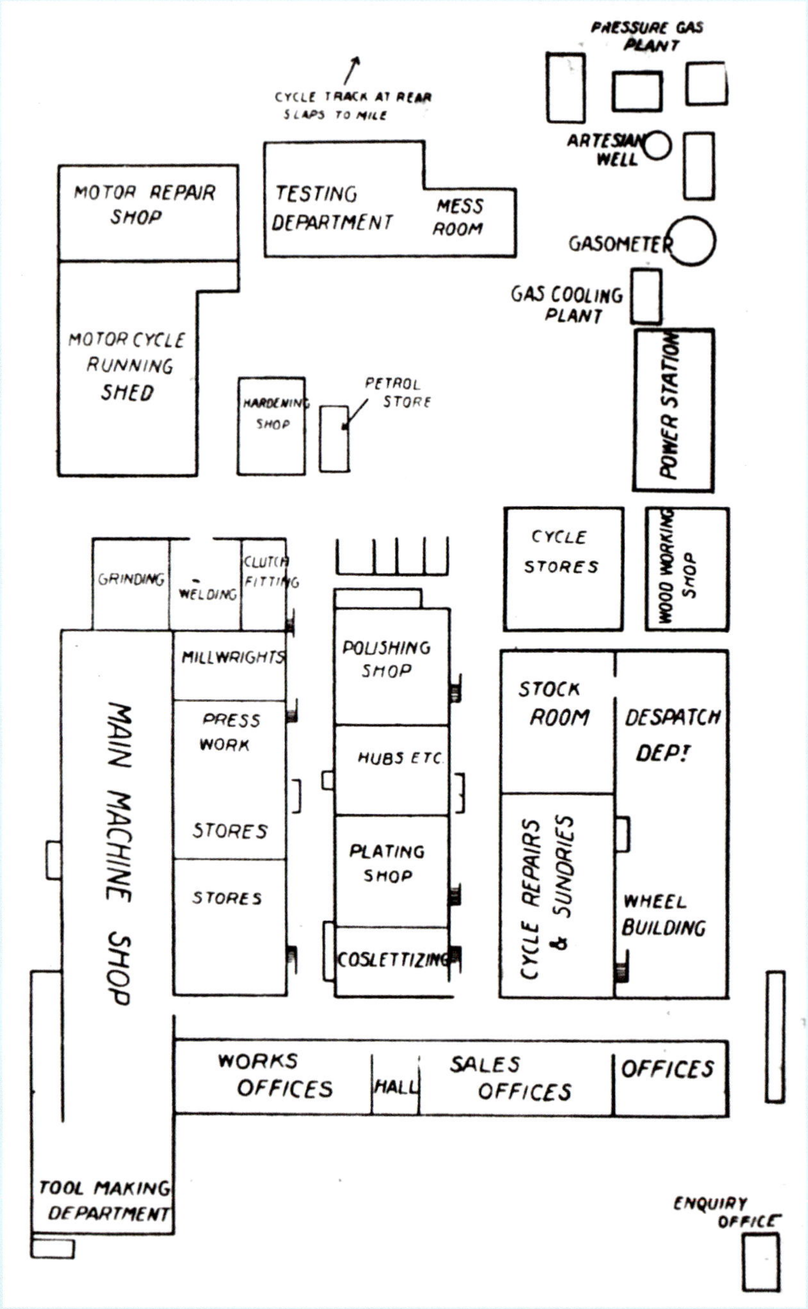

Plan of the Redditch Works. Save for a few exceptions such as tyres and electrical equipment, the Enfield factories were capable of carrying out all the operations inherent in the manufacture of cycles and motorcycles from raw material through to the finished machine

ABOVE: The photograph shows the various parts of the 3hp twin engines laid out for assembly.

RIGHT: The assembled 3hp twin engines undergoing testing.

LEFT: The sidecar body building shop at the Hunt End factory. The Enfield Cycle Company employed 40 coachbuilders in this shop.

RIGHT: Testing was carried out on the local roads, which included several severe hills, and on a specially constructed track at the rear of the works. The track had banked corners and, although only one fifth of a mile long, allowed high speeds to be attained. The illustration shows a 6hp Twin undergoing testing on the works track in 1913.

Two-stroke Debut

Since resuming motorcycle production the Enfield Cycle Company had concentrated efforts on lightweight V-twin motorcycles. These found favour with wealthy customers but there was a requirement for a cheaper model to appeal to the less well off. Under R.W.Smith's guidance, studies of alternative engine designs were carried out with particular emphasis on the two-stroke as being the most suitable design for the economy market. These had been underway for several years before the first prototype of the famous 225cc single-cylinder two-stroke was constructed in 1912. It would require a further eighteen months of intensive development to be carried out before the engine was considered suitable for production.

The first public appearance of a Royal Enfield two-stroke engine was at the Coventry and Warwickshire M.C. Hill Climb in Coventry at the end of August 1914. The Motor Cycle reported: 'It is interesting to note that the Enfield engine is in no way a copy of existing machines, but the designer has struck out boldly on lines of his own. Perhaps the most interesting feature is the slide valve employed to control the port openings. It is actuated from the crankshaft, and it is claimed that the petrol consumption is considerably improved by its use.' The slide valve would not be used on production machines. The engine was mounted in a loop frame although it was noted that production versions would use the standard Enfield divided frame arrangement.. A Enfield two-speed gear was fitted and The Motor Cycle understood that in the final form the engine and gear would be constructed as a unit.

Both the engine and frame would be considerably modified and improved before the two-stroke model was announced to the public and exhibited at the Motorcycle Show in November 1914. The new model would feature as part of the Royal Enfield model range for 1915.

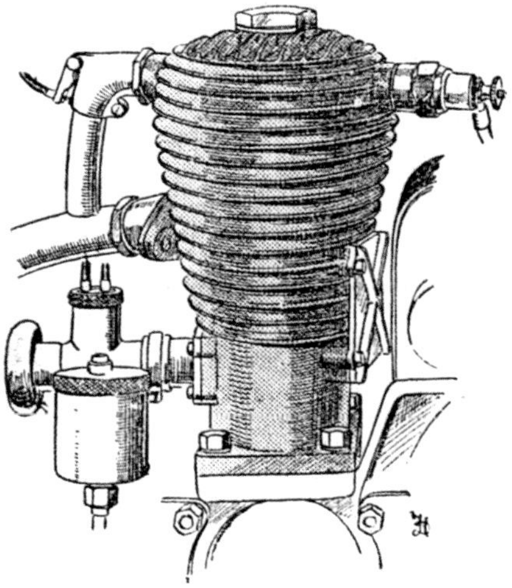

Detail of the Royal Enfield two-stroke engine.

The Royal Enfield Two-stroke which made its first appearance at a Coventry Hill Climb. Significant changes in design and appearance would be made before the machine was put into production.

The Successful Royal Enfield

The sporting successes of previous years had gained the Enfield Cycle Company valuable publicity and track usage had proven useful in testing the machines. As a consequence, 1914 saw the company commit more fully to racing and several more riders chose to ride Royal Enfield machiness. The successes achieved during the year were applauded under the banner 'The Successful ROYAL ENFIELD' in the following year's brochure.

The first appearance in 1914 of the Royal Enfield race team was at the BMCRC-organised Brooklands TT Races on Saturday 18 April. Bert Colver had by this time re-joined H.Collier & Son and the Royal Enfield riders were E.Keyte, H.Greaves and F.E. Wasling. The chosen race machines were again based upon the 424cc ioe models but with the stroke reduced to bring the capacity within the 350cc capacity limit for the Junior class. The race, which was over 26 laps of the circuit, appeared to be a close contest between Greaves and Les Bailey on his Douglas, but over the final laps Bailey pulled away to take the flag with a lead of just under a quarter of a mile. Greaves took second place, with an average speed of 57.89mph, and was followed home by Wasling in third place.

The BMCRC Brooklands Race Meeting on Saturday 2 May saw further success for Royal Enfield in the 350cc one-lap sprint race. The race was won by Keyte at a speed of 57.89mph with Greaves taking second place.

There was a very strong turnout of Enfield riders for the Isle of Man Junior TT on Tuesday 19 May. Nine of the 49 entries for the five-lap race were Royal Enfield twins. The race started well for Enfields with Irish rider F.J. (Fred) Walker taking an early lead with Douggie Alexander in second place and other Royal Enfield riders in third and fourth place. Alexander lost his second position when his seat loosened and fell off. He was forced to ride the remainder of the race seated on the tank and did well to finish in eighteenth position. His second position was soon occupied by Eric Williams riding an AJS. Walker retained a narrow lead over Williams until the third lap when a puncture caused him to crash. With the puncture repaired he rode flat out to catch Williams, but the delay over the puncture had been too great. He achieved a creditable third place at 43.72mph against Williams' winning speed of 45.48mph. Other Royal Enfield riders completed the race in positions 8 (E.Keyte), 15 (H.Greaves), 18 (D.S.Alexander), 20 (D.G. Prentice), 21 (B.A. Hill), 23 (D. Iron) and 26 (W. Sheard). Of the nine Royal Enfield

A photograph published in Motor Cycling of some of the Royal Enfield team in the Junior Tourist Trophy Race.

Douggie Alexander was lying second at the end of the first lap in the Junior TT when his saddle loosened and fell off. He was forced to complete the race sitting on the tank and achieved eighteenth place.

riders only Wasling failed to finish. F.J.Walker's 350 cc vee-twin machine, which achieved third place, survives with a replacement engine and is displayed at the National Motor Museum, Beaulieu.

This excellent performance by the Royal Enfield riders was marred by a tragedy at the end of the race. After his crash on lap 3 Walker had ridden flat out in pursuit of Williams and the AJS. He was reputed to have fallen at least twice during his mad pursuit and to have remounted in order to continue the chase. Such was his pace that he achieved third place. He crossed the finishing line seemingly unaware that the race had ended. It is not known whether he was in a dazed condition from his earlier crash or was physically exhausted, but he did not slow for the barriers denoting the stopping point and crashed headlong into them. Badly injured he was taken to hospital where he died five days later without having gained consciousness. It was a sad finale to a very creditable performance by the Royal Enfield riders.

The AJS of Cyril Williams again showed its supremacy over the Royal Enfields during the 56-lap Brooklands Junior TT Race on 13 June. In the early stages of the race the lead alternated between Les Bailey on his Douglas and Keyte (Royal Enfield), with Williams (AJS) and Prentice (Royal Enfield) vying for third. By lap thirteen Williams had established his dominance over Keyte and Prentice. He went on to win at 53.99mph with Keyte second at 53.18mph and Prentice third at 51.83mph. The other Royal Enfield riders were excluded from contention in the race by a back tyre blown off its rim (D. Iron) and petrol supply problems (F.E.Wasling).

The last BMCRC race meeting at Brooklands before the track was closed to motor sport for the duration of the war was held on July 25 1914. Royal Enfield were represented by E. Keyte and H. Greaves. The 350cc Solo Ten-mile Scratch Race was won by S. Mason on his twin-cylinder NUT-JAP with Greaves second and Keyte third. Keyte repeated his second place performance in the All-Comers Scratch Race. For the final race of the day, a 3-lap Passenger Handicap race, Greaves attached a lightweight racing sidecar to his machine. Starting from limit position, he won the race at an average speed of 47.58mph comfortably in excess of the existing 350cc class record for the distance. He was denied the record as the regulations at the time did not allow record breaking during the course of handicap events.

Departure of William H.Guillon

The outbreak of war, which caused a suspension in the design and production of civilian motorcycles, resulted in William H.Guillon leaving Enfields to return with his wife, a local Redditch girl, to his native Switzerland. He had joined Enfields in 1905 as works manager of the Enfield Autocar Company and was subsequently involved in motorcycle design.

Members of the Green Street Green (Kent) Fire Brigade.
It is a Standard 6 h.p. Royal Enfield Side-car Combination.
We do not advise this overloading but—
Royal Enfield reliability is equal to it.

An old Maxim gun, minus its breech and incapable of being fired, was purchased by the Enfield Cycle Company for advertising purposes. With the company's slogan 'Made Like a Gun', prominently displayed it was used for display in agents' shops. Whilst on shipment to the Belfast dealer, Mr W.H.Alexander, it was seized at Liverpool Docks owing to the prohibition on the importation of arms into Ireland.

A photograph from the Works album of Monsieur A Prudhomme of Marseilles, who travelled to Redditch in 1914 to collect his 3hp Model 140. Monsieur Prudhomme featured in the 1915 brochure.

A line-up of new 6hp Twins waiting shipment to New Zealand in 1914.

CHAPTER TEN

Model Years 1915 - 1916

The first appearance of a Royal Enfield lightweight two-stroke machine had been at the Coventry & Warwickshire Motor Cycle Club's Hill Climb on Saturday 29 August 1914. The machine on the company's stand at the Olympia Show in November 1914 had a similar specification, but in most respects was a totally revised machine. Identified as the Model 200, the machine was in the increasingly popular 225cc, 2¼hp, category. Its production meant that the Enfield Company was now able to offer a range to satisfy all tastes and requirements. Furthermore the company's machines were amongst the best of their type available and competitively priced.

The two-stroke, with its Enfield-built 225cc motor, provided economical transport for the solo rider. It fulfilled the role previously occupied by the Lightweight (2¾hp, MAG side valve twin) Model 160, which was by now of dated styling and dropped from the range. For the solo rider seeking a faster touring machine the choice was for the Guillon-designed 424cc, 3hp models, which remained available in standard form as the Model 140 and in sports form as the TT Roadster Model 150. The side-car Model 180 remained in production with a choice of body styles making it suitable for either the family man or the tradesman. The standard engine was still the 6hp side valve JAP, but for riders requiring more performance it could also be supplied with the larger 8hp 980cc side valve JAP engine. A new de luxe version of the passenger sidecar, designated the Model 190, was offered for the discerning rider and came equipped as standard with a Lucas dynamo lighting set.

Specifications of 1915 & 1916 Models						1915	1916
Model 200	2¼hp	225cc	ts	single	Solo	£39 18s	£42
Model 140	3hp	424cc	ioe	v-twin	solo	£52 10s	£52 10s
Model 150	3hp	424cc	ioe	v-twin	T.T.	£52 10s	£52 10s
Model 180	6hp	771cc	sv	v-twin	s/car	£84	£89 5s
Model 180	6hp	771cc	sv	v-twin	trade s/car	£84	£89 5s
Model 180	8hp	980cc	sv	v-twin	s/car	£86 2s	£91 7s
Model 190	6hp	771cc	sv	v-twin	de luxe s/c	£100	£105
Model 190	8hp	980cc	sv	v-twin	de luxe s/c	£102 2s	£107 7s

RIGHT: The lavish thirty-two page sales brochure, printed in colour on glossy paper, demonstrated the pride of the Enfield Cycle Company in its range of models for 1915.

BELOW: The Enfield works at Redditch and Hunt End.

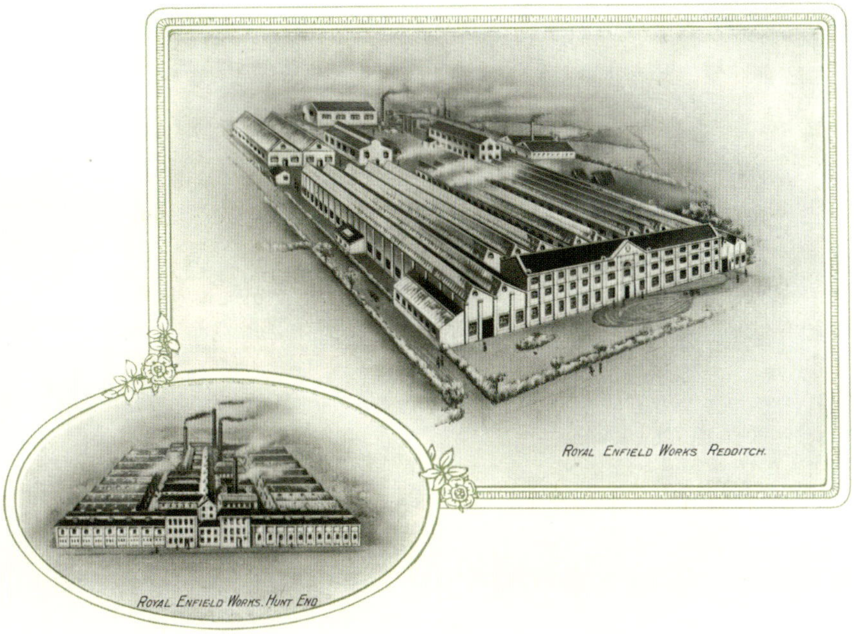

The 2¼hp Two-stroke Model 200

The Motor Cycle magazine was quick off the mark with a full description of the new Royal Enfield two-stroke on 5 November 1914, at the time of the Olympia Motor Cycle Show. This was followed up by a road test of the new model on 1 April 1915. The reviewers found much to praise: 'The new Enfield is of really practical design, and what is more, of attractive appearance – a combination difficult to obtain.'

The power unit had a 64mm bore and 70mm stroke, giving a capacity of 225cc, and was rated at 2¼hp. The cylinder, with 1¼in fins, was considered a fine specimen of the foundry worker's art. The piston was of the deflector type and was fitted with two rings. The crankshaft ran on roller bearings of Enfield design and the conrod on a plain bearing. The flywheel was mounted externally on the right side of the machine.

The Motor Cycle testers found much to praise, but also came up with some criticism, which was possibly harsh by the standards of the day: 'When we took the machine on the road its hill climbing ability was at

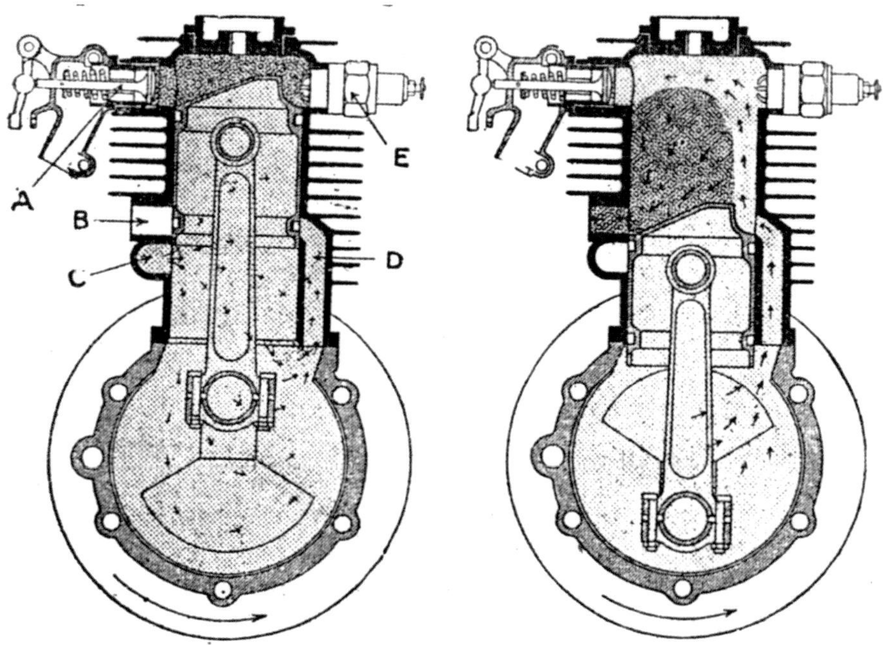

Enfield 2¼ h.p. two-stroke engine.
A. Release valve. B. Exhaust port. C. Inlet port.
D. Transfer port. E. Sparking plug.

ABOVE: The engine was of the three-port variety with the inlet and exhaust ports at the front and the transfer port in the rear. The diagrams show the operation of the engine.

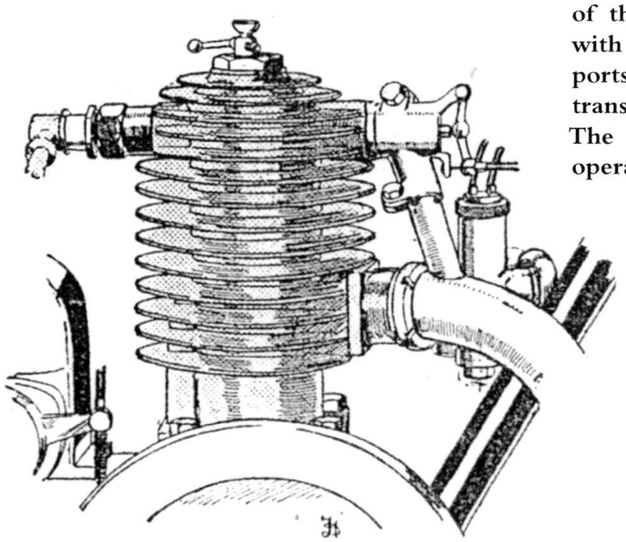

RIGHT: The inlet and exhaust pipes of the 2-stroke engine started from a common plate in front of the engine. The decompression valve was located just above the ports and was fitted with a pipe to convey the released gases into the exhaust.

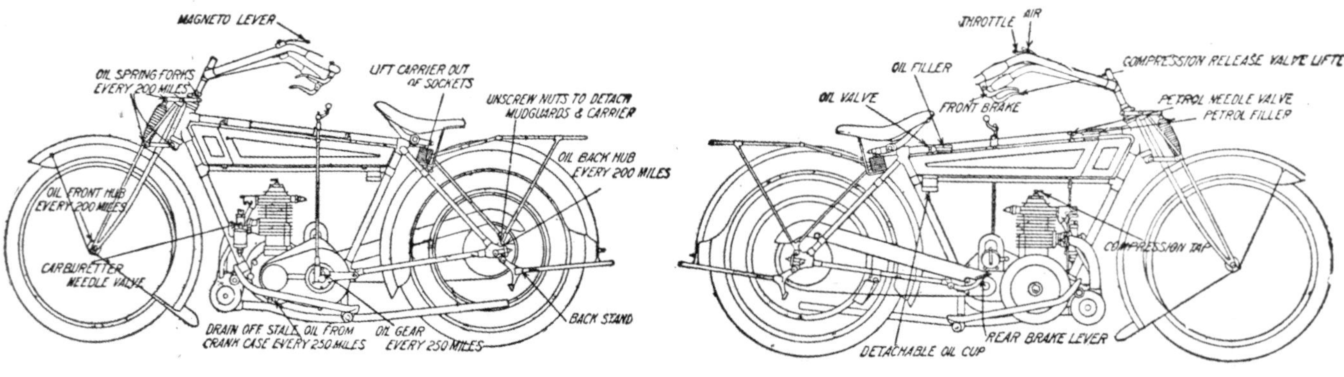

Illustrations from the booklet for the Royal Enfield 2¼hp, which was first produced by the Enfield Cycle Co Ltd in 1917. As well as giving instructions and useful hints for the operation of the machine, the handbook incorporated a full parts list.

Works photo dating from 1914. The oil pump is connected directly into the petrol compartment of the tank. This was discontinued on the production models

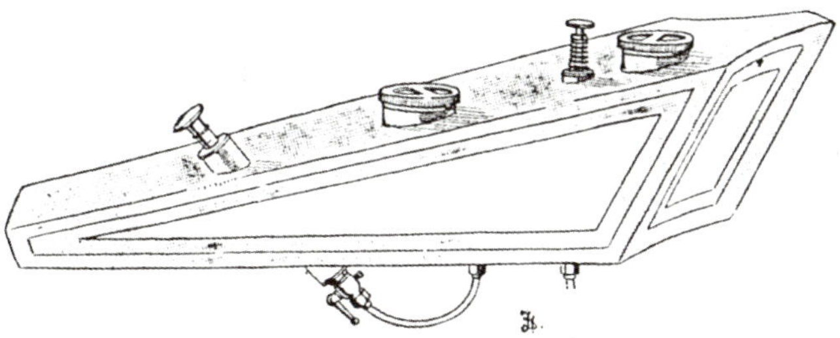

RIGHT: The handsome tapered tank of the Model 200. The front compartment contained the fuel, with oil contained in the rear compartment. Operation of the pump forced a measured quantity of oil through the connecting pipe into the petrol. The pump was replaced by a measuring cup on production models.

The power unit of the 2¼hp Two-stroke . The two-speed gear provided ratios of 5.4 and 9.3 to 1. The carburettor is a multiple jet Amac. The photograph is reproduced from the sales brochure and production models would be fitted with more stylish foot boards with rounded ends.

The oil pump arrangement of the prototype machine was replaced on early production machines by a small cup clipped under the tank, which was used to transfer oil from the rear compartment to mix with the petrol in the forward compartment. Flow into the cup was controlled by a needle valve and six cupfuls were needed to mix with a gallon of petrol.

The $2\frac{1}{4}$ h.p. ROYAL ENFIELD TWO-STROKE

Fitted with the Royal Enfield Patent Two-speed and Free-engine Gear, Cush Drive Hub, and Chain Transmission.

MANY years' experience of motor cycle construction has satisfied us that the two-stroke principle is undoubtedly the best adaptation of the internal combustion engine for the light-weight motor cycle. The simplicity of design, the comparatively few working parts, and the general handiness and compactness of build make it the *beau ideal* for both novice and experienced rider alike.

The $2\frac{1}{4}$ h.p. Royal Enfield includes, wherever possible, those highly efficient features which are embodied in each of the other Royal Enfield models. The Patent Two-speed Countershaft Gear, Cush Drive Hub, and Chain Transmission are all fitted to the standard model. The result is, that whilst being a genuine light-weight model, it is a motor cycle of exceptional strength of construction, great flexibility, and a splendid hill-climber; as a whole, it is a machine in which the utmost confidence can be placed.

SPECIFICATION

ENGINE. Royal Enfield two-stroke, bore 64 m/m; stroke 70 m/m; capacity 225 c/c.

TWO-SPEED AND FREE ENGINE GEAR. The patent Enfield two-speed and free engine gear, expanding clutch type. Direct drive for each gear.

TRANSMISSION. Best quality roller chains throughout.

CUSH DRIVE HUB. The Enfield patent cush drive hub fitted to the rear wheel.

IGNITION. Latest type of magneto, waterproof terminals.

CARBURETTER. AMAC multiple jet.

LUBRICATION. Oil mixed with petrol in proportion of six full measures to one gallon of petrol. Described on page 15.

BRAKES. Powerful rear wheel brake, operated by foot pedal. Inverted lever front rim brake.

TANK. Royal Enfield improved pattern, separate compartments for oil and petrol.

FRAME. Built throughout from best weldless steel tube; low and comfortable riding position.

TYRES. Palmer Motor Cycle tyres, 24in. × 2in.

HANDLEBAR. Royal Enfield latest design, special black finish.

MUDGUARDS. Back and front mudguards of ample width and strength.

CARRIER. Tubular steel carrier, with detachable rear guard.

SADDLE. Well sprung coil spring saddle, of best make and finish.

TOOL BAG. Strong leather, containing a complete set of tools.

FINISH. Enamelled in best black enamel. Tank artistically panelled and lined. Bright parts heavily nickel-plated.

Model No. 200 ... Price: **38** Guineas Net Cash.

A description of the more important Constructional Features appears on pages 14 and 15.

MADE LIKE A GUN

once apparent, and we found that the Coventry test hill at Stonleigh could be easily ascended on top gear with a hot engine. Starting was extremely easy; the machine could be paddled off, whether hot or cold, in a yard or so, and was pleasant and comfortable to ride at moderate or slow speeds, but when the pace was increased, the writer being accustomed to the 3 hp twin Enfield, which runs very smoothly, found the vibration in the handles and footboards more noticeable. The handle-bars were not the standard pattern, but were extra long. This possibly accentuated the vibration; also it is only natural that metal footboards should vibrate more than wooden ones sprung at the forward end.'

'We made no endeavour to reach the maximum speed, but the machine touched 30mph without being opened out fully. The exhaust was quiet, unusually so for a two-stroke engine which is often very noisy.'

The Motor Cycle test concluded by noting that the entire output of the machine had been booked for several months to come.

The Two-stroke proved easy to start by paddling away whilst seated astride the machine.

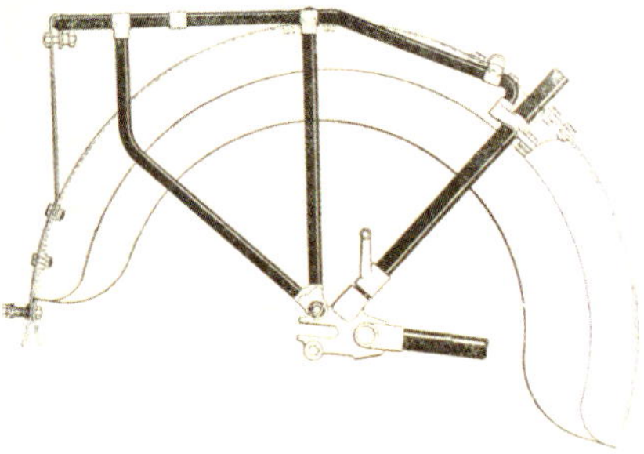

A particularly neat feature of the new 1915 models, including the Model 200, was the ease with which the rear carrier with rear section of the mudguard could be removed, by loosening two nuts, to facilitate rear wheel removal.

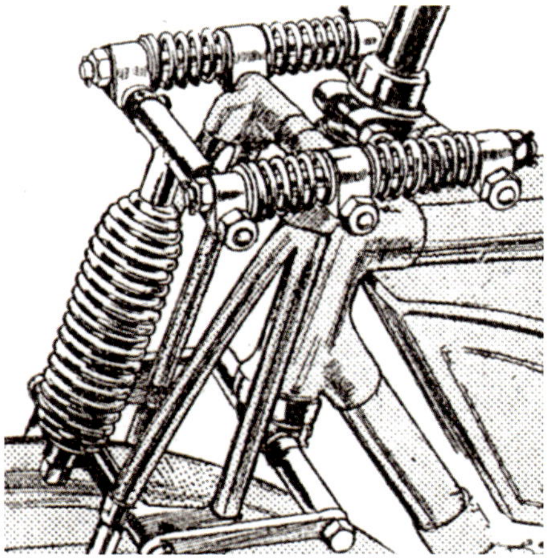

The front fork of the Motor Cycle road test machine was a lightweight version of the Royal Enfield design but with the addition of a horizontal springing. It combined the up and down movement of the linked forks with the backward and forward movement of the hinged variety of fork. This fork design does not appear to have been fitted to production models.

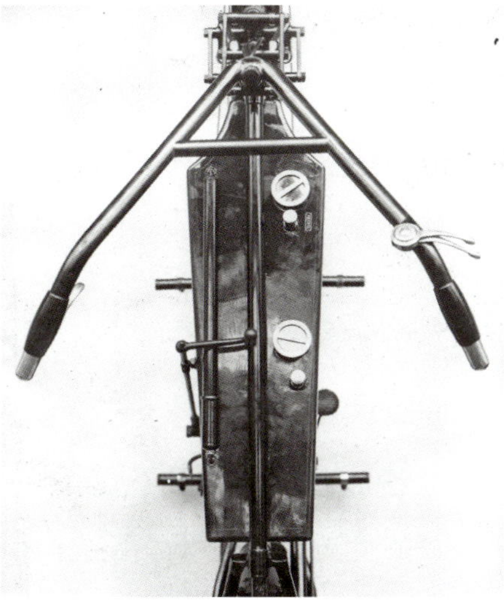

The slim lines of the 2-stroke Model 200 are apparent in this works photograph. The brace between the handlebars does not feature on the production machines.

3hp Twin, Model 140

Specification of Model 140

Engine: 60° ioe v-twin
Capacity: 424cc
Bore x stroke: 60mm x 75mm
Carburettor: AMAC multiple jet
Lubrication: Mechanical, dry sump
Ignition: Magneto
Gearbox: Two-speed, RE patent
Kick-starter: Direct chain to crankshaft
Transmission: Chain, RE cush drive
Tank: 1½ gallons petrol, 1 quart oil
Frame: RE loop frame
Suspension: RE spring front fork
Saddle: Well sprung leather
Tyres: Dunlop Cord 26 x 2¼in
Brakes: **F**ront stirrup, rear dummy rim
Finish: Black, panelled tank in two shades of green
Price: £52 10s

3hp TT Roadster, Model 150

Specification as Model 140 with following exceptions:
Semi TT style handlebars
Slim front mudguard and shortened footboards.
Kick-starter omitted

Mrs Riley of Birmingham on her new 1915 3hp Twin.

This photograph of Mrs Riley with a 6hp Twin was presumably taken when she visited the Works to purchase her 3hp Twin.

Model 140 Combination

The manufacturers considered the 3hp machines to be unsuitable for use with a sidecar attached. Nevertheless many owners were fitting sidecars and the decision was made to offer a special lightweight sidecar designed specifically for the Model 140. The sales brochure noted: 'It is built in our own coach-building shops from thoroughly seasoned three-ply wood, and is particularly light, but at the same time strong and reliable. The chassis is constructed of weldless steel tube throughout, and there is no other coach-built side-car in which lightness and strength are better embodied, or which is more suitable for this model.' No illustration of the sidecar was available at the time the brochure was printed, nor price quoted. There was no mention of the sidecar in subsequent sales literature.

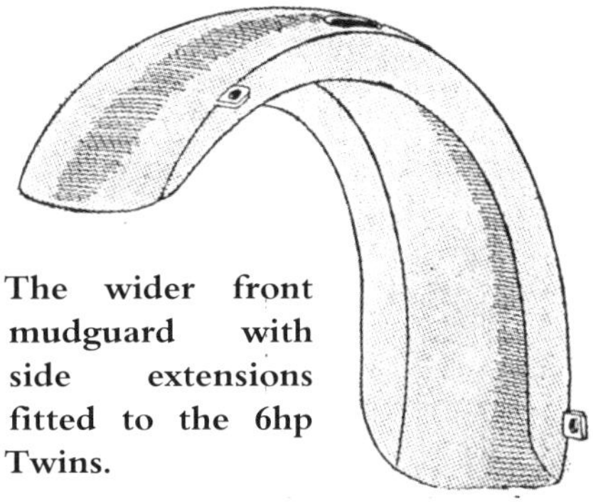

The wider front mudguard with side extensions fitted to the 6hp Twins.

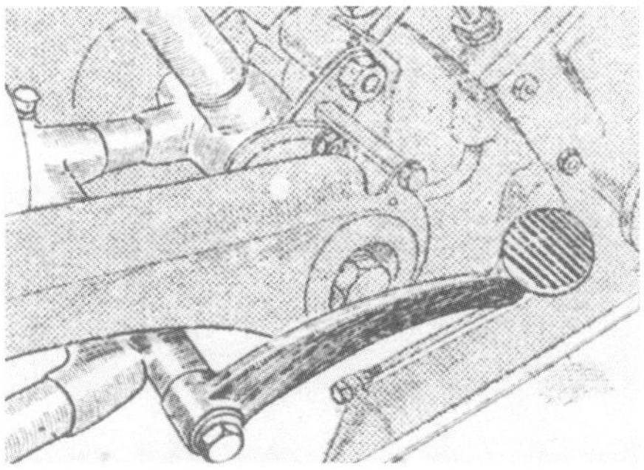

Rear brake pedal and cross-over shaft.

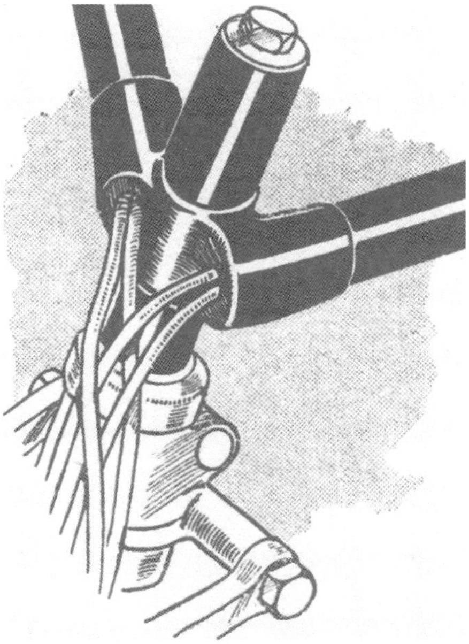

The revised steering head introduced on the 6hp Twins included an extension for mounting the headlight and acetylene generator.

6hp Side-car Combination Model 180

The most obvious design change to the 6hp model for the 1915 season was to the top tube and fuel tank. Both were now dropped at the rear after the style of the 3hp models. The machines also acquired a new wider front mudguard and the detachable rear carrier and guard similar to that fitted to the new two-stroke machine. To benefit from the more quickly detachable carrier and guard, which it was claimed could be removed in under half a minute, changes were made to the rear brake to allow the rear wheel to be more readily removed. The brake was arranged to operate inside the groove of the dummy belt rim on the left-hand-side of the machine; the pedal remained on the right-hand-side and a lug formed in the frame to transfer the connection.

8hp Twin Model 180

For owners wishing for greater power, the sidecar combination was also offered with an 8hp (85 x 85mm) JAP engine for an extra charge of 2 guineas.

De-luxe versions of both the 6hp and 8hp models were available with a Lucas llghting set and all fittings ready for the road. Designated the Model 190, they cost an additional £16.

6hp Side-car Model 180

Engine: 6hp JAP Side-valve v-twin
Capacity: 771cc
Bore x stroke: 76mm x 85mm
Carburettor: AMAC multiple jet
Lubrication: Hand pump with visible drip
Ignition: Magneto
Gearbox: Two-speed, RE patent
Starter: Hand crank and chain
Transmission: Chain, RE cush drive

Tank: 2 gallons petrol, 3 pints oil
Frame: RE loop frame
Suspension: RE spring front fork
Saddle: Leather covered pan
Tyres: Palmer Cord 'cycle car' 650 x 65mm
Brakes: Front stirrup, rear dummy rim
Finish: Black, panelled tank in two shades of green
Sidecar: RE coach built with luggage carrier
Price: £84

The Royal Enfield cane side-car body was lighter than the coach-built body, but provided less passenger comfort. It was already an old fashioned design and 1915 would be its last year of production.

A special frame and sidecar chassis giving improved ground clearance was offered for traversing rough roads; a full 7 inches of clearance being provided between the lowest part of the crankcase and the ground. The specification in other respects was unchanged. However the illustration in the sales catalogue shows the front stirrup brake replaced by a dummy rim brake. This improved front brake would subsequently be incorporated into the specification of the 6hp and 8hp models.

The 6hp Trade Delivery Combination Model 180. The dimensions of the carrier box were;:length, 3ft 6in; depth, 1ft 10in; width, 2ft 1in; capacity 14 cu ft. A square basket carrier could be substituted for the coach built box at no extra charge. For an extra £10 the combination would be supplied with an additional coach built passenger body to make a practical dual purpose vehicle.

The 6hp Twin fitted with Lucas dynamo electric lighting set.

The 6hp Twin with the sidecar seat removed to show the location of the battery for the Lucas dynamo lighting set.

The 6hp Twin Model with coach built sidecar body and full weather equipment. Particularly interesting is the additional skirt behind the front mudguard.

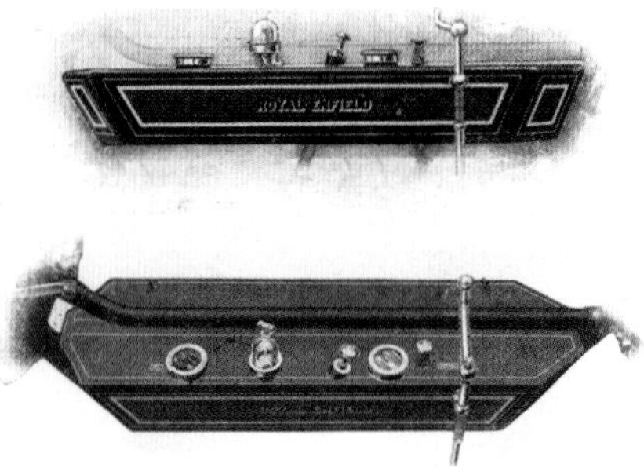

The pre-1915 style of tank fitted to the 6hp Twin.

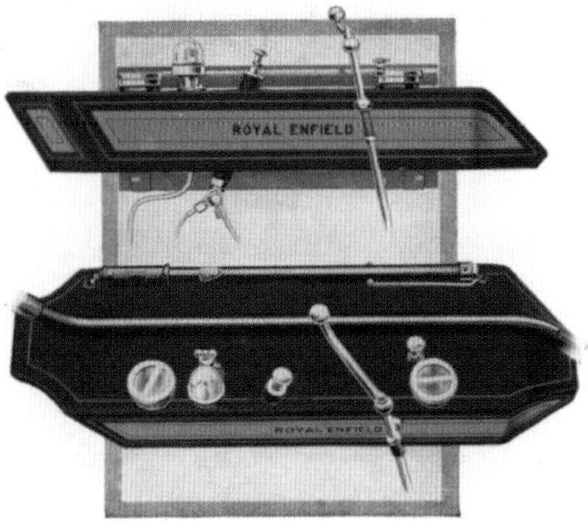

The later style of tank fitted to the 6hp Twin. The partial vacuum in the engine crankcase draws oil from the tank though the domed sight-glass to feed the engine; supply is adjusted by the adjacent turn-screw. A separate oil pump provides extra oil, as required, for negotiating severe gradients or for high speed riding. A two-way tap below the tank allows oil to be directed towards the 2-speed gear. The petrol supply is turned off using the knob adjacent to the petrol filler cap. The two-speed gear change lever passed through a clamp on the side of the tank.

A batch of 6hp Twins awaiting despatch in 1915.

An extension to the works being constructed in 1915.

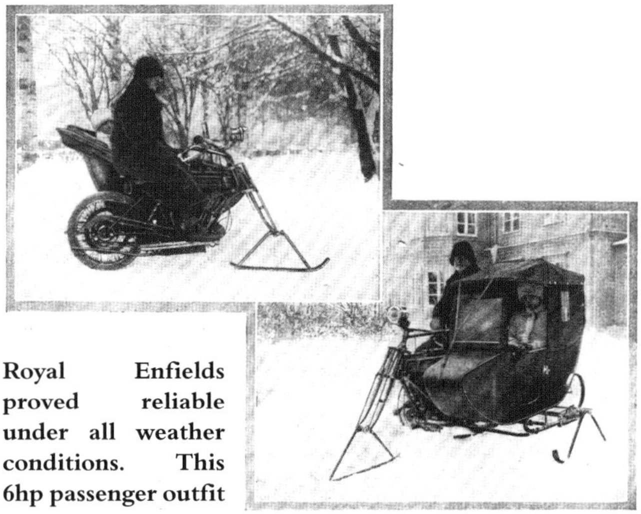

Royal Enfields proved reliable under all weather conditions. This 6hp passenger outfit was converted by its owner, Bertel Sandberg of Stockholm, to ride on skis in order to cope with the Swedish winter.

A works photgraph of a Model 190 De-luxe sidecar.

The 1916 Model Range

The 1916 range of Royal Enfield motorcycle models was unchanged from the previous year, although price adjustments were made. The 2¾hp two-stroke Model 200 was increased in price by two guineas to 40 guineas and both the passenger and trade delivery versions of the 6hp Twin Model 180 were increased in price by 5 guineas to 85 guineas. The price of the 3hp Twins, Models 140 and 150 were unchanged at £52 10s.

The inside covers of the brochure illustrated the dual roles of the Royal Enfield sidecar combinations at this time. The inside front cover, on the left, extolled the freedom of the open road and the ability to explore the beauty of the countryside. The inside rear cover depicted the military role of the machine gun sidecar combination. The first details of the machine gun carrier had been made public at the end of 1914. They are discussed in more detail in the following chapter.

Despite the restrictions and shortages due to the war, Royal Enfield continued to produce a lavish twenty page colour brochure.

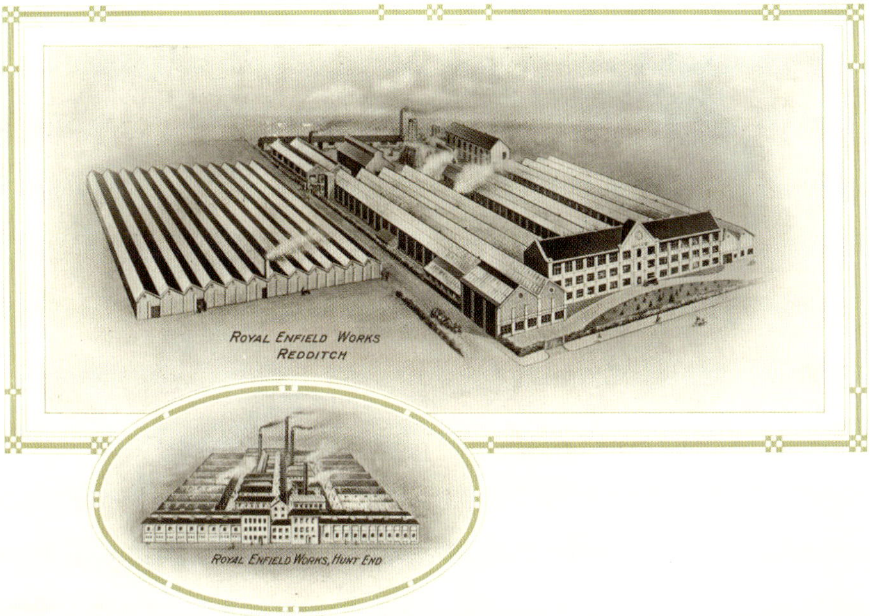

The illustration of the works in the 1916 sales brochure shows the extension to the Redditch factory made to satisfy wartime production demands. The Redditch works at this time extended over 9 acres and had an aggregate floor space of 200,000 square feet.

The two-stroke model had proved popular and no changes to the design were considered necessary. However, several modifications were made to the 3hp models. The chain and back stays were now of round section tubing whilst the whole of the frame was considerably strengthened, being built on exactly the same lines as the Colonial model, although it was noted that the earlier frame had never given any trouble from any structural point of view. Instances of poor lubrication of the front cylinder of this model were attended to by providing a direct gravity feed to the wall of this cylinder. Finally a revised induction pipe was adopted as in the previous arrangement the carburettor connection had been near the rear cylinder and required a rather sharp bend, which had caused the rear cylinder to run rich.

The front dummy rim brake, which was first used with the 1915 colonial version of the 6hp Twin, was now adopted for all 6hp and 8hp models. It was noted that wheel detachment was greatly improved with no blocks and shoes to get in the way. An alternate design of silencer with exit pipes at either end was also used on the big twins. A detailed change on all models was the support for the upright gear-changing rod being moved from the side of the tank to the top rail above the tank. This provided better support for the rod and also brought the gear knob more readily to hand.

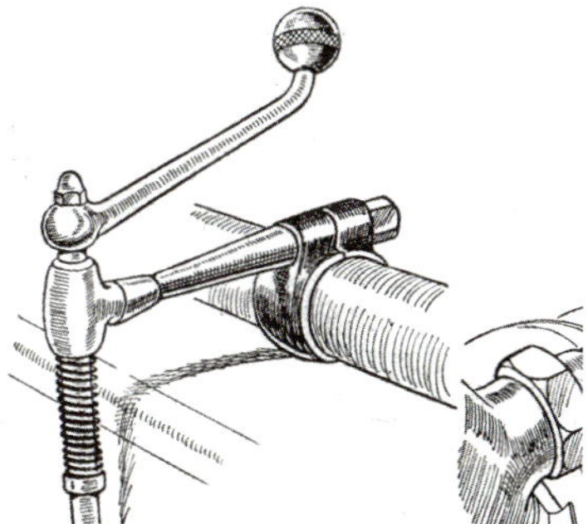

The revised method of supporting the gear-operating lever introduced with the 1916 models.

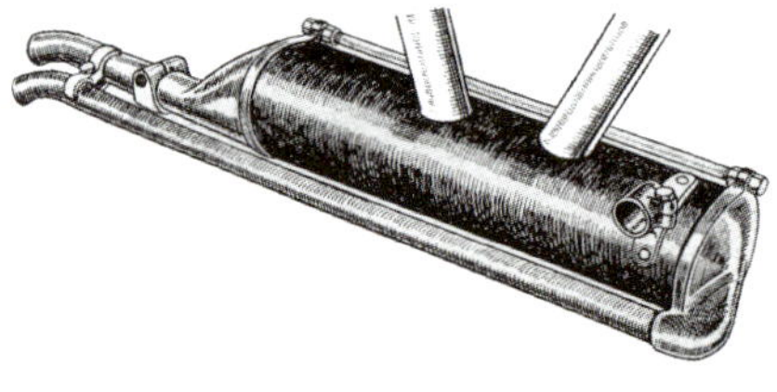

The new design of silencer fitted to the 6hp and 8hp twins.

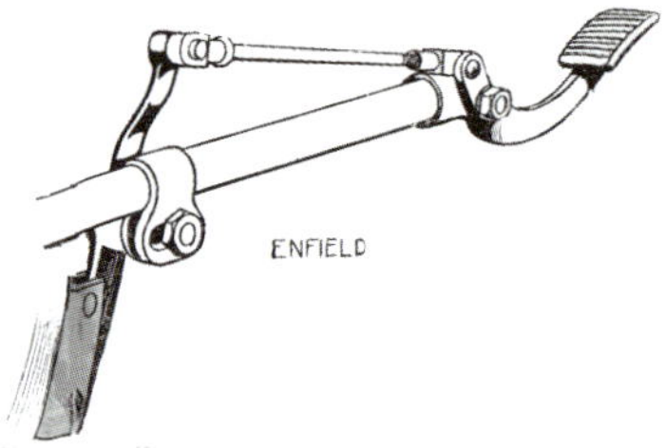

The foot-operated dummy rim brake fitted to the 2¼hp models. Dummy rim brakes were also now being fitted to the front wheel of the 6hp models.

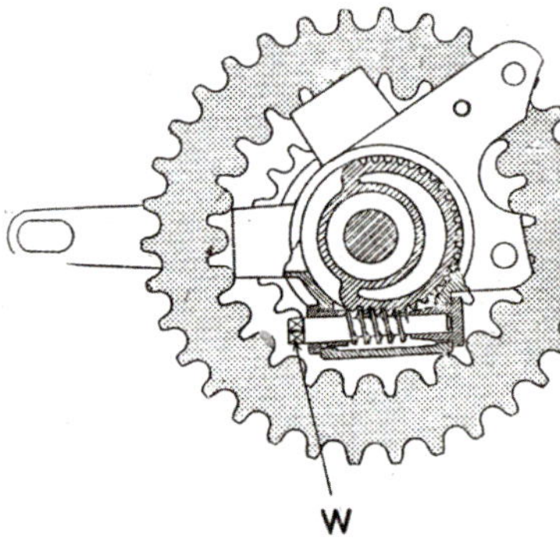

Improved adjustment of the gear chains on the 6hp models was provided by a worm gear operating on the bottom bracket eccentric.

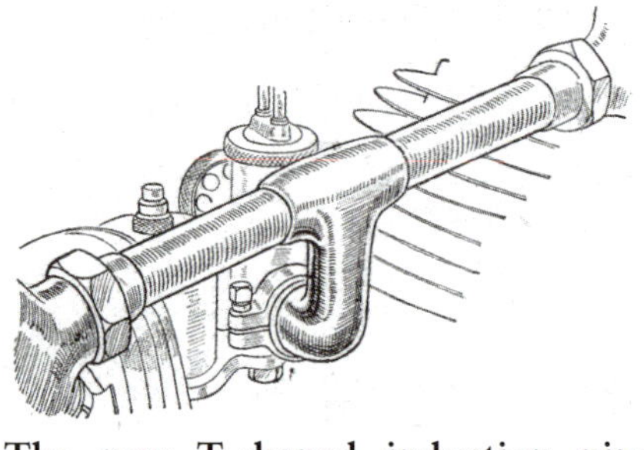

The new T-shaped induction pipe fitted to the 3hp Twins.

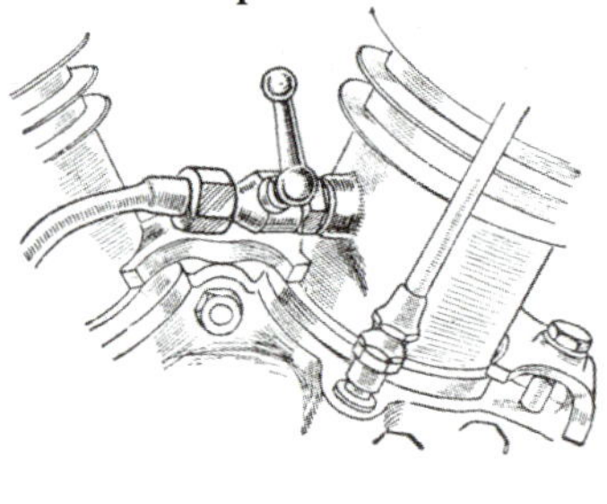

The extra oil feed to the front cylinder of the 3hp Twin.

2¼hp Two-stroke Model 200

3hp Model 2014

3hp TT Roadster Model 150

6hp Side-car Combination Model 180

6hp Dynamo Lighting Combination Model 190

6hp Combination Model 180

Colonial and Overseas Model

6hp Trade Delivery Combination Model 180

Royal Enfields for 1917

The Enfield Cycle Company was not allowed to sell motorcycles to the public during the war years. However, certain overseas and military sales were allowed and the Company continued to make improvements to the existing model range. In November 1916 a revised coach-built sidecar body for the 6hp Model 180 was announced. This provided greater leg room for the passenger and was fited with a screen of Triplex safety glass. The luggage carrier was now mounted on the sidecar body thus gaining benefit from the springing.

The new sidecar body for 1917. The sloping upward line of the nose of the body was designed to provide greater knee room with the screen in place.

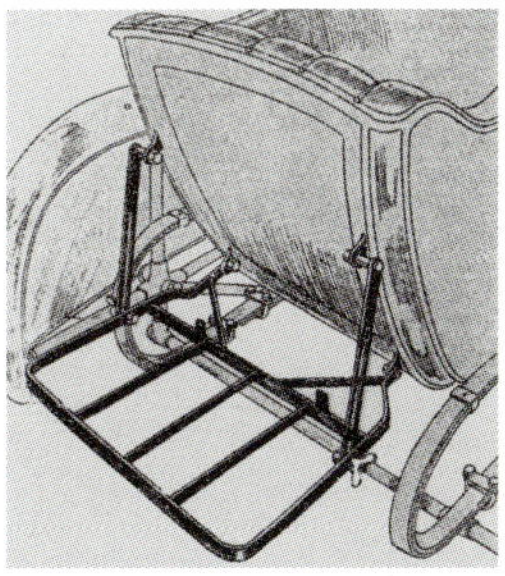

LEFT: The new design of luggage grid.

RIGHT: Wartime Petrol restrictions meant that fuel was not available for private motoring In October 1917, Motor Cycling ran a feature on a 6hp Enfield adapted to run on coal-gas. The photograph shows the Cox gas trailer being towed behind the sidecar.

The Lady Riders

The sale of cars and motorcycles for civilian use continued to be permitted during the early stages of the war. However, with so many men answering the call of duty by enrolling in the armed services, the public demand for motor vehicles was reduced and sales difficult to achieve. The Enfield Cycle Company realised that motorcycle sales would only be maintained by tapping into a new market, namely the women left at home whilst their husbands were away fighting the war. A new brochure titled 'The Lady Drives' was produced.

"The Lady Motor Cyclist."

THERE is something irresistibly fascinating in being one's own driver—the "captain of the craft"—instead of merely reposing as a passenger in the cosy comfort of a side-car. To have the means of journeying just when and where one's inclination leads; to take the high road or the by-road as fancy may dictate; to be as free as the air, and dependent on no one; all these delights—and many more—belong to the lady motorist.

¶ It is our purpose in the following pages to show that motor cycling with a side-car combination is a pastime eminently befitting the fair sex. The illustrations have been reproduced from photographs of lady motorists who drive their own Royal Enfields; whilst their letters, which we reprint by special permission, will best describe the pleasures motor cycling has conferred upon them.

¶ For the lady driver a reliable, easily-controlled motor cycle is a *sine qua non*. In the 6-h.p. Royal Enfield Side-car Combination we offer a model which experienced lady motorists endorse as the finest vehicle on three wheels ever produced. It is beautifully constructed—everything arranged with care and discrimination—and it is tastefully finished in a style which commands admiration from motorists the world over.

The ENFIELD CYCLE Co., Ltd.
Head Offices and Works: REDDITCH
London Office and Showrooms: 48, Holborn Viaduct, E.C.

Amongst the ladies featured was Miss Phyllis Relph (Mrs Lionel Atwill), 'the charming actress', who had been riding a Royal Enfield side-car combination for 15 months.

Testimonials from seven lady riders, together with photographs, were reproduced under headings such as:

'My Royal Enfield is simply delightful'
'Better in Health than I have ever been'
'The Maximum of Pleasure'
'An Ideal Machine for a Lady'
'Motor Cycling as a Sport is simply grand'
'Experienced no difficulty in learning to drive'
'Thousands of Miles without any difficulty'

LEFT: Mrs M.Evans of Wimbledon Park, S.W. found her 'Enfield was the most simple thing in the world to manage.' She considered it ideal for a lady and was sure that 'there would be many more lady riders if they realised its reliability, its low cost of upkeep, and the ease with which it can be managed.'

Mrs Mary Riley of Birmingham was one of the most experienced lady motorcyclists of the day and competed in many of the leading motorcycle trials. She had begun motorcycling on a 3½hp side-car machine, but quickly changed to a 6hp model. She frequently appeared in the Royal Enfield promotional material.

Mrs M.A.Whayman of Heathwood Gardens, Charlton, S.E. had never experienced the slightest trouble with her 1913 model, illustrated, which she had now changed for a 1914 model.

CHAPTER ELEVEN

Wartime Production 1914 - 1919

The years from 1910 through to the early years of the Great War had been a period of prosperity in Great Britain and this had resulted in an increased demand for motorised transport. The number of motorcycles registered for road use in 1910 had been 36,242 and the rate of increase had been only one or two percent a year. However, in 1911 the number of registrations increased by 31.3%, and further annual increases of 46.1% in 1912 and 40.7% in 1913 were recorded. The outbreak of war in August 1914 resulted in a reduced demand and cancelled orders, but the production of motor vehicles did not stop immediately with the announcement of hostilities. In 1914, the overall increase was 26.5% bringing to 123,678 the number of motorcycles registered for road use. Thereafter the increase for 1915 was 12.0% and for 1916 was 10.4%. By 1916 the number of motorcycles registered was 152,960, which compared with 141,621 cars registered for private use.

The commercial production of new vehicles finally came to a stop at the end of 1916 as a result of two measures imposed by the government. The first restricted the supply of petrol for private vehicles to those required for work of national importance. The second and more drastic measure introduced on 3 November 1916 by the Ministry of Munitions prohibited the manufacture or assembly of new motor vehicles without a military permit. From this date all factories were switched over completely to the production of materials for the war effort. Of course motorcycle production did not stop as a large number of machines were required by the military for use by dispatch riders and for other duties. Orders for solo machines were given to firms such as BSA, Douglas and Triumph. The Scott and Enfield Cycle Companies received contracts for the development and supply of a number of special sidecar outfits for use by the army.

The Enfield Armoured Sidecar

Within a few days of the commencement of the war, the Enfield Cycle Company had mounted an early type of Maxim gun on a 6hp chassis to advertise the fact that Royal Enfield machines 'Were Made Like a Gun'. The military combination was submitted to the military authorities and an order placed for further development and production. This first machine had the gun mounted at the rear of the sidecar. The gun could be swivelled and its elevation raised and lowered by a screw thread, but could only fire backwards. A substantial front leg, which was lowered to the ground to brace the machine when the gun was fired, supported a seat for the gunner. This arrangement afforded only a stationary platform for firing the gun. It was quickly appreciated that there was advantage in being able to fire the gun in either direction and with the machine in motion. The ability to fire rearwards allowed the machine to withdraw rapidly and maintain a hail of bullets during retreat.

First photographs of the Enfield sidecar machine gun were published in early December 1914. The announcement in the Motor Cycle

The photograph from the works album dates from late 1914 and is presumed to be the first Enfield combination adapted for use with a Maxim machine gun. The gun faced backwards and could only be fired once the leg at the front of the sidecar, which supported the gunners seat , and its support bracket was lowered to the ground.

The prototype machine gun combination, illustrated on the previous page, showing the riding positions with a three man crew. The front leg has been raised and is supported on a bracket.

noted that a large number of Scott 532cc and 6hp Enfield machines were now being delivered. The magazine noted that both machines had been selected for this very important work owing to being well tried and reliable. The published photographs of the Enfield machines showed the Vickers machine gun installed at the back of the sidecar, but the gun could be adapted for firing in any direction. When firing from the gun platform whilst stationary the two front legs of the gun tripod were released by a half turn of a simple lever. At the front of the platform a substantial stand was lowered which, combined with the two legs of the tripod, carried the weight of the gun and removed the recoil of the firing completely from the chassis springs.

The gun platform could be removed from the sidecar chassis in half a minute and replaced in a similar time. The ordinary complement of crew in charge of the machine consisted of the driver and gunner. A seat was mounted on the rear carrier of the motorcycle for the gunner and the entire sidecar platform was occupied

ABOVE and LEFT: The first photographs released of the Royal Enfield 6hp, 771cc JAP twin, combination adapted for military use. The sidecar chassis was unmodified from the standard road going machine with the exception of stronger springs fitted to take the weight of the gun and 2,000 round ammunition boxes. Longer connection bolts were fitted to give greater width between the motorcycle and sidecar, and larger than standard tyres fitted.

by the gun, ammunition boxes and equipment. The machine carried 2,000 rounds of ammunition.

Machine Gun Batteries

Details of the new machine gun batteries were also given at this time. A battery was to consist of sixty-four men, though an extra six men were to be kept in reserve. There were six gun teams, each having three motorcycles and sidecars. The first machine carried a gun, whilst the second was fitted with gun mountings, but without a gun, being intended to take the place of the former in the case of necessity. The third machine was adapted to carry ammunition and other equipment. In addition there were six riders with solo machines in close attendance to act mainly as scouts plus nine car drivers with further ammunition and other supplies.

ABOVE and RIGHT: The military combination adapted to carry ammunition and other supplies. The passenger sat facing to the rear.

Part of the first batch of Royal Enfield machine gun sidecar combinations outside the Redditch works prior to being sent to the new motor machine gun batteries.. These machines were allotted to the 5th Battery, the members of which were mainly Coventry motorcyclists. The machine guns were not fitted at this stage although a gun has been mounted on the machine in the foreground for the purpose of the photograph.

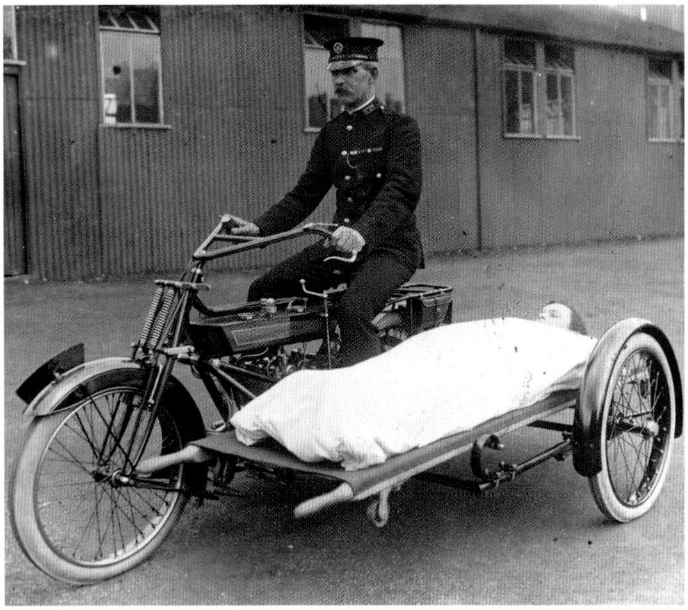

The photograph on the left is thought to be of the prototype stretcher combination as the stretcher and hapless patient are supported on the standard sidecar springs. It was quickly realised that more supple springing was required, as above, to improve the smoothness of the ride.

Further Military Adaptations

The adaptation of the 6hp combination as a stretcher carrier for the conveyance of wounded was also explored. The standard sidecar chassis was used and fitted with special springs to take the stretcher. The Motor Cycle and Cycle Trader noted: 'The 'riding' is wonderfully smooth and such is the flexibility of the engine and gears fitted to the Royal Enfield motor-cycle that the effort of taking up the drive becomes almost imperceptible and the inconvenience to the wounded occupant of the stretcher is thus minimised to the greatest possible extent.'

Revised Design of Machine Gun Sidecar

A revised design of machine gun sidecar, with separate gun mountings at the front and rear of the sidecar chassis, was announced in February 1915. With this arrangement the gun operator was able to adopt a low reclining position, resting against duplicated cushions and backrests, and fire either forwards or backwards whilst the combination travelled at full speed. In the forward mounting position, the gun could also be elevated to be used as a weapon against aircraft.

Capt. J.E.Alkin, commanding No. 5 Battery, M.M.G.S., demonstrates firing the gun backwards. The gun could be fitted to a similar mounting at the front of the sidecar chassis to allow it to be fired forward. The motorcycle is being driven by Sergt. J.Crocker, who was formerly a successful Brooklands rider.

With the gun fitted into the forward position it could be tilted upwards for attacking hostile aircraft. Sergeant Crocker has been called upon to demonstrate this position.

Requisition of the Enfield Works

The Enfield Cycle Company had been been fully committed to the war effort from the outset. On 17 August 1915 the company was requisitioned by the Government for war work. On 16 November the following year further wartime restrictions were introduced which brought an end to all non-essential vehicle production.

In addition to the production of bicycles and motorcycles for the War Department and essential services, the Enfield Cycle Company was involved in the production of artillery shells and in carrying out other engineering services for the war effort.

A batch of ten machines of the revised design lined up outside the works ready for despatch.

A batch of 8hp machines awaiting dispatch to the Imperial Russian Government.

An illustration published in The Motor Cycle in October 1914 depicting a 6hp Enfield in use by the South African Defence Force. By the time of the armistice in 1918, 161 military sidecars had been supplied to the British armed forces.

Experimental Machines

The development of new Royal Enfield models was continued during the early war years and a number of interesting and unorthodox engine configurations were studied and developed by the experimental department. However, because of their potential commercial value the new designs were kept a closely guarded secret. Indeed the studies remained largely unknown, even to those who had spent their working life at the Redditch factory, until the coming of the Second World War and a further curb on motorcycling. With neither new models to review nor competition riding to report, the motor cycle magazines were desperate for historical articles to fill their pages. Henry Laird, a staff reporter on Motor Cycling, visited the Redditch factory to explore the Enfield development stores and archives and he unearthed details of the design studies of the earlier war period. He wrote of his discoveries in May 1940. The engine configurations he described included a three-in-line two-stroke, a unit construction four-in-line four-stroke, a transverse v-twin with shaft drive, a 350cc ohv v-twin, and a 225cc two-stroke with a sleeve valve controlled inlet port. The Motor Cycling article recorded that only the four-in-line four-stroke existed as a complete machine at that time, whilst the others were no more than the engine units. The experimental engines, including the three-cylinder two-stroke, were subsequently donated to the Montagu Motor Museum (National Motor Museum) at Beaulieu.

The experimental 225cc sleeve-valve two-stroke. The sleeve controlling the inlet port also partly controlled the transfer port. The engine performed well, especially at low speeds, but the extra performance did not justify the added complexity and cost over the standard two-stroke.

Royal Enfield In-Line Triple

The only known picture of the Royal Enfield Triple, which was published in The Motor Cycle, shows the engine fitted into a largely unmodified version of the 1916-pattern 6hp v-twin frame complete with chain driven hand-starter. Although the engine has the appearance of three standard 225cc engines bolted together, the crankcases are specially cast and wider. The crankshafts are longitudinal with the frame and a single external flywheel is fitted at the front. A bevel box at the rear provides drive to the Enfield two-speed gear and magneto.

When the National Motor Museum reorganised its motorcycle display area the Royal Enfield experimental engines, donated by the Enfield Cycle Company, became superfluous and were offered for sale at a Beaulieu Autojumble. The triple engine has since been united with a 1915 Royal Enfield v-twin frame and running gear to recreate the complete machine. The standard frame had to be modified to accommodate the length of the engine by bending back the saddle tube by three inches and fitting longer chain-stays. This resulted in the luggage carrier sloping forward to an identical degree to that on the original machine suggesting the Enfield designers had used a similarly modified frame.

The engine was found to start easily and to be flexible. It produced good torque and would pull strongly; like many other two-stroke engines it performed best under load. The fuel consumption, however, was a disappointing 35mpg, which possibly explains why the design was not developed further. The recreated Royal Enfield Triple is now on display at Sammy Miller's Museum at New Milton.

Royal Enfield In-Line Four

Following on from the three-cylinder in-line two-stroke, the Enfield company went on to build a prototype four-cylinder in-line side-valve four-stroke. Each cylinder had a 60 mm bore and a 75mm stroke, the same as the 3hp models, giving a capacity of 848cc. The engine was in unit construction with a three-speed gearbox and had a horizontally split crankcase. It was air-cooled and used separate one piece cylinder/head castings. The lubrication oil was contained in a large capacity sump and distributed to the moving parts by

This illustration of the Two-stroke In-line Triple was published in The Motor Cycle in 1964. The seemingly anomalous registration is in fact a trade plate.

The Royal Enfield In-line Four.

The engine of the In-line Four.

splash. The magneto was mounted at the rear of the engine above the gearbox. Gear changes were made using a gate-change lever and relied upon an expanding band friction clutch similar in nature to the Enfield patented two-speed gear and dog clutches; first gear, and drive, being obtained using the expanding band clutch with further changes made using the dog clutches. Final drive was by chain and the top gear ratio was 3.5:1 for solo use and 4:1 for sidecar.

The foot boards were aluminium castings of a type which in coming years would be used across the Enfield range. The saddle was also a foretaste of things to come being the first Terry spring saddle ever made.

Wartime Production

The Enfield Cycle Company's works record books are held by the Vintage Motor Cycle Club. These indicate that the Enfield Company continued to make motorcycles throughout the war period. The machines are recorded by frame number and there are numerous gaps, which makes analysis difficult. The reason for the gaps is uncertain and possibly the WD machines were recorded separately. Numerous orders were being dispatched abroad.

The company produced over 500 motorcycles for the War Office. These represented only a small part of the Enfield Cycle Company's production in support of the war effort.

Private owners also adapted their Royal Enfields and made them available for essential war duties. This 6hp Enfield sidecar was converted to a Red Cross ambulance by an Exeter motorcyclist.

The attachment of sidecars to the Enfield lightweight was discouraged, but Capt Millard R.E. and his passenger Lt Ingham, together weighing 30½ stones, demonstrated that the 2¼hp Enfield was up to the task.

RIGHT: Production of machines for private use had been banned by the government but commercial vehicles could still be supplied. This promotional photograph was published in January 1918 in magazines such as The Field, 'The Country Gentleman's Newspaper' and The Queen, 'The Lady's Newspaper'.

The caption in The Queen read:
THE BUTCHER'S "BOY"
'This Royal Enfield side-car combination, used for the delivery of meat and other foodstuffs over a wide country area, is doing the work of two horsed vehicles and has covered some 14,000 miles. The girl driver had had no experience of motoring when she began her work, but mastered the little machine with ease.'

CHAPTER TWELVE

Early Post-War Years 1919 - 1920

The signing of the Armistice on 11 November 1918, which brought to an end the conflict in Europe, led to the immediate cancellation of all government contracts. Manufacturing companies were forced to reorganise quickly in order to resume civilian production. Enfield, as other manufacturers, had expanded their production capacity during the war and needed to increase bicycle and motorcycle production to greater than pre-war levels in order to maintain employment and fully utilise the factory. The demand for motorcycles was high. The Great War had been the first mechanized war and military service had provided many of the serving men with a first taste of motorized transport. They had survived the slaughter in their fight for freedom and now wished to enjoy the pleasures that freedom offered. For many that meant the acquisition of a motor vehicle. The Enfield Cycle Company was in a better position than many similar companies to satisfy this pent up demand as they had been able to maintain the manufacture of motorcycles and bicycles at a modest level throughout the war. However, the company suffered the same problems as the rest of the industry and was beset by shortages in the supply and the escalating costs of raw materials. It would be well over a year before the supply situation stabilised and the company was able to return to normal peacetime production.

1919 Model Range

The range of machines announced for 1919 was little changed from that of the early war years before civilian production came to an end. The prices of Enfield machines, as those of other manufacturers, would follow an upward spiral during the immediate post-war years reflecting the continued high demand for machines, shortage of supplies and inflation. The mainstays of the range were the v-twin combinations available with a 6hp or optionally an 8hp JAP engine, and the 2¼hp two stroke model which had first been introduced at the outbreak of war. The 6hp Model 180 was initially priced at £115 10s, compared with a pre-war price of £84, but few are likely to have been sold at this price. By February 1919 the price had risen to £126 and by the end of the year was further increased to £145. The extra sum for the de luxe model had been £16, but was now 20 guineas (£21). The two-stroke Model 200, which had been priced at £39 18s at the start of the war, was now priced at £52 10s and it would increase further in price to £61 for the 1920 season. The 3hp, 424cc ioe twins, which had first been introduced for the 1913 season, were initially identified as part of the range and offered at £69 6s; the prewar price was £52 10s. Presumably the factory was using up surplus stocks as these models would be dropped during the course of the year.

1919 Motorcycle Show

The Enfield Cycle Co Ltd decided to specialise on two models for the 1920 season and the stand at the 1919 Olympia Motorcycle Show displayed examples of the 6hp combinations and the 2¼hp solo machines. The 3hp 424cc solo machines were sadly missed by both visitors to the show and the press. The Motor Cycle commented: 'Many riders will miss the well-known

Specifications of 1919 & 1920 Models

						1919*	1920*
Model 200	2¼hp	225cc	ts	single	Solo	£52 10s	£63
Model 140	3hp	424cc	ioe	v-twin	solo	£69 6s	–
Model 150	3hp	424cc	ioe	v-twin	T.T.	£69 6s	–
Model 180	6hp	771cc	sv	v-twin	s/car	£126	£145
Model 180	6hp	771cc	sv	v-twin	trade s/car	£126	£145
Model 190	6hp	771cc	sv	v-twin	de luxe s/c	£148	

The 3hp Models 140 and 150 were dropped from the range for 1920.
The Models 180 and 190 were optionally available with an 8hp JAP engine at extra cost.
* Prices varied over review period - see text,

The Royal Enfield Stand at the 1919 Olympia Motor Cycle Show. In addition to the 2-stroke model, the company exhibited four examples of the combination and sixteen different models of pedal cycle.

3hp solo machine from the Olympia exhibit. This little machine, with its overhead inlet valves and mechanical lubrication, has long been an outstanding feature in the firm's production'.

The specifications of the two models were largely unchanged for 1920. The aluminium foot boards on the 2¼hp two-stroke were fitted with small but effective leg shields. In its review of the show, Motor Cycling considered this model the smartest little lightweight on the road. The fuel tank of the 6hp Twin gained larger diameter filling caps and a sight feed Benton and Stone lubricator. On the Model 190, fitted with Lucas Magdyno, the switch was neatly fitted in the offside of the tank. Both the two-stroke and four-stroke twin machines were now fitted with the new single-lever pattern AMAC carburettor and a Thomson-Bennett magneto. Motor Cycling sub-titled its review of the Enfield models at the Olympia Show: The New Enfield Models - machines for the Lady Rider. 'The low saddle position and long footboards give a very comfortable riding position, whatever the stature of the rider may be.' The review continued: 'For this reason and the tractability of the big twin the Enfield has always been popular with lady riders, and this year its increased ease of control due to the single lever carburettor should result in its being more popular than ever with the fair sex.'

Four examples of the 6hp Twin were exhibited: a standard Model 180, the tradesman's carrier with a large coach-built box carrier and two Model 190 machines with a Lucas Magdyno lighting set. One of the Model 190 machines was fitted with a windscreen, adjustable to any angle, and a hood of a new pattern, which when not in use folded back into the rear of the sidecar.

1920 Price Fluctuations

In most instances the quoted motorcycle prices have been derived from the annual sales catalogues. This was not possible for the 1920 model season, for whilst there is evidence that a combined cycle and motorcycle catalogue was produced a copy has not been found. The prices given in the table on page 157 are those quoted upon the announcement of the model range for the year and are based upon magazine articles and advertisements. That they were subject to early revision is borne out from a surviving Trade Price List for Royal Enfield Motor Cycles, which dates from May 10th 1920.

The trade price list indicates that the 2¼hp Model 200 increased in price

Experiences with a 3hp Twin

In May 1919 'FAS' writing in The Motor Cycle reported on his experiences with a 3hp Twin. The article was sub-titled 'An Appreciation and Some Criticism of an Interesting Mount, based upon a Year's All-weather Riding'. The machine in question was purchased early in 1917 and used to cover a distance of some 4,000 miles, which consisted of weekend journeys of about one hundred miles each way and some business trips. The machine was ridden at between 30 and 35 mph, which resulted in an average speed of 30 mph and a fuel consumption of 105 mpg; both of these figures were considered very commendable for a lightweight. The highest recorded speed was 46 mph and the tester felt confident that if tuned for speed it would be capable of exceeding 50 mph. The oil consumption was not accurately recorded, although estimated to be in the region of 1,800 mpg.

The machine was initially used on snow covered roads and when the snow turned to slush he began to wonder whether it was a 'mud-slinger' before blaming the high speed of 30 mph for the very dirty machine and rider. He would, nevertheless, have welcomed an efficient undershield and magneto cover. He recommended that these should be incorporated with leg-shields and extend backwards under the driving chain as the front wheel had an annoying way of spraying mud on the underside of the countershaft sprocket.

The Enfield two-speed gear and shock absorber were both praised. No attempt was made at 'freak hill-climbing' but most main road hills were ascended without recourse to second gear.

Starting on cold mornings proved extremely difficult, but once started no further difficulties were experienced. Part of the problem was put down to poor petrol and

ABOVE: The 3hp Twin was adapted for cold weather riding by arranging a warm air intake to the carburettor and through the use of handlebar muffs.

'vaporisers' fitted to the inlet manifold gave a modest improvement. It is likely that part of the problem was the kick-starter arrangement as 'FAS' found its forward position and direction of push to be distinctly awkward. He gave up on the kick-starter and preferred to start the machine by paddling-off in low gear.

He was also critical of the forward mounted magneto. He experienced the magneto platform working loose and considered it poorly supported by being mounted on an extension of the driving chain covers. He was enamoured of neither the Enfield method of enclosing the Bowden cables within the handlebars nor the fixed lugs to carry the control levers, although he did accept that both were neat. The filler cap for the glass oil reservoir was also criticised as it was impossible to fill using an ordinary can.

He felt he could not entirely recommend the machine for a tall rider but 'few lightweights are suitable for people over six feet' and noted that 'the best position I could find produced discomfort on a long run'. Nevertheless he concluded that he would still 'take an Enfield before many other makes, when I require a fast, reliable, solo lightweight'.

The test report was more critical than usual at a time when magazine editors, mindful of advertising revenue, were reluctant to publish criticism, except in the most guarded and oblique of terms. In this instance the editor

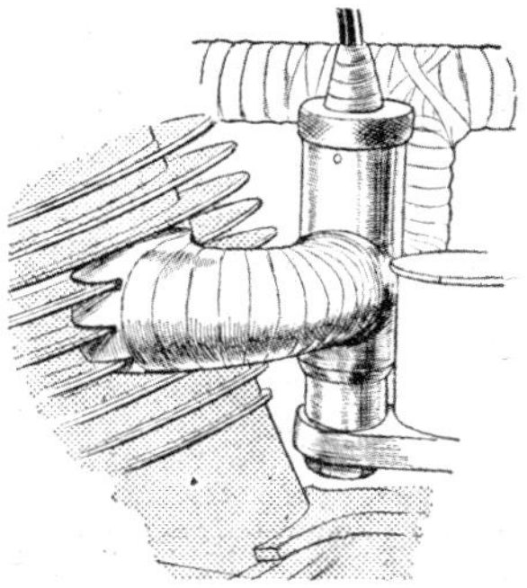

'FAS' devised a curved intake to draw hot air from between the fins of the rear cylinder. Insulation tape was also used to seal all joints and to prevent icing.

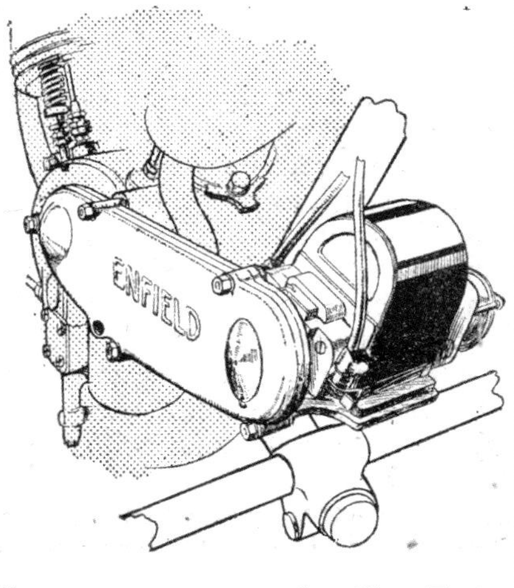

The magneto on the 3hp Twin, together with its platform and the driving chain cover, was held by three bolts on the timing case and a support at the front.

considered it appropriate to add a footnote. He recorded that a similar machine in their possession had suffered no trouble with the magneto fixing and he was of the opinion that the writer's trouble in starting was certainly curable.

2¼ hp Two-stroke, Model 200

Engine: 225cc two-stroke
Bore & stroke: 64 x 70 mm
Carburettor: AMAC, multiple jet
Magneto: Thomson-Bennett
Gears: Patent two-speed with free engine
Brakes: Stirrup front, dummy belt rim rear.
Tyres: Palmer 24 x 2 in
Tank capacities: 1⅝ gallons petrol; 1 quart oil.
Weight: 147 lbs

3 hp TT Roadster, Model 150

Engine type: ioe four-stroke
Capacity: 424cc.
Bore & stroke: 66 x 75 mm.
Carburettor: AMAC, multiple jet
Magneto: Thomson-Bennett
Gears: Patent two-speed with free engine
Brakes: Stirrup front, dummy belt rim rear.
Tyres: Palmer Cord 26 x 2 ¼in
Tank capacities: Petrol 1½ gallons.

3 hp Twin-cylinder, Model 140

The Model 140 was a touring version of the TT Roadster, Model 150. It was fitted with Royal Enfield registered design of handlebars.

6 hp Side-car Combination, Model 180

General Specification of 6 hp Machine
Engine type: Side-valve four-stroke
Capacity: 771cc.
Bore & stroke: 76 x 85 mm
Lubrication: Automatic drip-feed and auxiliary hand pump
Carburettor: AMAC multiple jet
Gears: Patent two-speed with free engine and handle starting
Brakes: Dummy rim front and rear
Tyres: Palmer Cord 700 x 80 mm
Tank capacities: Petrol 2 gallons
Oil 1½ quarts

6 hp Dynamo Lighting Side-car Combination Model 190.

Specification as Model 180 and fitted with Lucas dynamo lighting set consisting of a headlamp, tail lamp and sidecar lamp. The Model 190 also came complete with a bulb horn, which was an extra on the other models.

6 hp Trade Delivery Combination, Model 180.

Coach built box carrier body, with hinged top and folding doors at rear.
Dimensions of box:
Length 3ft 6in, Depth 1ft 6in, Width 2ft 1in.

ROYAL ENFIELD MOTOR CYCLES. **Net Trade Prices.**

	Model	Fitment	Net Trade Price £ s. d	Retail Price £ s. d.	Agent's Profit £ s. d.
2¼ h.p.	200	Chain Drive, Two-speed and Free Engine Gear, Palmer 24in. × 2in. Tyres ...	54. 0. 0 ~~61 0 0~~	65. 0. 0 ~~73 10 0~~	11. 0. 0 ~~12 10 0~~
6 h.p.	180	Chain Drive, Two-speed and Free Engine Gear, Handle starting, Palmer Cord Cycle-car Tyres 700 × 80 m/m, complete with Side Car, (with coach-built body)	133. 0. 0 ~~148 0 0~~	160. 0. 0 ~~175 0 0~~	27. 0. 0 ~~27 0 0~~
6 h.p.	190	As above, but fitted with Lucas "Magdyno" Lighting Set and Horn ...	151. 0. 0 ~~170 0 0~~	182. 0. 0 ~~200 0 0~~	31. 0. 0 ~~30 0 0~~

Models **180** and **190** ~~can be supplied with 8-h.p. engine instead of 6-h.p. at 30/- trade and 2 Guineas retail extra.~~

CAUTION.
¶ It is requested that every precaution be taken to prevent inspection of this List by unauthorised persons.

CONFIDENTIAL.

Royal Enfield Motor Cycles

TRADE PRICE LIST.

THE ENFIELD CYCLE Co., LTD., REDDITCH.

MAY 10th, 1920.

RETAIL PRICE MAINTENANCE.
Trade terms are allowed only on condition that no re-sale shall be made or offered at less or more than our current retail Catalogue price; that no "Royal Enfield" Motor Bicycle shall be offered for sale by auction; that no "Royal Enfield" Motor Bicycle shall be offered for sale as shop soiled unless it has been bona-fide in stock at least three months.

TERMS OF PAYMENT.
One third deposit and balance upon receipt of our advice that machine is ready for despatch. All Motor accounts are Strictly Net.

CARRIAGE AND PACKING.
All goods are delivered free on rail REDDITCH, in good condition, and are signed for as being so by the Carriers, who then become the Agents of the purchasers, the latter paying all charges for Carriage, etc. Purchasers should carefully examine goods before accepting delivery from the Carriers, and in case of damage should sign accordingly, and immediately claim upon the Carriers. Unless specially ordered otherwise, goods are consigned at Carriers' Risk.
Crates are charged at ~~20~~/- each, and an allowance of ~~10~~/- 15/- is made if they are returned carriage paid and in good condition.
Crates and Cases for shipment abroad are charged at cost price.

CONDITIONS OF SALE.
All goods are sold subject to the guarantee printed in our current Catalogue.
The prices and specifications in this list are subject to alteration at any time without previous notice.
We cannot hold ourselves responsible for Side Cars of other manufacture which may be attached to our Machines.
Wherever possible, Sole Agencies are granted to Agents for special districts, on the basis of mutually satisfactory contracts, all enquiries being referred to Sole Agents. We are not responsible in any way for Machines bought in one district and taken or sent into others.
Agencies for Motor Bicycles are separate and distinct from those of ordinary cycles; but Sole Agents for ordinary cycles will have preferential consideration when applying for Motor Bicycle Agencies.

ABOVE and RIGHT: The Trade Price List issued by the Enfield Cycle Company Ltd on May 10th 1920. The list, consisting of three pages, was a confidential document issued to the company's agents setting down the terms of sale.

to as much as £73 10s, compared with the pre-war price of £39 18s. The 6hp Models 180 and 190 increased in price to £175 and £200 respectively, compared with pre-war prices of £84 and £100. The date of the handwritten corrections to the trade price list is unknown, but it is likely the corrections were made just prior to the announcement of the 1921 model range as the adjusted prices remained current for 1921. By this time, as we shall see in the following chapter, the 6hp JAP engine fitted to the Models 180 and 190 would be replaced by the Enfield designed 8hp Wolseley engine. The standardisation of the 8hp Wolseley engine for the Models 180 and 190 removed the need to provide an 8hp alternative to the 6hp engine.

The trade list also provides informative on the profit margins available to the dealers. The May 1920 figures indicate the dealers mark-up was 20.5% on the Model 200, but this was reduced to 18.2% and 17.7% on the 6hp Models 180 and 190, respectively. With the introduction of the 8hp Wolseley engine the dealers mark-up was a more consistent figure of just over 20% across the range.

The final page of the trade list sets down the terms of business. The

Enfield Cycle Company was concerned about maintaining the retail price structure of its products. With long waiting lists for motor vehicles during the early post war years there was as much concern over dealers making excess profits by selling to the highest bidders as in gaining advantage by undercutting competing dealers.

The debut of the 1920 Model 190 was in the Six Days Trial. The machine was in the hands of J.Dosher.

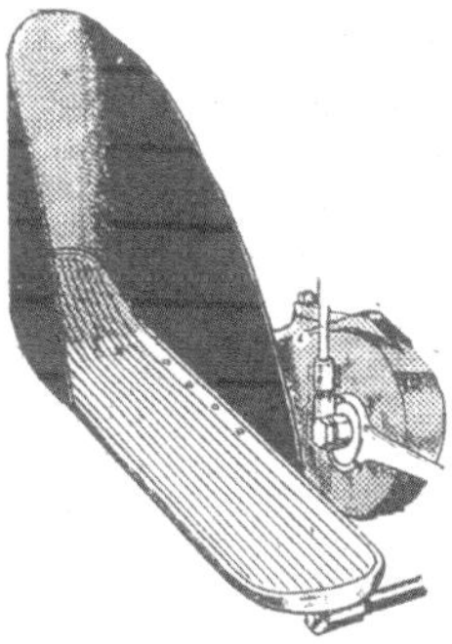

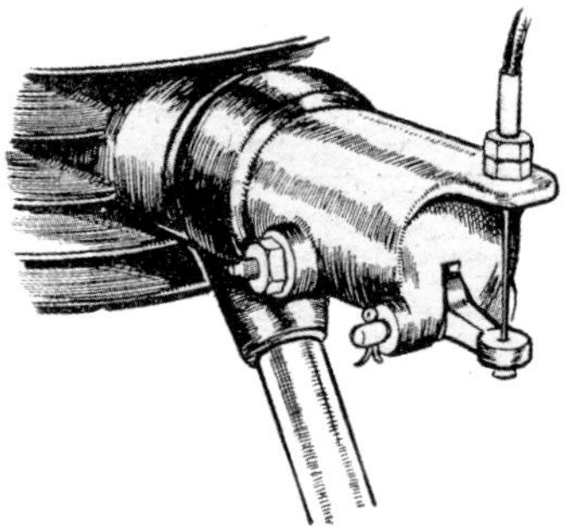

Noteworthy features on the 2¼hp two-stroke models for 1920 were the neat mud shields fitted to the front of the footboards, UPPER, and the decompressor arranged so that escaping gas was released into the silencer, LOWER.

For the 1920 season, the 6hp sidecar body was revised with higher sides to provide better passenger protection.

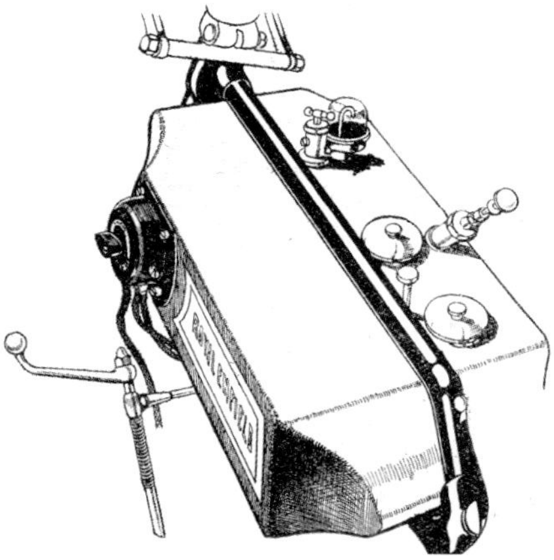

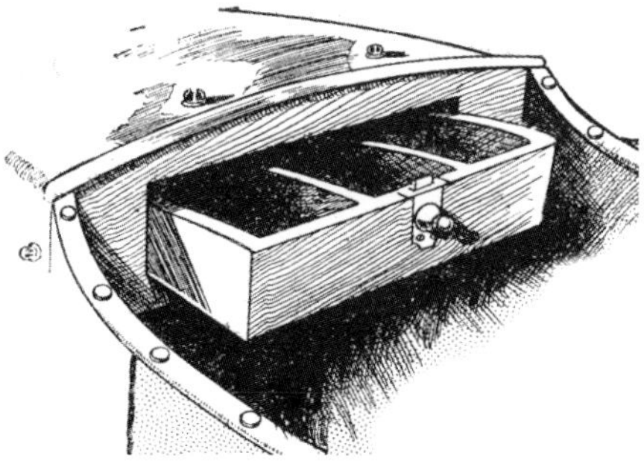

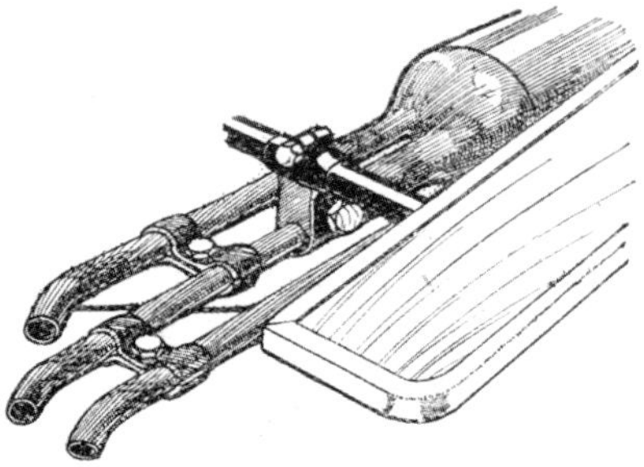

Motor Cycling, in its edition of 24 September 1919, carried illustrations of new mechanical details seen on machines participating in the Six Days' Trial. Of the eleven items considered worthy of note, four were seen on Royal Enfield 6hp twin machines. TOP LEFT: the new front brake operating within a dummy rim. TOP RIGHT: the extra-wide Enfield tank fitted to the Model 190 with lighting switch mounted on the side. (On production models this would be moved to the offside) BOTTOM LEFT: the tool locker incorporated in the Enfield Sidecar. BOTTOM RIGHT: an unusual three tail pipe arrangement fitted to the Enfield silencer.

CHAPTER THIRTEEN

Two Model Policy 1921 - 1923

November 29th, 1920.

MOTOR CYCLE PRICES.

8 h.p. Royal Enfield Side-car Combination Model 180 ...	£160 0 0
8 h.p. Ditto ditto but fitted with the Lucas "Magdyno" Lighting Set Model 190	£182 0 0
8 h.p. Royal Enfield Trade Delivery Combination Model 180	£160 0 0
2¼h.p. Royal Enfield Two-stroke Lightweight Model 200 ...	£65 0 0
Extra for Kick Starter on Lightweight Model	£5 0 0
,, ,, M. & L. "Maglita" Lighting Set on Lightweight Model	£12 0 0

We reserve the right to alter prices whenever alteration may be rendered necessary through increased costs of labour or material.

The above price list was inserted in the Royal Enfield brochure issued at the time of the 1920 Motor Cycle Show. Despite the rather plain appearance of the cover, shown on the right, the brochure was a high quality publication of sixteen pages and printed in colour. The aerial photograph was reproduced in the brochure

An Aeroplane Photograph of The Royal Enfield Works.

The post-war period of austerity had gradually come to an end during 1920 as the raw materials for motor vehicle manufacture became increasingly available. At last manufacturers had no restriction on the number of machines they could build. Also there was no shortage of customers from amongst the men returning from war service with experience of mechanical transport and in receipt of a service gratuity sufficient for the purchase of a motorcycle or small car. For a period it became a seller's market with manufacturers struggling to meet the pent-up demand. As a consequence production of motorcycles increased from an annual figure of 65,000 machines in 1919 to 100,000 machines in 1920. However, rising material and labour costs forced manufacturers to increase prices and the boom conditions soon passed. By 1921 all manufacturers were struggling to find sales. British motorcycle production fell by 20,000 to 80,000 machines in 1921 and by a further 20,000 in 1922. The Enfield Company had been in a good position to benefit from the boom years and had maximised production by concentrating on the well respected 2¼hp two-stroke solo and large capacity v-twin combinations. The Company wisely chose to maintain this two model policy and keep production costs low during the depression years. It was 1923, with the demand for motorcycles again increasing and the British motorcycle industry showing signs of recovery, before the Enfield Company began the introduction of new models. This proved a sound decision as the national production figure would increase back to 80,000 machines in 1923 and by 1927 would rise to a peak figure of 160,000.

The 2¼hp two-stroke lightweight model remained in production and was unchanged, except for the optional fitment of a kick-starter. The large capacity v-twin models were also continued, but the engine options of either a 6hp or 8hp JAP v-twin were now replaced by the single option of an Enfield-designed 8hp v-twin side-valve engine manufactured by the car manufacturing company Wolseley Motors. Wolseley at this time was owned by the Vickers armament group and the engine was produced at the Vickers plant at Weybridge, Surrey. The v-twin models were available only

4

Royal Enfield

The 2¼ h.p. ROYAL ENFIELD Two-Stroke

STANDARD MODEL 200

Fitted with the Royal Enfield Patent Two-speed and Free-Engine Gear, Cush Drive Hub and Chain Transmission.

SPECIFICATION.

ENGINE.—Royal Enfield Two-Stroke; bore and stroke 64×70 m/m.; capacity 225 c.c.

TWO-SPEED AND FREE ENGINE GEAR. Royal Enfield Patent; expanding clutch type. Direct drive on each gear.

TRANSMISSION.—Superfine Roller Chains throughout.

CUSH DRIVE.—Patent Cush Drive in rear hub.

IGNITION. British made high tension magneto.

CARBURETTER.—Special Two-lever type.

LUBRICATION.—Oil mixed with petrol in the proportion of eight measures to one gallon of petrol. Measure combined with filler cap on tank.

FRAME.—Low and comfortable riding position. Aluminium Footboards and protective Legshields. Royal Enfield Front Forks.

TANK.—Sloping Tank of pleasing design; Combined Filler Cap and Oil Measure.

HANDLEBAR.—Registered design; concealed cables through handlebar.

BRAKES.—Powerful Rear Wheel Brake, operated by foot pedal. Inverted Lever Front Brake.

TYRES.—Dunlop, 24×2¼in.

MUDGUARDS.—Graceful design; of ample width and strength.

CARRIER.—Detachable Carrier. By unfastening two nuts, the Carrier and Rear Mudguard are lifted clear of the back wheel and tyre.

SADDLE.—Well-sprung Saddle of best make and finish.

TOOL BAG.—Strong Leather, containing a complete set of Tools.

FINISH.—Enamelled in Best Black Enamel. Tank artistically panelled and lined. Bright parts heavily plated.

For CURRENT PRICES, see slip affixed inside the front cover of this List.

This Model can be supplied fitted with an Electric Lighting Set at an extra charge.

The specification is reproduced from the 1922 Sales Brochure and was largely unchanged during the period 1921 to 1923. The early tanks had a separate compartment of 2 pints capacity for storing oil. By 1923 a single compartment tank for the petroil mixture was used.

as combinations with both passenger-carrying and trade delivery bodies being offered. The Enfield Cycle Company would continue with this two model policy from 1921 through to 1923 and the introduction of new solo machines.

2¼ hp Two-stroke

The two-stroke model was largely unchanged from the model which had gone out of production as the factory moved over to war work in 1915. Footboards were still fitted but these were now aluminium castings and fitted with leg shields. The tool container had previously been a box attached to the top of the carrier, but was now the more usual leather fronted metal box attached to the side of the carrier adjacent to the rear mudguard. The carrier itself was made readily detachable. The saddle nose on production machines, as on the entire Royal Enfield range, was hinged at a bracket attached to the frame top tube and could be lifted together with the rear mudguard and carrier in order to expose the rear wheel.

The other innovation on the two-stroke model was the option for a kick-starter. The two-stroke had proved an easy starter and the machine was popular with lady riders but the action of 'running and jumping' or 'paddling away' was not considered ladylike and impractical when wearing a long skirt. The simple external mechanism added four pounds to the weight of the machine and a hefty £5 to the the £65 basic cost.

The illustration is of the 1921 Model 200 included in the 1921 Sales Brochure. It shows the saddle mounted on an extension of the frame top tube although a hinged saddle and carrier appear to have been used on production machines. The saddle height was 28½in and the weight of the machine ready for the road was 147 pounds.

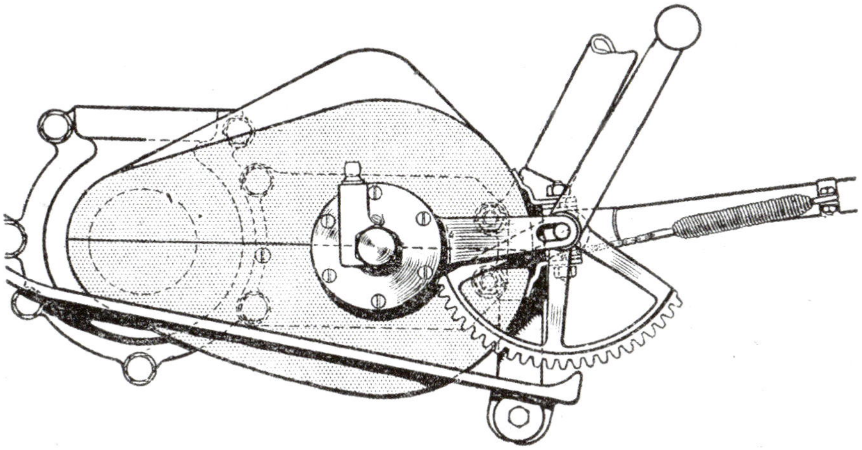

Kick-starters had become standard fitments on larger capacity machines and customers were demanding similar fitments on lightweights. The innovation with the 2¼hp two-stroke model for 1921 was the option of a kick-starter. The kick-starter mechanism was fitted to the near side of the machine and consisted of a geared quadrant, attached to the foot pedal, which engaged with a gear on the end of the countershaft.

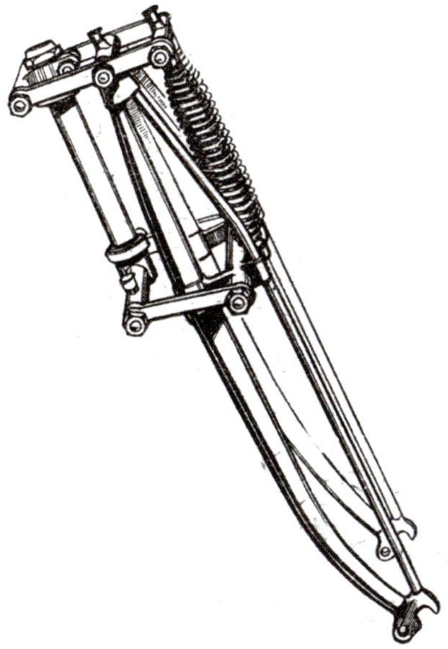

The Enfield front fork design consisted of a single spring in tension.

LEFT: The 1922 Catalogue illustration of the 2¼ hp Royal Enfield Two-stroke Model 200; the Standard Model.

BELOW: The illustration from the same 1922 catalogue is of the 2¼ hp Two-stroke Model 201; the Kick-starter Model. The two machines were identical except for the external kick-starter mechanism on the near-side of the machine.

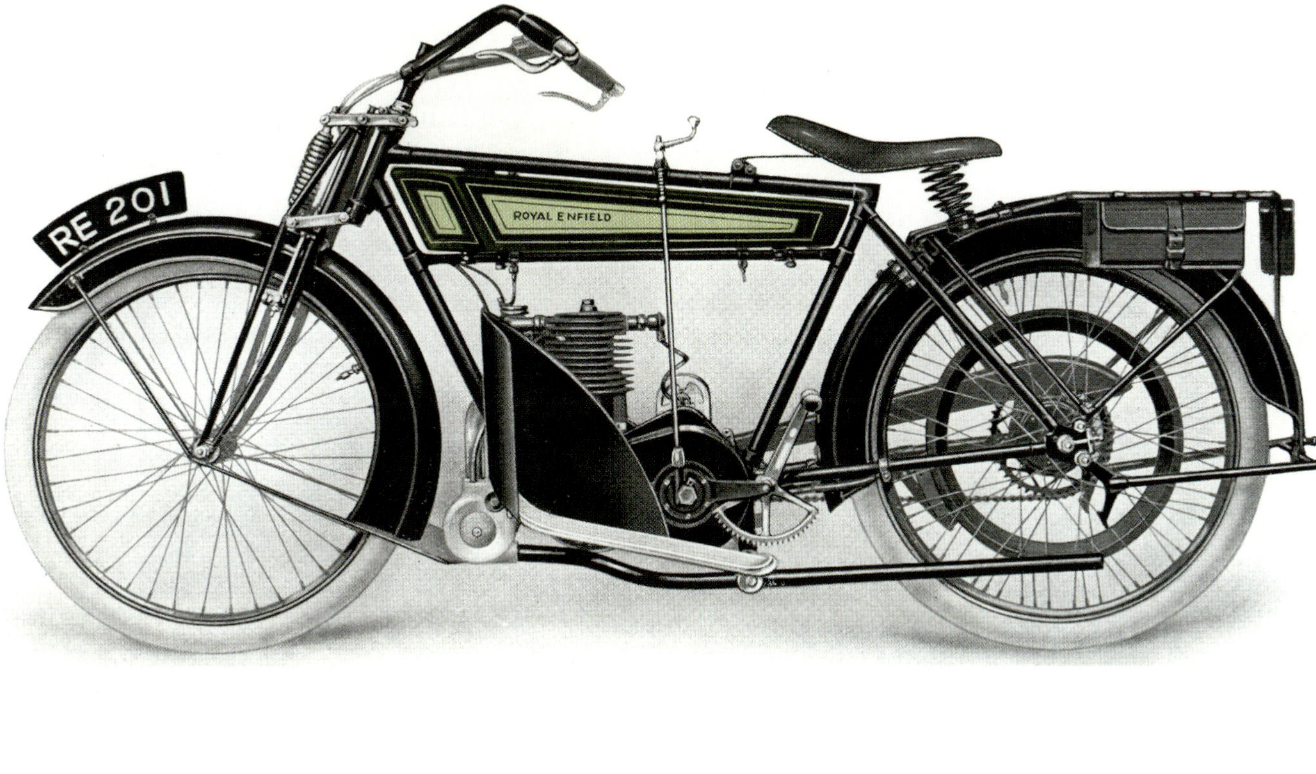

Sporting 225cc Two-stroke Enfield

A sporting version of the 2¼hp two-stroke was announced in April 1922. The most obvious changes to the standard model were the use of semi TT handlebars, in place of the Enfield design of swept back bars, a long straight-through large bore exhaust pipe and the replacement of the footboards and leg-shields of the standard model by adjustable footrests. The changes were considered by Motor Cycling to give the machine a racy appearance. Improved performance was achieved through the use of a larger than normal exhaust port, a Senspray carburettor and a Lucas magneto with variable ignition; hitherto magnetos with fixed ignition advance had been used. The quoted price was £55, the same as the standard model.

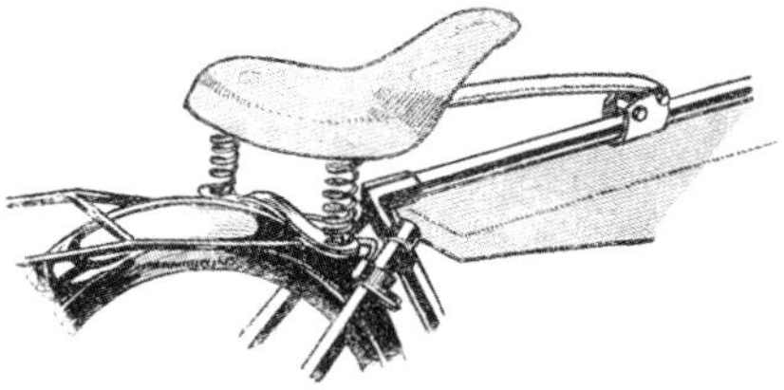

TOP: The seats of all the Royal Enfield models during this period were designed to hinge forward once two attachment screws at the base of the carrier had been loosened. The action of swinging the seat forward raised the carrier and mudguard and permitted easy removal of the rear wheel.
RIGHT: A works photograph providing a practical demonstration.

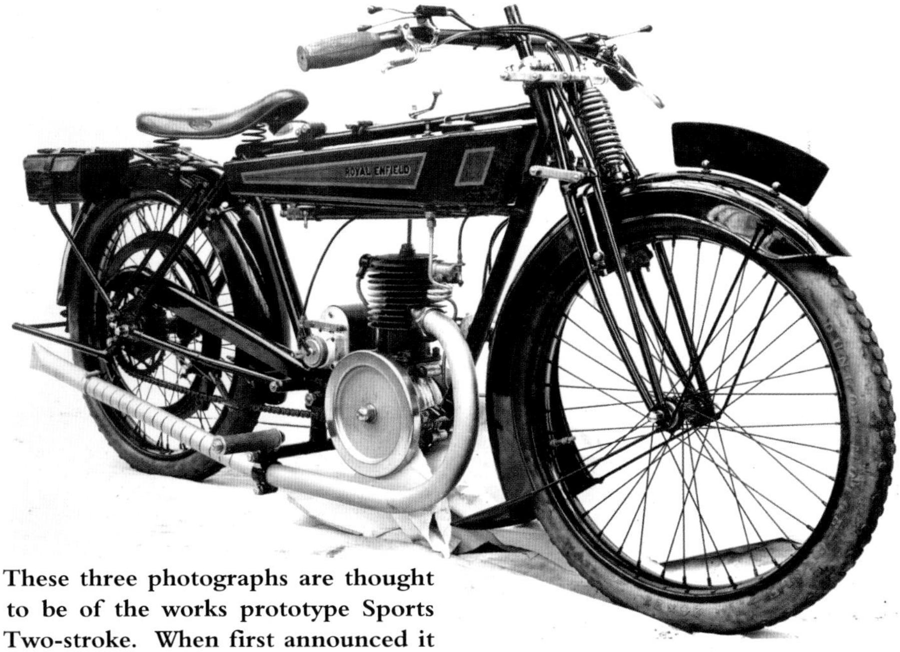

These three photographs are thought to be of the works prototype Sports Two-stroke. When first announced it was not fitted with the carrier toolbox.

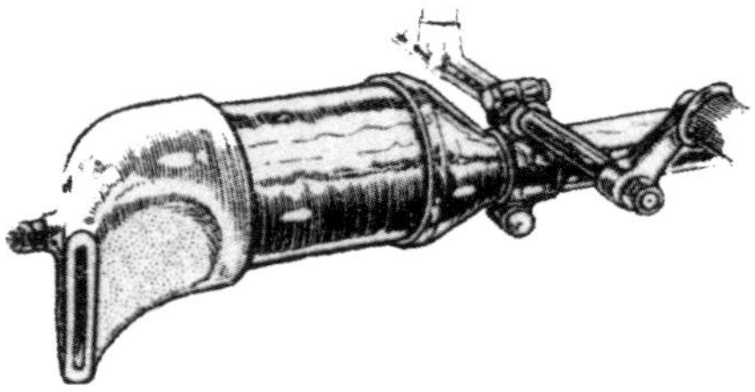

The Two-stroke Sports Model 200, illustrated above, was available from June 1922, by which time it was fitted with the silencer illustrated on the left. It was first catalogued in the 1923 brochure.

Prices of 2¼hp Two-stroke Models over years 1921 - 1923

1921	
Model 200	£65
Model 201 with kick-starter	£70
ML Maglita lighting set extra	£12
1922	
Model 200	£55
Model 201 with kick-starter	£57
Lighting set extra	
1923	
Model 200 Sports	£50
Model 201 includes kick-starter	£50
Lighting set extra	

New V-twin Engine

The sidecar machines exhibited at the Olympia motorcycle show in December 1920 were fitted with a new 8hp Wolseley v-twin engine built by Vickers Ltd. The engine was an Enfield design, but based upon the 6hp and 8hp JAP engines used on the previous year's models. The same 85.5mm bore and 85mm stroke as the JAP 8hp engine was adopted giving a capacity of 976cc, although 965cc was originally quoted. Cast iron pistons were used and fitted with three piston rings at the top. The gudgeon pins were considered to be of liberal size; they were a taper fit in the pistons and fitted with brass end caps to avoid scoring the cylinder. The substantial connecting rods were of the forked variety and fitted with steel sleeves at the big-ends to accept the three cages of replaceable roller bearings. A double ball bearing was used for the drive side main bearing. The four valves were fitted in the usual side-by-side position. They had recessed heads and were considered to be extremely light. Separate cam wheels, with integral cams to operate the inlet and exhaust rockers, were used for each cylinder and driven by the half-time pinion.

A mechanical lubrication system was used. The pump, fitted at the forward end of the the timing case, was operated by a driving shaft rotating at one-fortieth engine speed by a spiral gear from the front cam wheel. An eccentically located stud at the top of the drive shaft operated a small diameter piston working on the reciprocating principle to draw in oil and force it to the front cylinder wall, which was apt to be starved of oil on a v-twin, before lubricating the remaining moving parts by splash. A hand pump in the tank provided a supplementary supply as required.

The Woseley engine remained in the range until the end of 1924 when Enfield switched to manufacturing their own engine, which like the earlier JAP, had a single camshaft.

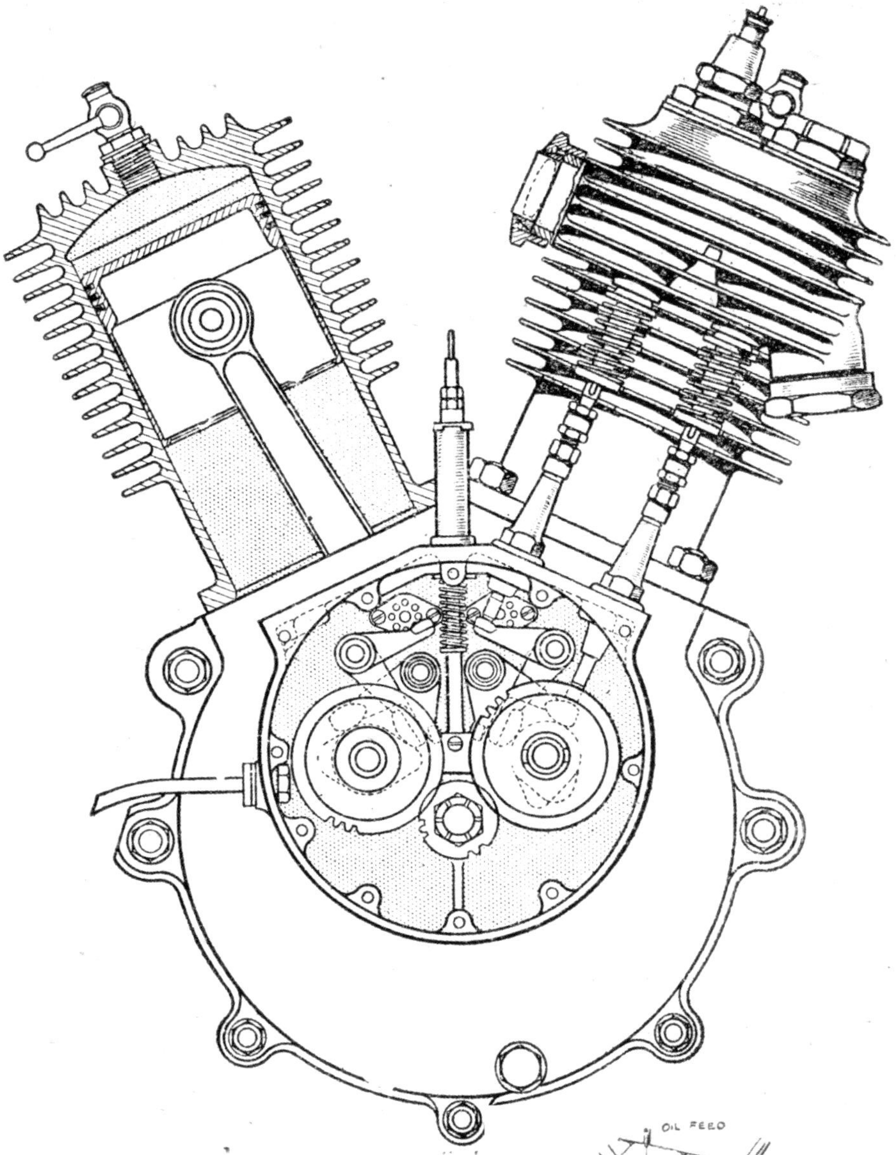

ABOVE: The new 976cc Wolseley engine, manufactured by Vickers, used to power the Royal Enfield big-twin sidecar outfits from 1921 through to the 1925 season.

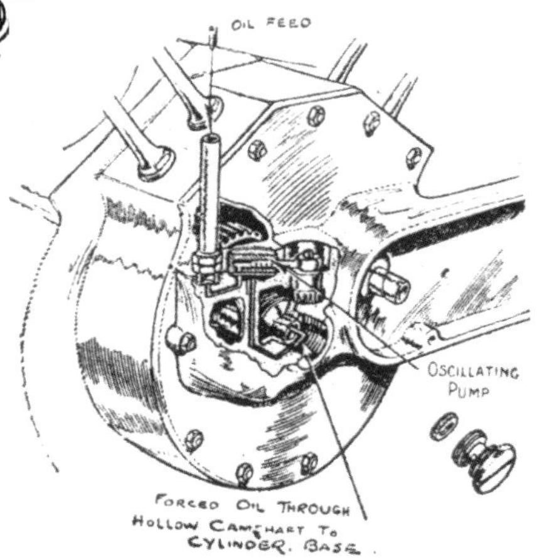

The mechanical oil pump fitted to the Enfield-Vickers engine. Oil from the oil compartment in the fuel tank was metered by a simple plunger pump operating at one-fortieth engine speed and pumped to the front cylinder wall and from there by splash to all moving parts.

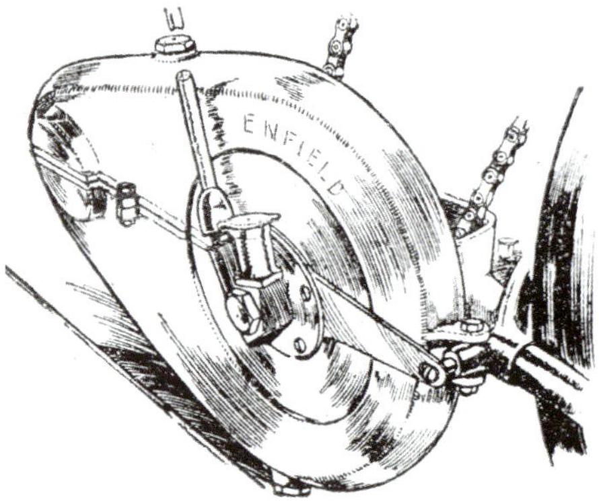

The primary chaincase of the 8hp model showing the two-speed gear.

8hp Sidecar Combinations

Apart from the use of the new Vickers-built engine, the new Enfield combinations for 1921, announced in November 1920, were little changed from those of the previous year. Such improvements as were made were of a minor nature. The well established Enfield two-speed gear was retained but was now lubricated using a large lubricator fitted above the operating mechanism. The frame design was slightly modified to fit wider mudguards with improved valances to improve access to the rear wheel for cleaning and tyre repairs. Large ball races were also fitted to the rear wheel and steering head. An improved front wheel stand and mudguard stay was also fitted.

The range of sidecar combinations exhibited at the Motorcycle Show at Olympia in December 1920 consisted of the sidecar combination Model 180, or Model 190 with lighting set, and the Model 180 Trade Delivery Combination. Few changes had been made to the design although experience gained in testing and during competition had suggested detailed improvements. The sidecar mudguard was made wider and attached directly to the body to reduce vibration. All tool boxes on the motorcycle had been dispensed with and instead tools and spares were carried in a special tool drawer, with different sized compartments in the sidecar nose, and in a special locker between the sidecar body and wheel.

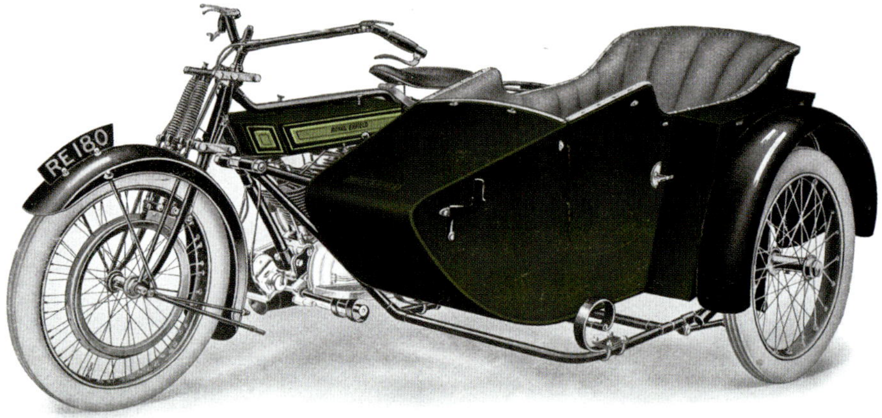

The 1921 8hp Sidecar Combination Standard Model 180. The machine was little changed from the previous model. The sidecar mudguard had been widened and was attached to the body to reduce vibration.

The 8 H.P. ROYAL ENFIELD SIDE-CAR COMBINATION MODEL 180.

Fitted with the Royal Enfield Patent Two-speed and Free-engine Gear, Cush Drive Hub and Chain Transmission.

THE ROYAL ENFIELD SIDE-CAR COMBINATION has been universally acknowledged as being the most perfect three-wheeled vehicle on the road. Built solely for side-car work it represents the skill and experience gained through many years of close association with motor cycle practice. The motor cycle and side-car are made interdependent while the connecting lugs are actually built into the motor cycle frame, a practice which ensures perfect alignment and safety. The adoption of the 8 h.p. **Royal Enfield Twin Cylinder engine,** together with newly designed and thoroughly protective mudguards, the roomy and comfortably upholstered side-car, will further enhance the reputation of this already famous outfit.

SPECIFICATION.

ENGINE . . Royal Enfield twin cylinder, bore and stroke 85·5 x 85 m/m (8 h.p.) mechanically operated valves.

TWO-SPEED AND FREE ENGINE GEAR.—Royal Enfield patent. Direct Drive on each gear.

TRANSMISSION. Best quality roller chains; with the Royal Enfield patent cush drive hub fitted to the rear wheel.

IGNITION . . Latest type waterproof magneto.

CARBURETTER Two lever.

LUBRICATION . Automatic engine pump and auxiliary hand-pump.

TANK . . . Two compartments; petrol capacity, 2 gallons; oil capacity, 3 pints, fitted with large fillers and needle valve.

FRAME . . . Royal Enfield design, with side-car attachments built into frame. Fitted with footboards and Royal Enfield spring fork.

TYRES . . . Dunlop tyres, 700 X 80 m/m on all threc wheels.

MUDGUARDS . Strong and wide; thoroughly protective.

CARRIER . . The improved Royal Enfield detachable carrier. By unfastening two nuts the carrier and rear mudguard are instantly removed.

STANDS . . Stands for front rear, and side-car wheel are fitted.

HANDLE-BAR . Royal Enfield registered design; concealed cables through handle-bar.

BRAKES . . . Inverted lever front brake; powerful rear brake operated by pedal.

SADDLE . . . Best make and finish, with large and comfortable seat.

TOOLS . . . Complete set contained in side-car tool locker.

FINISH . . . Enamelled in best black enamel: Tank artistically panelled in green, and lined. Bright parts heavily plated.

SIDE-CAR . . Royal Enfield chassis, exceptionally strong and well sprung; coach-built body; luggage grid fitted at rear of body.

Model No. 180 . . . . . { For Current Price see slip affixed inside the front cover of this list.

We also supply this model equipped with the Lucas "Magdyno" Lighting Set. Full particulars will be found on page 12.

Specification of the 8hp Sidecar Combination Model 180 reproduced from the 1921 sales brochure.

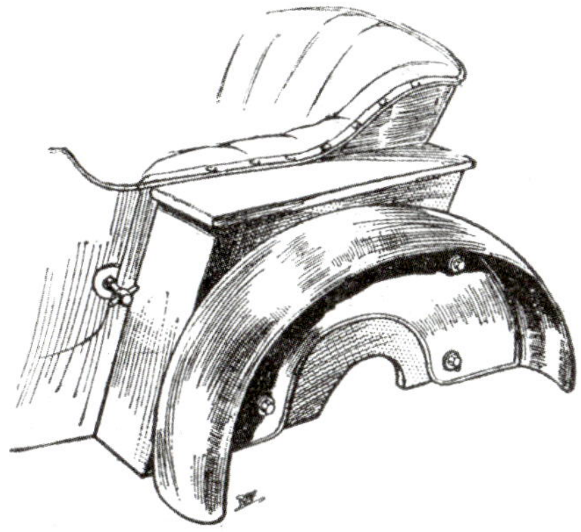

The 8hp combination was fitted with a handy locker between the body and mudguard of the sidecar together with additional storage in the nose of the sidecar.

The Model 180 with specially designed collapsible hood and adjustable windscreen.

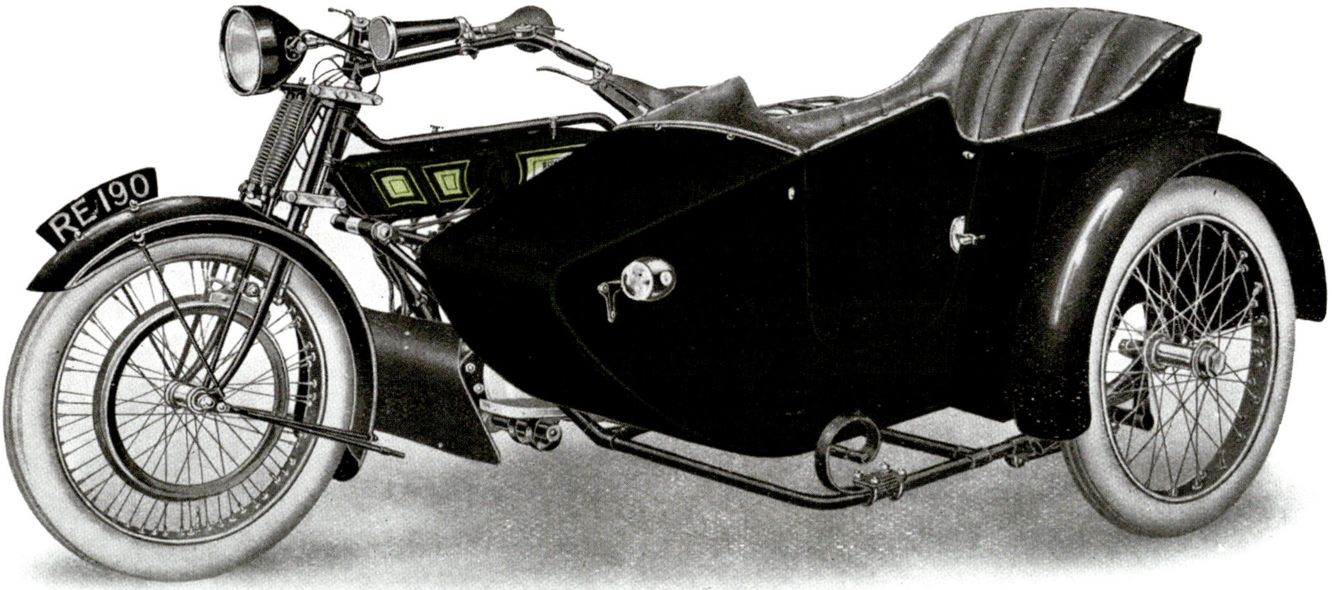

1921 8hp Model 190 Motorcycle Combination. The general specification of the Model 190 was as the Model 180 but with the addition of a Lucas Magdyno Lighting Set. This consisted of a dynamo, battery (carried in the sidecar), head lamp, sidecar lamp and rear lamp together with a best quality Lucas bulb horn.

Combinations ordered for use abroad would be supplied with a specially designed Colonial frame and sidecar chassis giving 7 inch ground clearance.

1922 Model 180 Improvements

A slight improvement in the mechanical lubrication of the Vickers-Enfield engine was noted when the new season Royal Enfields were announced in November 1921, but in other respects the 8hp machines were unchanged. However improvements were made to the sidecars. The folding luggage grid on the standard model was discarded and a roomy locker incorporated into the rear of the sidecar body. The lid of the locker was made flat in order to serve as a platform for extra luggage. This replaced the locker between the body and mudguard. Extra storage for tools was available under the seat, and the compartment at the front of the sidecar for maps and the like was retained.

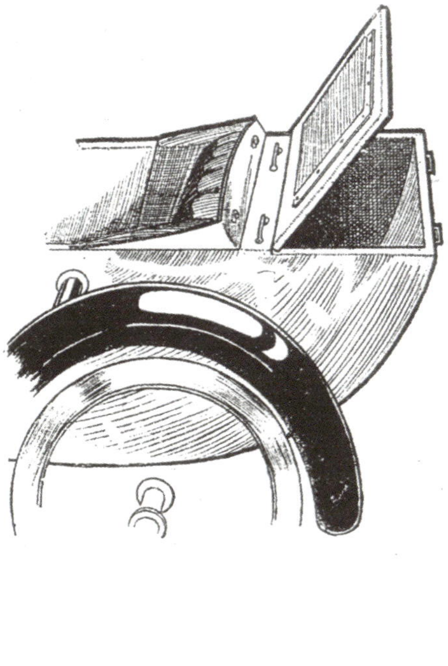

RIGHT: The roomy locker fitted to the Models 180 and 190 was considered suitable for carrying several tins of petrol together with a quantity of luggage.

TOP: 1922 Catalogue illustration of the Standard Model 180. Rather unusually it depicts the motorcycle rather than the sidecar. The saddle is a Terry patent spring saddle rather than the leather covered pan saddle usually depicted. This was made standard for 1922.
RIGHT: 1922 Catalogue illustration of the Lighting Set Model 190.

Two-seater Sidecar Body

A new design of two-seater sidecar body was introduced for 1922. Of similar lines to the Standard Body, it could be fitted to either the Model 180 or Model 190 for a small extra charge. The sidecar was fitted with a folding front seat, to allow access into the rear, and a folding rear luggage grid. The green upholstery previously used for sidecar bodies was replaced in 1922 by black grained leather cloth.

RIGHT: The new Enfield sidecar was designed to accommodate two passengers in comfort.

ABOVE and RIGHT: The 1922 Model 180 showing the arrangement of the folding windscreen and hood.

1923 Model 180 Improvements

Two modifications were introduced to improve the ease of use and practicality of the braking system used on the 8hp combinations. The front brake, previously operated by the right hand, was changed to foot operation with the pedal conveniently placed at the front of the left hand footboard. The pedal was linked to the front, dummy belt rim, brake by Bowden cable. The Ferodo brake block was eight inches long and made interchangeable with the rear. In the view of Motor Cycling it gave 'an immensely powerful and smooth action'.

The other braking innovation was a simple hand-operated rod device to hold the rear brake pedal down to provide a parking brake.

Other improvements to the 1923 models included the fitment of spring footboards for greater rider comfort. Tools were now carried in a wallet in the sidecar door where they were both readily accessible and secure.

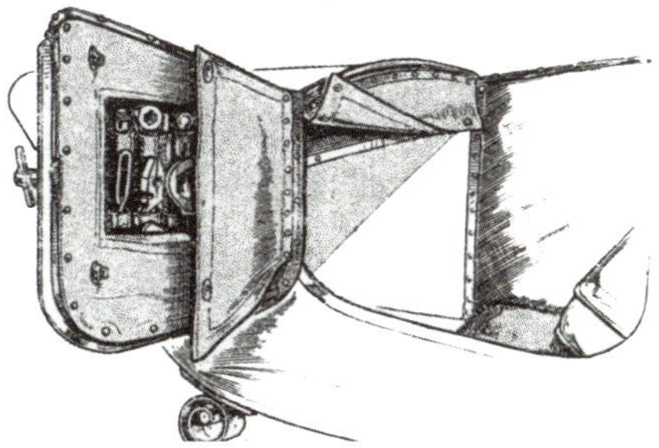

For 1923 tools were carried in a compartment, behind a flap, in the sidecar door.

The merits of the Royal Enfield Two-speed Gear.

The Royal Enfield was by now the only twin-cylinder sidecar outfit on the market having only two gear ratios. When the new model was announced, The Motor Cycle noted: 'These, however, must not be regarded as inadequate, for the reasons that render a three-speed gear necessary on most machines do not exist with the Royal Enfield, i.e., friction losses in the gear box and excessive overall weight. It is not generally realised that the Royal Enfield is the lightest machine of its power on the road.' The magazine continued: 'Weight-power ratio plays a very important part in the performance of a machine, and in this respect the Royal Enfield is certainly at an advantage. The performance of this machine in the classic reliability trials have proved its hill climbing abilities, while there are but few private owners who consider that a third ratio is necessary. The majority who have given this matter consideration probably affirm that the advantages obtained by the use of the simple Enfield gear outweigh any inconvenience caused by the lack of a middle ratio, since all friction losses are eliminated by having both gears direct.'

The Motor Cycle testers, responsible for putting the machine through its paces, were less certain. During the road test, published in April 1921, the testers took a trip to Malvern to attempt an ascent of the Old Wyche road. They wished to form an idea of the climbing powers of the machine on its 9 to 1 low gear bearing in mind that most outfits used three-speed gearboxes with a 14 or 15 to 1 bottom ratio. They set out with some trepidation and, perhaps as a consequence, failed before the peak gradient of 2.9 to 1 was reached. A second attempt was made after gaining more experience with the machine and this time a successful ascent was achieved. The success was attributed to changing down early and 'the driver's morale'.

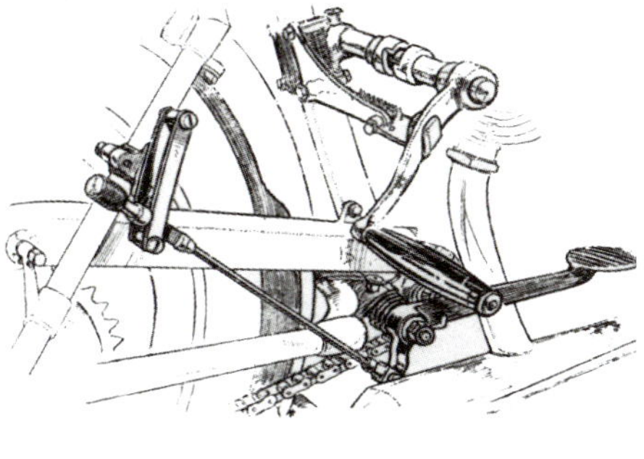

The rear brake locking device introduced for the 1923 season. With the brake pedal depressed, the rod is lifted by hand to the top of the arc where it is held by pressure until released. The starting handle retaining clip is also illustrated.

1923 8hp Standard Model 180.

A photograph from the works photograph album of a family enjoying an outing in a Model 190 with two-seater sidecar.

Leg shields had been fitted as standard to the 2¼hp Two-stroke for several years before being offered in 1923 as extras on the 8hp models.

LEFT: The boy in the sidecar above would have been seated on a folding pedestal seat of the type illustrated. It was used with the two-seater bodywork and positioned as required.

Weights of 8hp Motorcycles and Sidecars

Model 180 Standard	2cwt.	2qrs.	0lbs.	(128kg)
Sidecar, chassis, etc	1cwt.	2qrs.	14lbs.	(77kg)
Model 190 Lighting Set	2cwt.	3qrs.	4lbs.	(141kg)
Sidecar, chassis, etc	1cwt.	3qrs.	7lbs.	(90kg)
Trade Delivery Box, chassis, etc	1cwt.	1qrs.	7lbs.	(64kg)

Prices of the 8hp Sidecar Combinations over years 1921 - 1923

1921	
Model 180 Standard	£160
Model 190 Lighting Set	£182
1922	
Model 180 Standard	£140
Model 190 Lighting Set	£158
Extra for two-seater body	£5
Extra for hood & windscreen	£8
Extra for windscreen & one side-curtain	£4
1923	
Model 180 Standard	£120
Model 190 Lighting set	£135
Extra for two-seater body	£5
Extra for hood & windscreen	£8
Extra for windscreen & one side-curtain	£4
Extra for legshields	£1 5s.

Fred Bladon, the Royal Enfield publicity manager, demonstrates the 1922 Model 180.

Trade Delivery Combinations.

The Trade Delivery Combination consisted of the Standard 8hp Royal Enfield motorcycle fitted with a large coach-built box carrier of approximately 13 cubic feet capacity. The dimensions of the box were: length 4ft 6in, depth 1ft 9in, width 1ft 8in. It was painted green and black to match the motorcycle and lettering on the panels would be inscribed to customers requirements at nominal charge.

A claimed advantage of the 8hp Royal Enfield Combination was that it was a 'dual-purpose' vehicle and the Standard Passenger Body could be fitted to the chassis of the Trade Delivery Combination with very little trouble; a separate set of springs and shackles only being necessary.

RIGHT: Catalogue illustrations of the Trade Delivery Combinations. The upper illustration is taken from the 1921 catalogue and the lower from the 1922 catalogue, which was unchanged for 1923. The only identifiable difference is the use of a 'C' section sidecar mudguard in 1921 replaced by a 'D' section in 1922.

Model 155, Special Trade Delivery Combination

The merits of the 8hp Sidecar Combination were appreciated by commercial operators as an economic method of transporting goods up to a weight of 3-4 cwt. During 1922 the GPO placed a contract for a batch of machines for evaluation. These were fitted with a special sidecar body specifically adapted for parcel carriage. The machines were painted GPO red. The prototype machine carried the initials GR (George Rex) on the near-side. The evaluation was successful and further orders were placed. The production machines were more clearly identified on a front panel as belonging to the Royal Mail. It was identified as Model 155 GPO Type Special Delivery Combination.

The combination came with a complete tool kit and an easily operated jack. It was priced at £140

The box was provided with a lid in front and a double door at the back. The dimensions are:
length 4ft 9in; width 2ft 9½in; height 3ft 3in. The capacity was 31 cubic feet and the recommended loading was 3 – 4 cwt.

In 1923 when the Model 155 first became available it was priced at £140.

The bodywork was mounted on a strongly made chassis with quarter elliptic springs at the front and semi-elliptic springs at the rear.

Three photographs from the works photograph album of the prototype Model 155 purchased for evaluation for postal delivery services by the GPO Authorities. The claimed operating cost was 2¾d per mile.

A consignment lined up ready for despatch, October 1922.

The works electro-plating shop.

Two photographs taken in the Despatch Department at the Hewell Road Works during the 1921/22 period.

The Enfield Cycle Company had experienced difficulty in meeting demand for their machines at the start of the 1921 season and had taken steps to ensure there would not be similar delays in the future.

The bottom picture dates from from just prior to the 1921 Motor Cycle Show in November. It shows the new Enfield Despatch Department building. The building measured 60 feet wide and was 250 feet long. The machines are 1922 model Royal Enfield sidecar combinations which had been produced in anticipation of increased demand brought on by the show. Most of the machines shown were despatched before the show.

Wooing the Lady Rider

Motorcycling remained a predominantly male activity and whilst wives and girl friends were content to ride as pillion or sidecar passenger relatively few considered riding their own machines. Enfield, as other manufacturers, hoped to extend their market by selling to the lady rider and promoted the 2¼hp Two-stroke for this purpose. The Model 200 was well suited to lady motorcyclists being of light weight, easy to ride and with a low saddle height. For the 1922 season the company produced a brochure extolling the virtues of the machine. Titled 'Testimonials from Lady Riders' it contained the illustration above and the following two letters of endorsement:

Dear Sirs,

I have had no trouble with my Enfield Two-stroke, it is a most comfortable and practical mount for a lady. It has taken every hill I have tried easily so far. Altogether, I feel I could not have chosen a more suitable machine.

Yours sincerely,
Nora Burton.

Gentlemen,

I have had very many enjoyable rides on the 2¼-h.p. Two-Stroke Motor Bicycle and have not yet found a Devonshire hill which it was not able to climb.

It has taken me up Peak Hill, Sidmouth, and to the top of Haldon, and I think these are good tests of its hill-climbing ability.

The arrangement of controls and brakes is very simple and can be learnt in a few minutes. The Bicycle starts easily, runs smoothly, and the engine gives no trouble.

I remain,
Yours truly,
(Miss) Elsie M. Toone

A New Ladies Model

A new ladies' machine was introduced to the public at the end of July 1923. It was based upon the mechanics of the Model 201, including the two-speed and free engine gear, and featured a new dropped frame and redesigned fuel tank. The excellent weather protection of the standard model was retained and a metal strip was fixed to the frame tube at the rear of the tank to prevent the rider's skirt coming into contact with the cylinder. The company announced that orders were to be accepted for early delivery although it is uncertain as to whether the machine entered production as a model with the revised frame and cylindrical fuel tank featured at the Motor Cycle Show in October.

Current Prices of the 2¼-h.p. ROYAL ENFIELD Two-Stroke Lightweight.

Standard Model - £55.
Kick-starter Model £57.

An illustration from the 1922 brochure extolling the virtues of the Model 200 for lady riders.

Prototype of the Ladies' Model two-stroke announced in July 1922. The two-speed gear selection lever protruded above the frame and looks vulnerable.

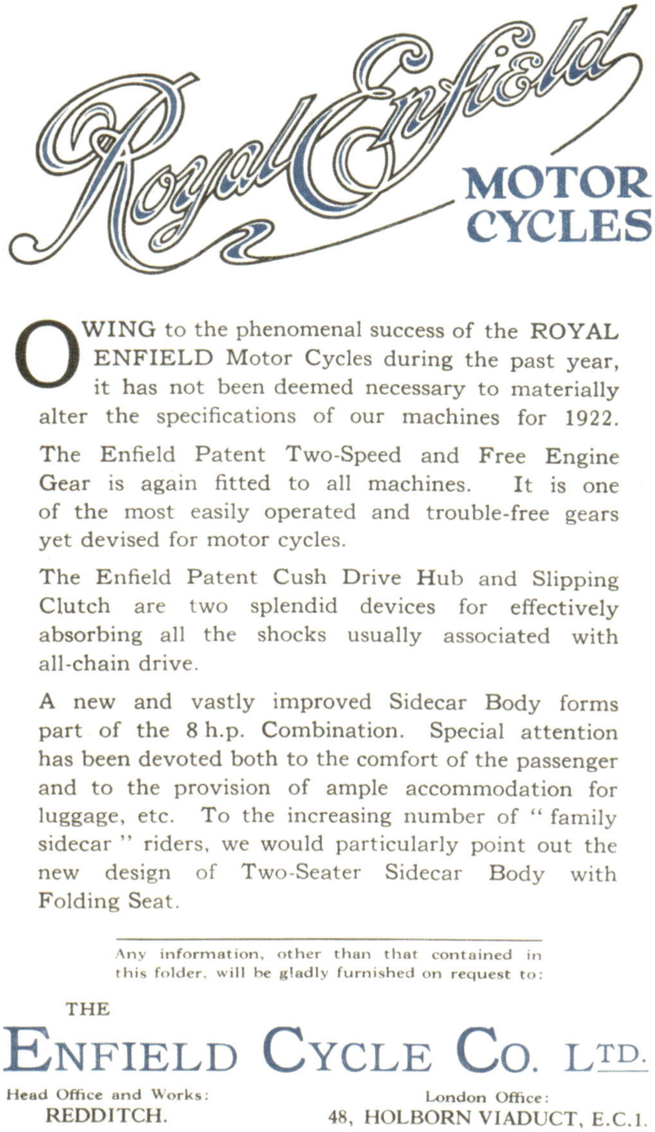

Royal Enfield MOTOR CYCLES

OWING to the phenomenal success of the ROYAL ENFIELD Motor Cycles during the past year, it has not been deemed necessary to materially alter the specifications of our machines for 1922.

The Enfield Patent Two-Speed and Free Engine Gear is again fitted to all machines. It is one of the most easily operated and trouble-free gears yet devised for motor cycles.

The Enfield Patent Cush Drive Hub and Slipping Clutch are two splendid devices for effectively absorbing all the shocks usually associated with all-chain drive.

A new and vastly improved Sidecar Body forms part of the 8 h.p. Combination. Special attention has been devoted both to the comfort of the passenger and to the provision of ample accommodation for luggage, etc. To the increasing number of " family sidecar " riders, we would particularly point out the new design of Two-Seater Sidecar Body with Folding Seat.

Any information, other than that contained in this folder, will be gladly furnished on request to:

THE
ENFIELD CYCLE CO. LTD.

Head Office and Works:
REDDITCH.

London Office:
48, HOLBORN VIADUCT, E.C.1.

A recognised way of promoting companies in the twenties was by fitting delivery vehicles with novelty bodywork to draw attention to the nature of a firm's business. This bodywork was made to special order by the Sandham Engineering Company of Greys Inn Road, London WC.

Death of Thomas Evans

In his Directors' Report to the Annual General Meeting the chairman, Lord Seymour, recorded the death of Thomas Evans, who had been a director of the Company since its formation. His place on the board was filled by appointing Mr T.C.Patchett as Secretary and Commercial Manager.

Falling Profits

The Company recorded a significant reduction in earnings during the financial year ending 31st August 1923; profit having fallen over the year from £52,627 to £9,779. The directors offered no explanation in the annual report although clearly the problems had arisen from the sharp fall in selling prices which had faced all cycle manufacturers. The dividend paid to shareholders was reduced to 7.5%, half that paid in each of the previous three years but still a significant amount at the time when other companies were struggling to survive.

General Specification of 8 hp Machines

Engine type: Side-valve, four-stroke
Capacity: 976cc.
Bore & stroke: 85.5 x 85mm.
Lubrication: Mechanical oil pump
Carburettor: AMAC, multiple jet
Gears: Patent two-speed with free engine and handle starting
Frame: Royal Enfield design
Front fork: Royal Enfield design
Brakes: Dummy rim front and rear.
Tyres: Dunlop 700 x 80 mm
Tank capacities: Petrol 2 gallons, oil 1½ quarts

A photograph from the Enfield Cycle Company's photograph album of Reg Steele on his Model 200 Two-stroke.

Summary of Balance Sheet of the Enfield Cycle Co Ltd at 31st August 1923

LIABILITIES		ASSETS	
Issued capital	£225,000	Fixed assets	£114,962
Creditors	£40,308	Stocks	£144,127
Divs proposed	£15,674	Investments	£17,890
Reserve	£70,000	Debtors	£49,264
Carried forward	£37,645	Cash	£62,384
	£388,627		£388,627

Works photograph dated 7 December 1921 which is believed to be of the prototype Sports model.

CHAPTER FOURTEEN

Model Year 1924

Despite the poor trading conditions of the previous years, the demand for Royal Enfield motorcycles had remained strong. Indeed at some stages the company had struggled to satisfy demand. The Royal Enfield two model policy, a 225cc lightweight two-stroke for the solo rider and 8 hp twin sidecar outfits for the tradesman or family man, had proved successful. However, the policy had been maintained for three years with only minor alteration and the designs were showing signs of stagnation. The Enfield Cycle Co, was falling behind its competitors and it was clear that new up-to-date models were required if sales were to be maintained. The most obvious gap in the range was for a medium weight four-stroke machine aimed at the solo rider, who having gained experience with the economical two-stroke now wished to own a less basic and more powerful four-stroke. Without a machine in the 2¾hp or 3½hp (350 to 500cc) category the company was likely to lose existing customers to rivals offering a more comprehensive range of machines. Lacking a suitable in-house engine the company would need to rely upon the use of JAP proprietary engines.

The Royal Enfield patent two-speed gear had served the company well over many years. It provided direct drive on both gears and with its self contained clutch was simple for the rider to master. Arguably two-speed gears were perfectly adequate for the low-tuned engines of the period, but by this time most manufacturers were offering machines with 3-speed gearboxes. For Enfield to remain competitive it was necessary to follow the trend and the Royal Enfield two-speed gear would be phased out over the next two years. The new mid-range machines were offered with the Royal Enfield two-speed gear as standard, but with an option for a three-speed Sturmey-Archer Three-speed gearbox and hand change.

Three or more gears are required to extract the best performance from an engine and the Company's desire to return to racing following a gap of some eleven years must also have been an influencing factor hastening the demise of the patent two-speed gear.

1924 Model Range

Solo Models				
Model 201	2¼ hp	Two-stroke	225cc	£42
Model 201A	2¼ hp	Open Frame Two-stroke	225cc	£42
Model 200	2¼ hp	Sports Two-stroke	225cc	£42
Model 202	2¼ hp	Two-stroke	225cc	£35
Model 350	2¾ hp	Standard Side-valve	347cc	£49
Model 351	2¾ hp	Sports OHV	347cc	£57
Side-valve Twin Combinations				
Model 180	8 hp	Standard	976cc	£90
Model 190	8 hp	De Luxe	976cc	£110
Model 185	8 hp	Trade Box	976cc	£90
Model 155	8 hp	Trade GPO Type	976cc	£120

A Ten Model Range

The Royal Enfield machines for 1924 were announced at the Motor Cycle Show held at Olympia in October 1923. The 2¼hp two-stroke was retained but was now available in four versions. The two new four-stroke solo machines were powered by 347cc JAP engines and offered the choice of overhead-valve sports or side-valve touring performance The range was completed by the well established 8hp Wolseley-Vickers combinations. The new models doubled the size of the Royal Enfield range on offer to the public.

The 1924 Sales Brochure extended over twenty pages and provided full details of the Royal Enfield motorcycle range and listed accessories. Details were also provided of a Gradual Payment Plan and of the cost of a Comprehensive Insurance Policy through London Assurance.

The Two-stroke Models

The Model 201 and Sports, Model 200 were continued and identical in specification to the 1923 machines except for the adoption of a foot gear change, rather than the hand change used previously. The prices, however, were considerably reduced with models available for £42, £8 less than in 1923.

The new Model 202 was a stripped-down economy version. It retained the hand gear change of the earlier models, lacked a kick-starter and was fitted with pedals rather than the footboards and legshields of the Model 201. It was priced at £35.

The fourth 2¼hp lightweight, Model 201A, was a drop frame ladies model. The preliminary version of this machine, with a lozenge shaped fuel tank, had been announced in July 1923. The 1924 season model featured an unusual cylindrical fuel tank. It was priced as the Model 201.

The two works photographs above are dated October 1923 and appear to show an experimental version of the 2¼hp Model 201. The machine is fitted with an aluminium primary chain drive cover, which does not appear to have gone into production.

Foot Gear Change

Hand changing of gears would remain in use for many more years, but the tram-handle type of lever was becoming antiquated. Also it was impractical on drop frame models due to the lack of an upper anchorage point. R W Smith's solution was to adapt the Enfield Patent two-speed gear to foot operation. His revised design is presented in Patent application number 23,132/23 which was titled 'Improvements in or relating to variable speed gear mechanism, suitable for motor cycles and the like'.

The operation of the Enfield patent two-speed gear with hand operated lever has already been described in Chapter 6. The foot operation was similar. A rotating sleeve attached to the foot pedal incorporated a spiral slot, which engaged with a pin at the end of the cam member. Rocking the foot pedal, using toe and heel, caused rotation of the sleeve and thereby axial displacement of the cam lever to

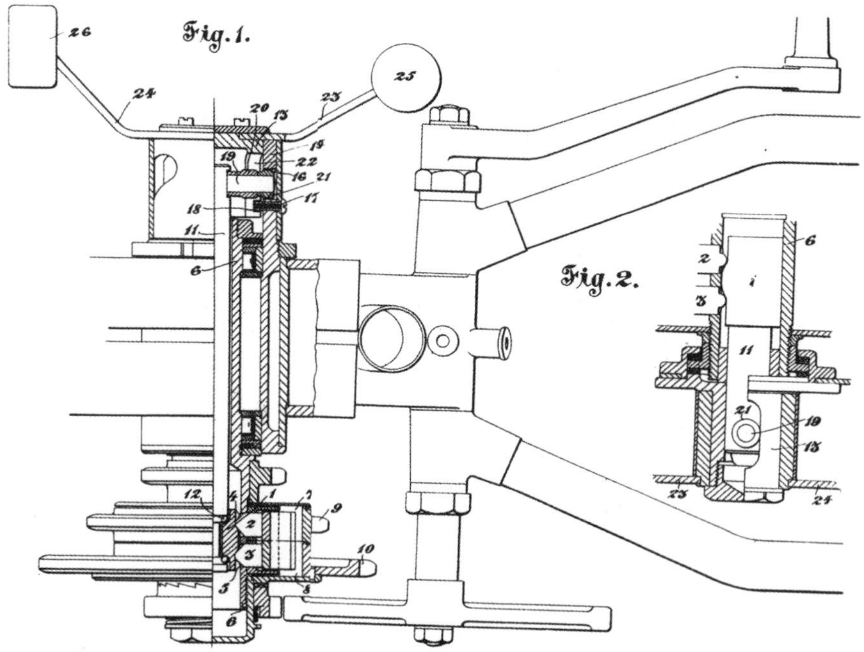

The illustration, from Patent GB 216,759 granted to Enfield Cycle Co Ltd and Robert Walker Smith, shows the Enfield Patent Two-speed gear adapted for foot operation. Also shown is the kick-starter mechanism which has been adapted to move the pedal to the off-side of the machine.

expand the clutch bands and obtain drive.

The improved gear change mechanism was standardised on the company's single cylinder models with the exception of the cheapest two-stroke, the Model 202.

The toe and heel pedals of the new foot change gear fitted to several of the 1924 2 ¼hp and 2¾hp models.

2¼hp Open Frame Model 201A

Open frame cycles had proved popular from the early years of the safety cycle and open frame motorcycles were offered by several manufacturers for a brief period immediately prior to the Great War. The first Open Frame Royal Enfield machine was announced for the 1912 season, but proved unpopular and was listed for just the one year. Royal Enfield's second attempt at marketing an open frame machine was able to exploit the popularity of the existing 225 cc two-stroke models and was very much more successful. It would continue to be available through till the 1929 season.

The Open Frame model was promoted as suitable for lady or gentleman and apart from the frame, which allowed the rider more easily to become seated, the model was identical to the Model 201. The patented Royal Enfield two-speed gear was retained and, as with the other models of the 1924 range, with the exception of the 2¼ hp Lightweight, it was operated by

2¼ h.p. TWO-STROKE LIGHTWEIGHT

OPEN FRAME MODEL **201 A**

(SUITABLE FOR LADY OR GENTLEMAN)

Fitted with Enfield Patent Foot-operated Gear, Cush Drive Hub, Chain Transmission and Kickstarter

SPECIFICATION

Engine.—Royal Enfield Two-stroke; bore and stroke, 64 × 70 m/m.; capacity, 225 c.c.

Lubrication.—Oil mixed with petrol.

Ignition.—High tension Magneto.

Carburetter.—Special two-lever type.

Gear.—Royal Enfield Patent; expanding clutch type. Direct drive on each gear. Kickstarter.

Transmission. — Roller Chains; Patent Cush Drive in rear hub.

Frame.—Low and comfortable riding position. Aluminium Footboards and protective Legshields. Royal Enfield Front Forks.

Tank.—New shape Tank of pleasing design; Combined Filler Cap and Oil Measure.

Handlebar.—Registered design; concealed cables.

Brakes. — Powerful Rear Wheel Brake, operated by foot pedal. Inverted Lever Front Brake.

Tyres.—Dunlop, 24 × 2¼ in.

Carrier. — Detachable Carrier. By unfastening two nuts, the Carrier and Rear Mudguard are lifted clear of the back wheel and tyre.

Tool Bag.—Strong Leather, containing a complete set of Tools.

Finish. — Enamelled in Best Black Enamel. Tank artistically panelled and lined. Bright parts heavily plated.

Price **£42** Net Cash

or **£11 12s. 6d.** down and twelve monthly instalments of **£3 1s. 1d.,** including Insurance.

This model can be supplied fitted with Electric Lighting Set at an extra charge of £8.

— 4 —

ROYAL ENFIELD

Specification of the Open Frame Model 201A reproduced from the 1924 sales brochure.

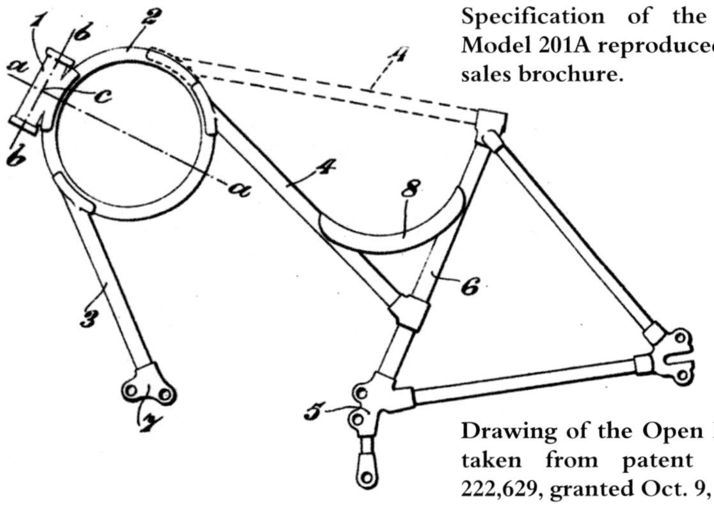

Drawing of the Open Frame design taken from patent number GB 222,629, granted Oct. 9, 1924.

a rocking pedal arrangement. This arrangement did not require the rider's hand to be removed from the handlebars when starting off or changing gear. It also obviated the problem of where to mount the lever on an open frame design.

Right: 'The new Open Frame Model illustrated is designed for the Lady or Gentleman who prefers not to wear special clothing while riding to the golf links or tennis courts, into town on shopping expeditions, to and from business, etc'. Extract from a brochure produced during 1924 specifically to promote the Open Frame model.

General Specification of 2¼hp Models
Engine type: Two-stroke, single
Capacity: 225cc
Bore and stroke: 64 x 70mm.
Lubrication: Oil mixed with petrol
Ignition: Magneto
Gearbox: Royal Enfield Patent Two-speed
Drive: Chain, with cush drive in rear hub
Front forks: Royal Enfield
Tyres: Dunlop 24 x 2¼ in
Brakes: Front stirrup, Rear dummy rim
Carrier: Detachable with rear mudguard section
Finish: Black enamel
Weight: 155 lbs

Open Frame Model 201A
Kick-starter: Fitted
Carburettor: Two-lever type
Tyres: Dunlop 24 x 2¼ in
Handlebars: RE registered design
Tank: Finished in black enamel
Equipment: Footboards and leg shields
Price: £42

Model 201
Kick-starter: Fitted
Carburettor: Two-lever type
Tyres: Dunlop 24 x 2¼ in
Handlebars: RE registered design.
Tank: Finished in black enamel, panelled and lined.
Equipment: Footboards and leg shields

2¼hp Sports Model 200
Kick-starter: Fitted
Exhaust: Large diameter pipe
Carburettor: Senspray Model B
Tyres: Dunlop 24 x 2¼ in
Handlebars: Semi TT
Tank: Panelled and lined
Equipment: Adjustable footrests
Price: £42

2¼hp Model 202
Handlebars: RE registered design
Carburettor: Two lever type
Tyres: Clipper 24 x 2 in
Handlebars: Sports bars optional
Tank: Finished in black enamel and lined
Equipment: Adjustable footrests
Price: £35

The New 350 cc Models

As already noted the adherence to a programme of producing only lightweight two-stroke and large capacity v-twin machines meant that the Enfield Cycle Co lacked design experience with the four-stroke single cylinder engine. In consequence it was decided to fit the new 350cc machines with the best available proprietary engines. The new 350 Standard Model used the JAP 2¾ hp (347cc) side-valve unit, which had been announced in August 1923. It weighed a modest 35 lbs. The new 351 Sports Model was fitted with the JAP 2¾ hp (347cc) overhead-valve unit. Both engines were of modern design and claimed to be the most efficient 350cc engines of

RIGHT: The first 2¾hp Standard Model 350. The machine used the Enfield countershaft two-speed gear and kick-starter pedal on the left. The machine at this stage was fitted with foot pedals.

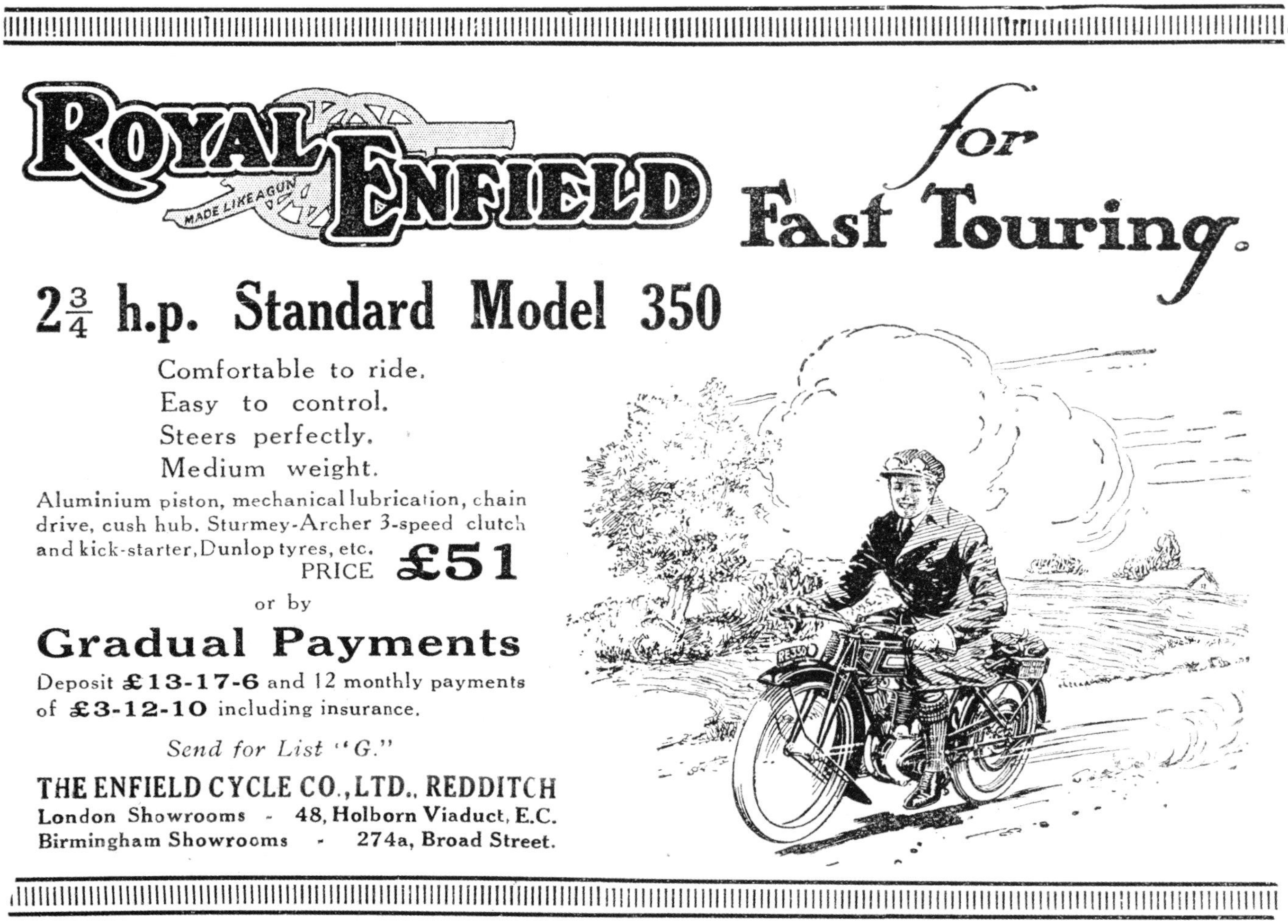

This advertisement dates from July 1924

their type on the market. Both were fitted with an aluminium alloy piston, roller bearing big-end and a roller bearing drive side main bearing. Lubrication relied upon an automatic mechanical pump although an auxiliary hand pump was also fitted. Both machines were fitted with the Royal Enfield patent two-speed gear although the option was made available to fit a Sturmey-Archer three-speed gearbox, with handlebar controlled clutch and kick-starter, at an extra charge of £2 for the Model 350 or £3 for the Model 351. The extra charge with the Sports model resulted due to the gearbox option including the fitment of a hub front brake. The standard option on both models was a stirrup brake operating on the front wheel rim and a dummy rim brake on the rear wheel.

LEFT: The photograph shows the on-going development of the prototype 2¾hp Standard Model 350 shown on the opposite page. The kick-starter pedal has been moved to the off-side. The Enfield patent design of handlebars, fitted to the 8hp models would be offered as an alternative to the semi TT handlebars.. The foot pedals have been replaced by footboards. On production machines aluminium footboards with rubber studs and legshields would be offered as an alternative to the footrests at a cost of £1. The production machines were fitted with two pannier toolboxes.

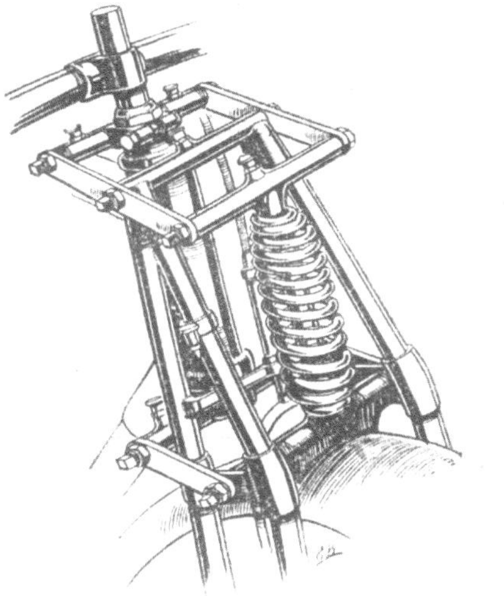

The front fork fitted to the new Models 350 and 351 used a single barrel spring.

The 2¾hp Side-valve model showing the optional Touring Handlebars, footboards and leg shields which cost £1 when supplied with a new machine.

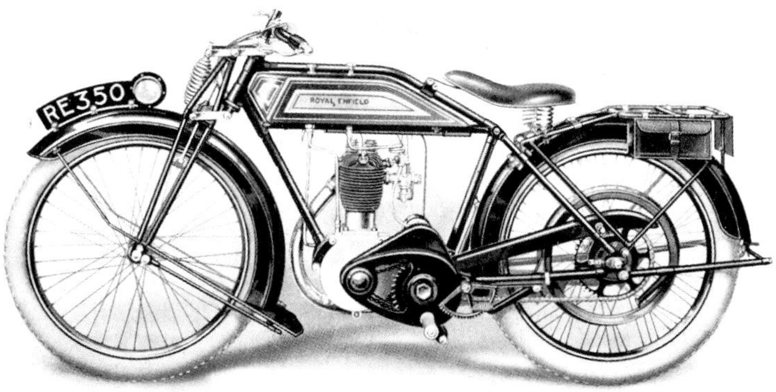

2¾ h.p. STANDARD MODEL 350

Fitted with Enfield Patent Foot-operated Gear, Cush Drive Hub, Chain Transmission and Kickstarter

SPECIFICATION

Engine.—2¾ h.p. (547 c.c.) Single Cylinder: bore and stroke, 70 × 90 m/m. Aluminium Alloy Piston.

Lubrication.—Automatic Mechanical Pump and Auxiliary Hand Pump.

Ignition.—High tension Magneto.

Carburetter.—Two-lever type.

Gear.—Royal Enfield patent; direct drive on each gear; foot-operated. Kickstarter.

Transmission—Roller Chains; Cush Drive in rear hub.

Frame.—Low saddle position; Adjustable Footrests. Royal Enfield Front Forks.

Brakes.—Foot-operated Rear Brake; Stirrup Front Brake.

Handlebar.—Semi-T.T.

Tyres.—Dunlop, 26 × 2¼ in.

Tools.—Complete set in two pannier bags on carrier.

Finish.—Best quality Black Enamel. Tank panelled and lined. Bright parts heavily plated.

See page 15 for particulars of Touring Handlebars, Footboards, and Legshields.

Price **£49** Net Cash

or **£13 7s. 6d.** down, and twelve monthly instalments of **£3 10s. 3d.,** including Insurance.

This Model can be supplied with Sturmey-Archer Three-speed Gear, Handlebar-controlled Clutch, and Kickstarter, at an extra charge of £2. Electric Lighting Set, £8 extra.

— 9 —

ROYAL ENFIELD

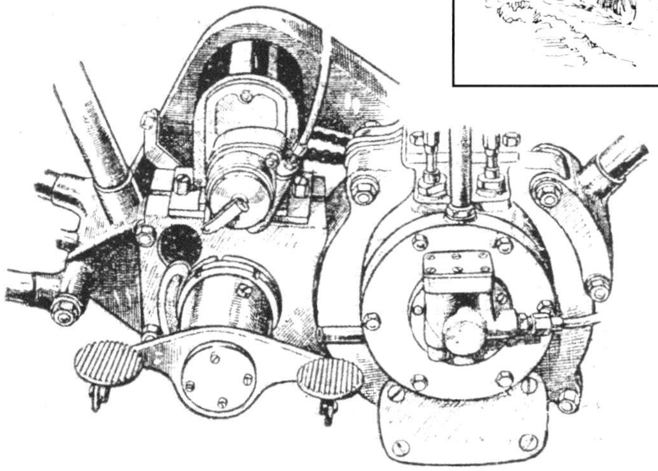

LEFT: The mechanical oil pump fitted to the 2¾hp models was mounted on the outside of the timing cover and driven off the half-time shaft. It was similar to the that fitted to the 8hp machines. An auxiliary hand pump, fitted into the tank, allowed a charge of oil to be given as required. The illustration also shows the heel and toe pedal of the two-speed gear and the magneto mounted on a platform above the two-speed mechanism.

2¾ hp Standard Model 350

Engine: Side-valve 2¾ hp (347 cc) single
Bore & stroke: 70 x 90 mm
Lubrication: Mechanical pump and auxiliary hand pump
Ignition: High tension magneto
Carburettor: Two-lever type
Gears: RE patented 2-speed. Kick starter
Transmission: Chain, with cush drive in rear hub
Frame: Adjustable footrests. RE front forks
Handlebars: Semi TT type

Brakes: Front stirrup on wheel rim,
Rear dummy rim
Tyres: 26 x 2¼ in Dunlop
Finish: Best quality black enamel
Tank: Panelled green and gold
Optional: Touring handlebars, footboards and leg shields
Weight: 189 lbs
Price: £57
Price with electric lighting set: £65

2¾ hp Standard Model 350 with optional Sturmey-Archer Three-speed Gearbox

Specification
Sturmey-Archer Three-speed Gear box
Otherwise as Two-speed model above

Price: £51
Price with Electric Lighting Set: £72.

2¾ hp OHV Sports Model 351

Engine: Overhead-valve 2¾ hp (347 cc) single
Bore & stroke: 70 x 90 mm
Lubrication: Mechanical pump and auxiliary hand pump
Ignition: High tension magneto
Carburettor: Racing type
Gears: RE patented 2-speed. Kick-starter
Transmission: Chain, with cush drive in rear hub
Frame: Low saddle height, adjustable footrests. RE front forks

Brakes: Front, stirrup on wheel rim
Rear, dummy rim. Hub brake optional
Handlebars: TT type
Tyres: 26 x 2½ in Dunlop
Finish; Best quality black enamel
Tank: Panelled green and Gold
Weight: 206 lbs
Price: £57

The prototype Model 351 was fitted with the rocking pedal gear change on the left hand side of the machine.

Illustration from the JAP catalogue of the Standard OHV Three-fifty. This was a single port design with 70 x 90mm bore & stroke. It weighed 60lb.

LEFT: The 2¾hp OHV Sports Model 351 was available fitted with a Three-speed Sturmey-Archer gearbox with hand change and kick-start. This version of the Sports Model 351 also came with a 5in drum brake, which replaced the front stirrup brake of the Two-speed model. The extra cost of the machine with gearbox and drum brake was £3.

LEFT: The photograph is of a prototype Open Frame model fitted with a JAP 347cc Side-valve engine and Sturmey-Archer 3-speed gearbox. The Open Frame two-stroke model had proved popular and the company explored the possibility of providing a similar mid-range machine with 350cc engine.

Return to Competition

Royal Enfield participation in racing had ended with the outbreak of World War 1. The company would have been aware of the boost in sales that followed sporting successes and keen to participate upon the return to peace. However, the company's post-war machines were functional, rather than sporting, and competition performance was restricted owing to the use of the Royal Enfield patent two-speed gear. This changed with the introduction of the 347cc JAP-powered Models 350 and 351 and the subsequent announcement of the JAP 344cc race engine.

JAP's new 344cc, 74 x 80mm, engine which was announced in April 1923, had been designed to be suitable for sporting and fast touring machines. A version of the engine with a twin-port cylinder head and other improvements was announced in time for the TT and subsequently made available for sports and racing singles. It very quickly became the dominant racing engine. A version of the Model 351 with the new JAP race engine, which would subsequently be identified as the Model 352, became available during 1924, but did not feature in the sales catalogues.

The first recorded appearance of the new Royal Enfield in competition

The above two works photographs are of the new two-port Sports Model 352 fitted with the JAP 344cc, 74 x 80mm race engine.

Illustration from the JAP catalogue of the OHV Three-fifty Sports engine. This was a twin-port design with 74 x 80mm bore & stroke. It weighed 66lb.

was at the Kop Hill Climb in March 1924. Stan Greening, who worked in the experimental department at J.A.Prestwich, the manufacturer of the JAP engines, won Class 3 (250cc - 350cc) and Class 7 (350cc sidecar) with his 344cc Royal Enfield-JAP. The machine was non-standard in respect of its duplex racing frame and small fuel tank. Perhaps as a result of his initial success at Kop Hill, Greening was penalised by the handicappers for his debut with the Royal Enfield at the Surbiton Motor Club's July meeting at Brooklands and was unplaced. The machine's next outing was at the Ealing Club's 200-Mile Sidecar Races in August. Greening took the lead on lap 39, but could only hold it for one lap, and retired on lap 67 of the 73 lap race with lubrication problems.

The weeks before the Olympia Motor Cycle Show were a popular time for record breaking as successes provided valuable publicity at the show. Stan Greening, accompanied by Teddy Prestwich, took the 344cc Royal Enfield-JAP outfit to Brooklands where they were successful in setting a total of twenty four long distance World's records in Classes B (350cc Sidecar) and F (600cc Sidecar); the 700 kilometres records were taken at 90.89kph (56.5mph)

8hp Sidecar Combinations

The 8hp sidecar machines for 1924 retained the Wolseley 976 cc twin-cylinder engine. The machines also retained the Royal Enfield patented two-speed gear although the Sturmey-Archer Three-speed gear box, with handlebar-controlled clutch and kick-start was offered as an alternative for an additional charge of £5.

As in previous years a de-luxe version of the standard machine was offered to the customer seeking passenger comfort and the convenience of an electric lighting system.

Both the Standard Model 180 and the De Luxe Model 190 could be purchased with two-seater bodywork for an additional £4. The passengers

The perfect combination; an Open Frame Two-stroke Model 201A for her and a 2¾hp Standard Model 350 for him.

8hp Sidecar Combination Standard Model 180

Type: Side-valve v-twin
Capacity: 976cc
Bore & stroke: 85.5 x 85 mm
Carburettor: Two-lever
Gearbox: Royal Enfield Patent Two-speed, Handle starter
Brakes: Foot operated dummy rim front & rear
Handlebar: Registered design with concealed cables
Tyres: Dunlop 700mm x 80mm
Saddle: Terry Patent Spring Saddle
Weight: 476 lbs. Solo weight; 296 lbs
Finish: Black enamel, tank panelled in green
Price: £90 without lights

were seated in tandem and the front seat folded forward to give access to the rear. The Model 190 was fitted with a single front windscreen to the sidecar. The front screen on the Model 180, the rear screens, the hood and side curtains were extra. The windscreens were priced at £4, The hood, and side curtains cost an additional £3 for the Model 180 and £3 10s for the Model 190.

Trade Delivery Combinations

Royal Enfield offered two trade delivery combinations based upon the Model 180 combination. The Model 185 was a conventional box carrier with a hinged top. It was painted green to match the motor cycle and lettering on the panels of the box was available for a nominal extra charge. The Model 185 was identified as a dual purpose vehicle; a standard passenger body could be fitted in place of the box. It was however necessary also to change the springs and shackles.

The Trade Delivery Combination, Model 155, had originally been designed for postal deliveries, but had found more general application and was being offered to a wider market.

Colonial Models

It had become the practice for manufacturers seeking sales in overseas markets to offer colonial versions of their larger capacity machines. Royal Enfield offered a Colonial Model of the 8hp Sidecar Combination with a specially designed frame, sidecar chassis, tank, etc., to provide extra ground clearance. The sales brochure noted: 'Our machines are giving the utmost satisfaction in all corners of the world, where the roads are extremely rough.'

RIGHT: The list of accessories available was listed in the sales catalogues. With the exception of the Model 190, the prices quoted did not include items such as a speedometer , lights or horn. Acetylene lights were most commonly used and were purchased with an acetylene generator.

The 8hp Standard Model 180 was listed as being available fitted with a Sturmey-Archer three-speed gearbox, with kickstarter and hand clutch for an extra £5.

8hp Sidecar Combination De Luxe Model 190

Specification as Model 180, but inclusive of a Lucas magdyno lighting set with headlamp, sidecar lamp and rear lamp. Mud shields were added to protect the dynamo and a Lucas bulb horn came as standard. The sidecar was more luxurious and fitted with a windscreen. The Model 190 was priced at £110.

ACCESSORIES

The following Accessories only can be included above :—

ANY MODEL.—Cowey Speedometer, **£4 15 0** ; Electric Lighting as per Catalogue.

ANY COMBINATION.—Legshields, **£1 1 0** ; Hood, Windscreen and Side Curtains at Catalogue prices.

MODELS 200, 201, 201a, 202.—

Lucas No. 331 Head Light and Generator - -	**£1**	**7**	**6**
Lucas No. 344 Rear Light and Generator - -		**13**	**0**
Lucas No. 63 E.B. Bulb Horn - - - -		**10**	**6**

MODELS 180, 185, 155.—

Lucas No. 462 Head Lamp and Generator - -	**3**	**10**	**0**
Lucas No. 354 Sidecar Lamp and Generator - -	**1**	**2**	**6**
Lucas No. 344 Rear Light and Generator - -		**13**	**0**
Lucas No. 60 E.B. Bulb Horn - - - -		**15**	**6**

The single-seater sidecar body fitted with the optional hood and side curtains.

The Two-seater sidecar body was available with either the Standard Model 180 or the De Luxe Model 190 at extra cost.

LEFT: The two-seater sidecar body is shown fitted with optional hood and side curtains. These offered excellent weather protection for the sidecar passengers but added little to the styling.

Trade Delivery Combination Model 185

Motorcycle specification as Model 180
Box dimensions: Length 4ft 6in
Depth 1ft 9in
Width 1ft 8in
Capacity: 13 cubic feet approximately
Finish: green to match the motorcycle
Price: £90

8 h.p. TRADE DELIVERY COMBINATION

MODEL 155

MODEL 155, with specially large Box Carrier, was originally designed for the G.P.O. Authorities, and after exhaustive tests the latter have placed orders for and taken delivery of a number of these machines. They have emerged successfully from strenuous trials as the most efficient and economical means of transporting parcels, etc., **up to 3 to 4 cwt.**

The 8 h.p. motor cycle is as previously described, but the chassis is a specially wide one to accommodate the large Box Carrier, and has a five-point attachment. Quarter-elliptic springs support the front of the Box, and semi-elliptic springs the rear.

The Box is strongly made from finest quality timber, and has a lid at the front, and double doors at the back. It can be painted any colour, and be lettered at a small charge, similar to the outfit described on page 12.

Dimensions of the Box Carrier:—

Length, 4 ft. 9 in. Depth, 3 ft. 3 in. Width, 2 ft. 9½ in.
Capacity, approximately 31 cubic feet.

A complete kit of tools is provided, and an easily-operated jack accompanies each outfit.

Dunlop Magnum Cord, 26 in. × 3 in. Light Car Tyres are fitted on all three wheels.

Price **£120** Net Cash

or **£32 8s. 9d.** down and twelve monthly instalments of ~~**£4 17s. 9d.**~~, including insurance.
£8 10 4

This Model can be supplied with Sturmey-Archer Three-speed Gear, Handlebar-controlled Clutch, and Kickstarter, at an extra charge of £5.

— 13 —

ROYAL ENFIELD
MADE LIKE A GUN

CHAPTER FIFTEEN

Model Year 1925

Building on from the extended 1924 range, Enfield offered for 1925 two additional sporting models, the Model 352 and Model 182. The former was a twin-port version of the 350cc OHV machine introduced for 1924. The second was a sports solo version of the 8hp Twin, which previously had only been offered as a combination. In addition two lightweight sidecars were offered for fitting to the 350cc machines. In other respects a casual glance at the new season's offerings might suggest that little development had occurred, but such was not the case.

The Enfield Patent Two-speed Gear, which had proved popular with riders owing to its simplicity and ease of operation, was by now dated and would be gradually phased out. For the 1924 range, purchasers had been offered the choice of the Enfield Two-speed Gear or a three-speed Sturmey-Archer gearbox at extra cost. For 1925 the Sturmey-Archer box was the standard fitment on all except the two-stroke models. The fitment of the Sturmey-Archer box necessitated redesign of the frame to allow use of either the Enfield or Sturmey-Archer gear change. Cycle-type stirrup front brakes and dummy rim brakes were also gradually being phased out in favour of internal expanding drum brakes front and rear. The early system was retained on the two-stroke machines until the models were extensively redesigned for the next season. All the four strokes gained mechanical lubrication and the riders of these machines, with the exception of the 350cc side-valve model, enjoyed the extra comfort provided by a Terry saddle.

1925 Model Range

Model No.	Model Name	Engine Size	Type	Price
201	2¼ hp	225cc	Two-stroke	£35
201A	2¼ hp Open Frame	225cc	Two-stroke	£36
200	2¼ hp Sports	225cc	Two-stroke	£35
202	2¼ hp	225cc	Two-stroke	£32 10s
350	2¾ hp Standard	350cc	Side-valve single	£47
351	2¾ hp Sports	350cc	OHV single	£55
352	2¾hp Sports	350cc	OHV twin-port single	£75
182	8 hp Sports Solo	976cc	Side-valve twin	£75
180	8 hp Sidecar	976cc	Side-valve twin	£95
190	8 hp Sidecar	976cc	Side-valve twin	£105
185	8 hp Trade comb.	976cc	Side-valve twin	
155	8 hp Trade comb	976cc	Side-valve twin	

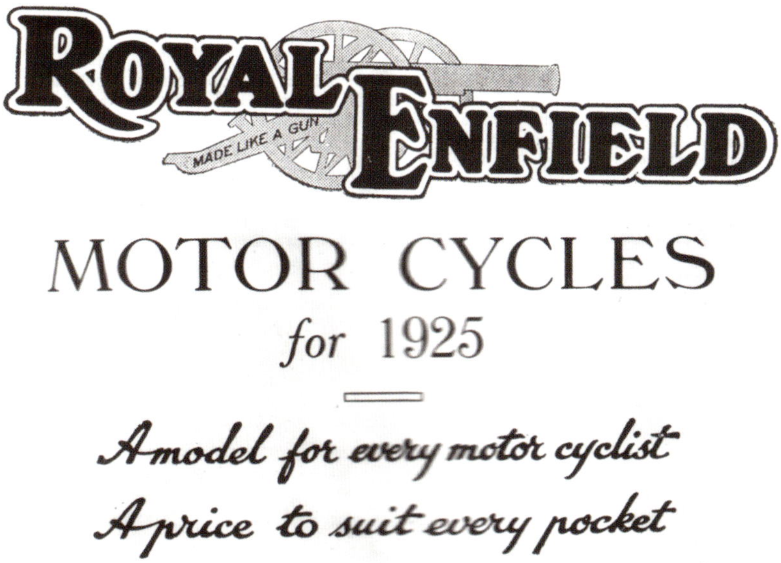

General specification of 2¼hp Two-stroke Lightweights
Engine: 225cc single cylinder two-stroke.
Bore and stroke: 64 x 70mm.
Lubrication: Oil mixed with petrol
Ignition: Magneto.
Gearbox: Royal Enfield patent, two speed gear.
Transmission: Chain, with cush drive in rear hub
Front forks: Royal Enfield
Brakes: Front stirrup, rear dummy rim
Carrier: Detachable with rear mudguard section
Finish: Black enamel
Weight: 155 lbs

Model 201
Tyres: Dunlop 24 x 2¼in
Handlebars: Royal Enfield design, concealed cables
Kick starter: Fitted
Fittings: Aluminium footboards & leg shields

Model 201A
Tyres: Dunlop 24 x 2¼in
Frame: Open design
Handlebars: Royal Enfield design, concealed cables
Kick starter: Fitted
Footrests: Aluminium footboards & leg shields
Price: £36
Extra for electric lighting set £8

Model 202
Tyres: Dunlop 24 x 2in
Footrests: Adjustable
Kick starter: Not fitted
Price: £32 10s
Available with 'Sports' handlebar if required.

Model 200
Sports model fitted with TT handlebars
Ignition timing: Variable
Carburettor: Senspray
Footrests: Adjustable
Kick starter: Not fitted
Price: £35

2¾hp Four-strokes

The side-valve Model 350 and overhead-valve Model 351 were retained in the range. The Model 350 was now fitted with a four-inch drum brake on the front wheel whilst retaining the dummy rim brake at the rear. The Model 351 was fitted with five-inch internal expanding brakes, both front and rear. Both machines were fitted with a three-speed Sturmey-Archer gearbox, which necessitated a slightly modified frame. The OHV Sports Model 351 had new duplex saddle tubes bolted at their extremities to the main frame. These were designed to carry an aluminium bracket, mounted approximately at mid height, for the gear change lever. It was claimed that this lowered the riding position by 2 inches. Both machines had a new design of flat-ended silencer.

The new Model 352 was similar to Model 351 but fitted with the double-port Super Sports version of the JAP overhead-valve 350. It was fitted with long exhaust pipes with removable tail pieces and racing style handlebars; it was priced at £20 more than the standard model.

Lubrication on all models relied upon a gravity-fed mechanical oil pump, of Enfield's own manufacture, mounted externally on the timing cover and driven from the crankshaft. Oil from the pump was delivered directly to the timing chest, from where it was sucked through a non-return valve into the crankcase; surplus oil being returned to the timing gears through three release valves. The pump was of the low speed oscillating plunger type and its flow was preset at the works.

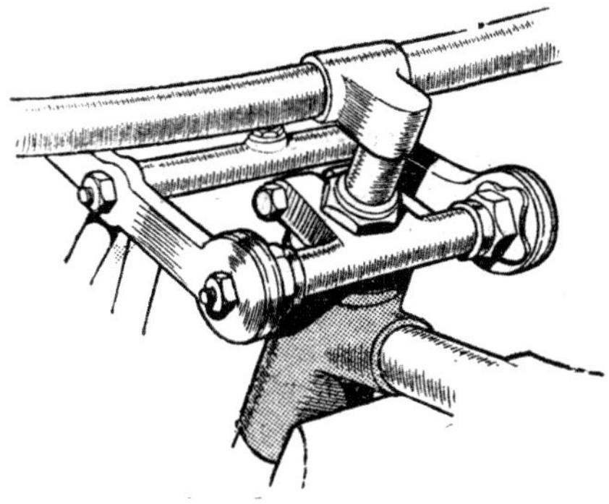

The damper incorporated into the front forks of the two overhead-valve 350 models.

Brochure illustration of the 2¾hp Side-valve Model 350.

2¾hp Side-valve Model 350

Engine: Side-valve 2¾ hp (346 cc) single
Bore & stroke: 70 x 90 mm.
Lubrication: Mechanical pump and auxiliary hand pump.
Ignition: High tension magneto.
Carburettor: Two-lever type.
Gears: Sturmey-Archer 3-speed, Kick-starter
Transmission: Chain, with cush drive in rear hub
Frame: Adjustable footrests. RE front forks
Brakes: Front 4in drum, Rear dummy rim
Handlebars: Semi TT type.
Tyres: 26 x 2¼ in Dunlop.
Finish: Best quality black enamel.
Tank: Panelled green and Gold.
Price: £47

A nearside view of the 2¾hp Side-valve Model 350 showing the primary chain cover fitted to the 2¾hp machines. The machine has been fitted with the touring handlebars, aluminium footboards with rubber studs and legshields. The footboard and legshield assembly cost an additional £1. A Lucas lighting set cost a further £8.

2¾hp OHV Sports Model 351

Engine: ohv 2¾ hp (347 cc) single
Bore & stroke: 70 x 90 mm.
Lubrication: Mechanical pump and auxiliary hand pump.
Ignition: High tension magneto.
Carburettor: Racing type.
Gears: Sturmey-Archer 3-speed , close ratio, kick-starter
Transmission: Chain, with cush drive in rear hub
Frame: Adjustable footrests. RE front forks
Brakes: 5in drum front and rear
Handlebars: TT type.
Tyres: 26 x 2½in Dunlop.
Finish; Best quality black enamel.
Tank: Panelled green and Gold.
Price: £55

2¾hp Super Sports Model 352

Engine: ohv twin-port (344cc) single
Bore & stroke: 74 x 80 mm.
Lubrication: Mechanical pump and auxiliary hand pump.
Ignition: High tension magneto.
Carburettor: Two-lever type.
Gears: Sturmey-Archer 3-speed Close ratio, kick-starter
Transmission: Chain, with cush drive in rear hub
Frame: Adjustable footrests. RE front forks
Brakes: 5in drum front and rear
Handlebars: Racing type.
Tyres: 26 x 2¼ in Dunlop.
Finish; Best quality black enamel.
Tank: Panelled green and Gold.
Price: £47

Sidecars for 2¾ hp Motorcycles

Two lightweight sidecars of attractive design were offered for use with the 2¾hp machines. The option of either a sporting or touring version was available. Both used the same chassis and were coach built with polished aluminium panel work and upholstered in the finest quality black leathercloth. The touring model boasted a door.

The Sports Model was priced at £13 10s, whilst the Touring Model cost an additional £1.

Touring Model

Sporting Model

The lightweight sidecars available for use with the 2¾hp models.

The Model 350 with Sporting sidecar body was priced at £60. This works photograph shows a sidecar body with a more rounded nose than was featured in the sales brochure.

New 8hp Enfield Twin

First announcement had been made in August 1924 of the new three-speed version of the 8hp Twin to be sold alongside the well established two-speed models. The new machine was fitted with a new engine which, although outwardly similar in appearance to the Enfield-Vickers unit, was a totally new design built in the Enfield works. Its quoted capacity was 976cc (85.5 x 85mm). The cylinders were interchangeable and the combustion space designed to be of the semi-turbulent type with sloping head. The ports were increased in size and of modified design. The pistons were aluminium alloy die castings and were fitted with two rings. The two connecting rods were fitted with double row roller bearings and each big end located side-by-side. The timing gear consisted of two pinions, the back of the larger one being machined with three cams set in line. The central cam operated both exhaust valves whilst the outer ones the two inlet valves.

A small oscillating plunger pump, operated by a reduction gear, delivered oil via a duct in the crankcase to the back of the front cylinder. The oil dripped from here into a well at the front of the engine from whence it was circulated by piston splash. The timing gear was lubricated by oil mist drawn into the timing chest via two disc valves.

The engine had a compression ratio of 4.7 to 1 and was claimed to be capable of revving to about 6,000rpm; the quoted output was 18hp at 2,800-3,000rpm. The gear ratios were 4.7, 7.1 and 13.7 to 1.

The new machine was fitted with a Sturmey-Archer gearbox, located below the bottom bracket, with the hand gear change lever attached to the saddle down tube. The magneto was moved from its earlier position ahead of the front down tube to a less vulnerable position behind the engine. It was bolted to a platform on top of the bottom bracket. The rear half of the magneto chaincase formed part of the timing case.

The new Enfield designed and manufactured 8hp (976cc) V-twin, which was announced in August 1924.

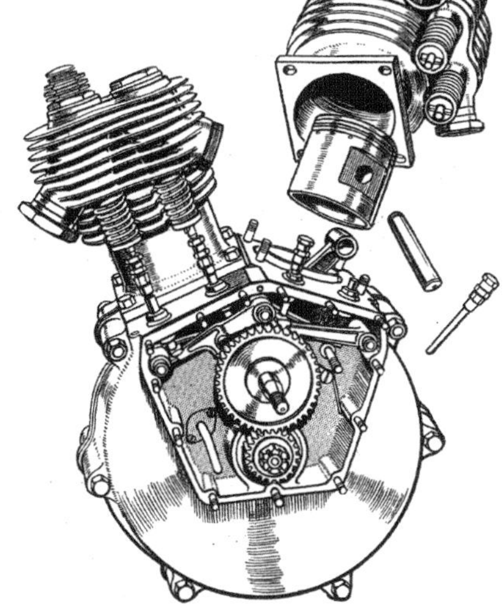

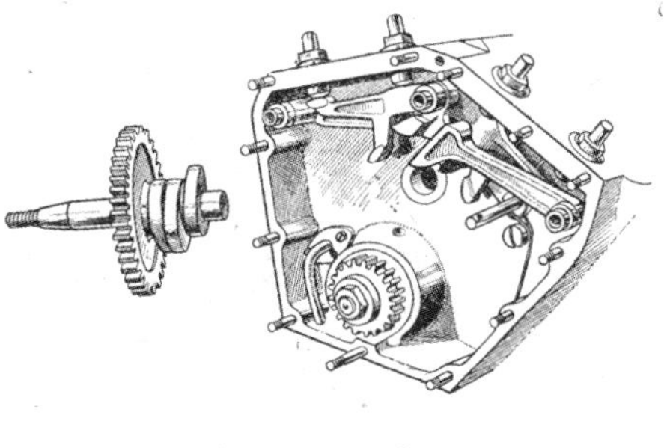

ABOVE and LEFT The two cutaway drawings show the general arrangement of the new 8hp Enfield engine. The valve operation relied on a single camshaft carrying three cams, two inlets and a single exhaust, acting on the tappets through hinged rockers.

The timing cover of the new engine proudly proclaimed it to be of Royal Enfield manufacture. The drawing indicates the position of the magneto, gearbox and the mechanical oil pump.

The works photographs are thought to be of the prototype 8hp (976cc) Twin announced in August 1924, for the 1925 season, and used for press demonstrations. The machine is fitted with a three-speed Sturmey-Archer gearbox and the earlier type of Terry saddle. The dummy rim brakes would be replaced by 8 inch drum brakes on the production models.

Minor modifications were made to the frame of the new model. It was fitted with a straight top tube, which was dropped considerably at the rear to achieve a low saddle height. The special Terry saddle was bolted at its nose directly to the top tube and suspended on tension springs. The rear portion of the frame was bolted to the main frame behind the saddle tube and to the back of the engine plates.

The tank capacity was increased to over two gallons of fuel and three pints of oil. Despite the fitment of a mechanical pump, an auxiliary hand pump was retained for emergencies.

Dummy belt rim brakes, with the friction material operating within the v of the rim, were fitted to both wheels of the machine announced in August 1924, but would be replaced by 8 inch drum brakes for the 1925 season. Both brakes were foot operated; the front brake was controlled by the right toe and the rear brake by the left heel.

The complete outfit was priced at £95 compared with £90 for the standard two-speed outfit, a relatively modest increase which would indicate the demise of the Enfield patent two-speed mechanism.

8hp Sports Solo

The 8hp twin models had long been considered the workhorses of the Enfield range and were supplied with either passenger or commercial sidecars. The solo rider, whether ride-to-work or sporting had been catered for by the 350cc singles. This had left a considerable gap in the range as by

The prototype 8hp Sports Solo. The machine retains the dummy belt rim brakes and the hand lever for the gear change is attached to the frame saddle tube.

8 h.p. SPORTS SOLO

MODEL 182

Fitted with Sturmey-Archer Three-speed Gear, Enfield Cush Drive Hub, Chain Transmission, and Kickstarter.

SPECIFICATION

Engine.—Royal Enfield Twin Cylinder; 85·5 × 85 m/m; capacity 976 c.c. Aluminium Alloy Pistons.

Lubrication.—Automatic Mechanical Pump and Auxiliary Hand Pump.

Frame.—Strong and rigid; Enfield Front Forks; detachable Rear Mudguard and Carrier.

Brakes.—Internal Expanding Front and Rear, with 8 in. drums (both foot operated).

Handlebar.—Semi-T.T., rubber and celluloid grips.

Saddle.—Terry Patent Spring Seat.

Tyres.—Dunlop, 700 × 80 m/m.

Stands.—Front and Rear Stands fitted.

Tools.—Complete kit in two strong pannier bags.

Finish.—Best Black Enamel; Tank panelled in green and gold. Bright parts heavily plated.

Price **£75** Net Cash

or **£19 15s. 0d.** down and twelve monthly instalments of **£5 3s. 8d.,** including insurance.

The De Luxe version of the Sports Twin Combination with Lucas Magdyno lighting set. The switch for the lights is on the left of the tank. On this pre-production model the dummy belt rim brakes have been replaced by internal expanding drums, identical to the five inch units fitted to the 2¾hp ohv machines, and the early type front forks retained.

this time most manufacturers were offering larger capacity singles and twins for the solo rider seeking a more sporting and faster mount. The Enfield answer to this was to offer the 8hp twin as a solo mount for the rider seeking effortless high speed. The mid-range gap in the Enfield range would not be plugged before the 1927 model season.

The specification was little different to the other twin-cylinder models, Differences included moving the mount for the gear-change hand lever from the saddle tube to the frame tube below the tank, a change which the other models would soon adopt, and the use of semi-TT handlebars. It was also fitted with internal expanding brakes, with 8in drums from the outset, although the other models would soon follow.

8hp Sports Combination

Although offered as a solo machine, the Sports model was also available as a Sports Combination. It used the standard 8hp sidecar chassis with a lightweight polished aluminium body having a folding windscreen.

The 8hp Range

The range of 8hp models, both passenger and commercial were little changed for the 1925 season. The range of bodies illustrated in the sales brochure included a Dairyman's Open Box version of the Trade Delivery Combination, two versions of the Trade Delivery bodywork, including the GPO version, and what appears to be a de luxe closed passenger coachwork body. The sidecars were manufactured in the Company's own coach building works and bodywork could be supplied to meet the customer's requirements. Details of the Dairyman's Open Box body are given in the next chapter.

Dunlop balloon tyres were available for the de luxe combination for an extra £5, but would soon

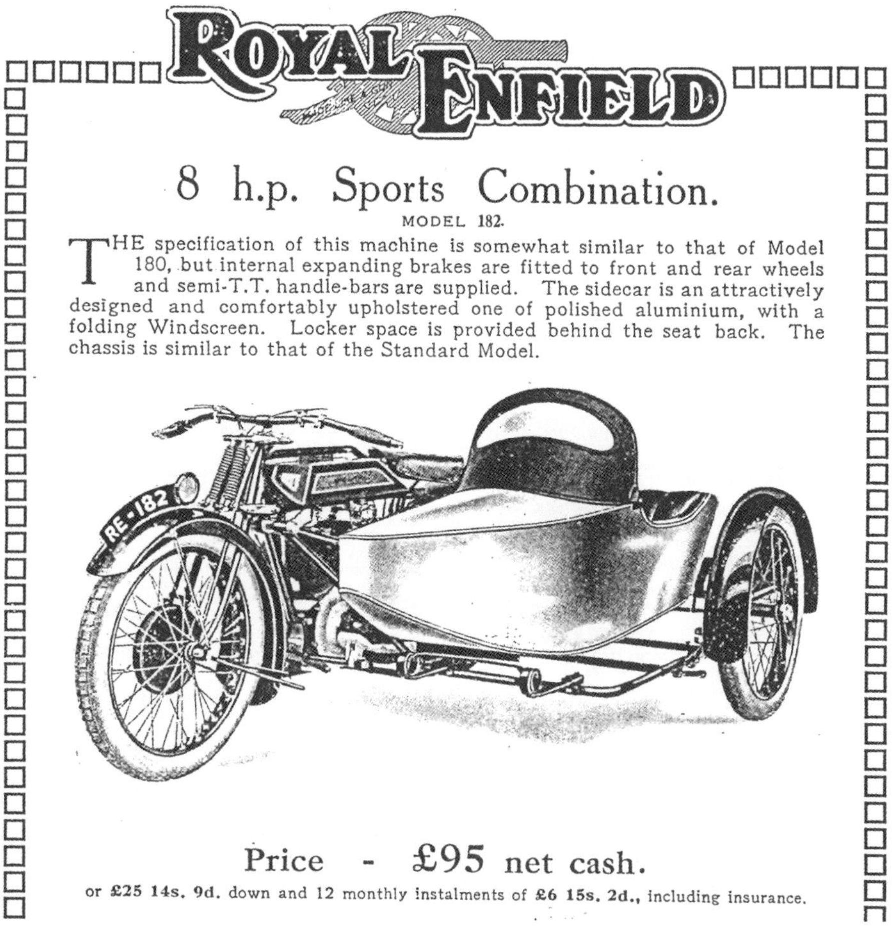

Catalogue illustration of the 8hp Sports Combination. By this time the 5-inch drums of the prototype model have been replaced by 8-inch drums, but the early 'narrow' type of front forks has been retained.

become a standard fitting. To accommodate these, a new widened and strengthened front fork was introduced in which the bottom links were brought inside the girders.

Factory illustration of the 8hp Sports Model Combination fitted with the wider front forks.

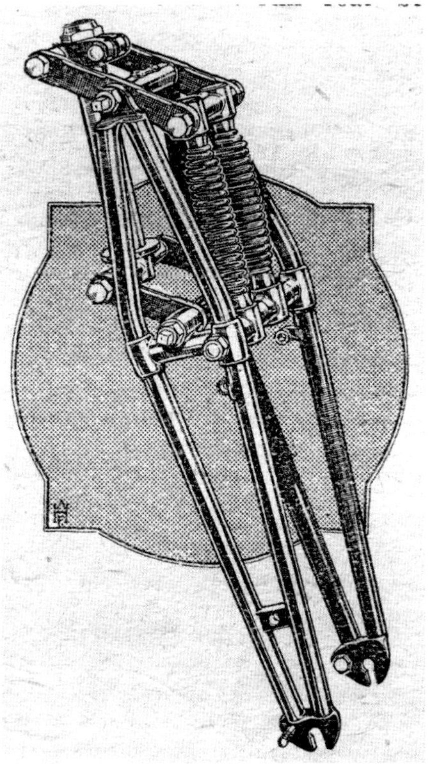

The more substaantial and wider design of front forks was introduced on the 8hp twin models to accommodate the new balloon tyres.

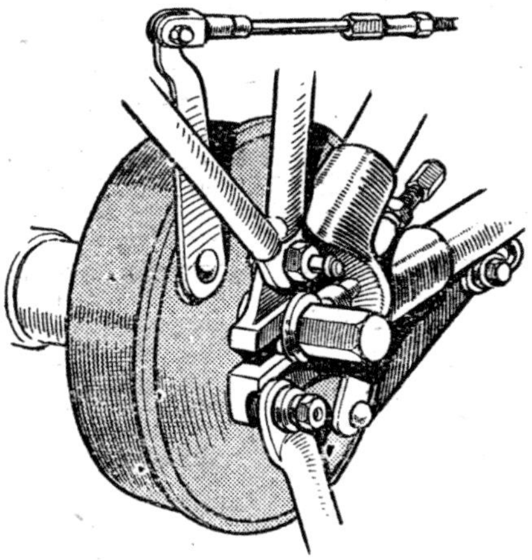

Internal expanding brakes were gradually being introduced on all models across the range. Those fitted to the 8hp Twins were 8 inch diameter and 1 inch in width.

ROYAL ENFIELD

8 h.p. Sidecar Combination.

DE LUXE MODEL 190

THE general specification of this Model is somewhat similar to that of the Standard Combination, but is supplied ready for the road, complete with Lucas Magdyno, Switchbox, Battery (in the Sidecar), Head Lamp, Sidecar Lamp, Rear Lamp and Ebony Black Bulb Horn.

The sidecar on this Model is fitted with Windscreen, side-curtain, etc. Internal expanding Brakes are fitted, front and rear, both operated by foot-pedals.

Enfield two-speed Gear or Sturmey-Archer three-speed Gear, Clutch and Kickstarter.

Price - £105 net cash,

or **£28 5s. 3d.** down and 12 monthly instalments of **£7 8s. 5d.,** including insurance.

Two-seater Sidecar, one Windscreen, extra	£4	0	0
Hood - - - -	£3	0	0
,, (Two-seater) -	£3	10	0
Legshields - - -	£1	1	0

MODEL 192, similar to Model 190, but fitted with special front forks and **Dunlop Cord Balloon Tyres,** Pressure Gauge and Legshields (not illustrated).

Price - £110 net cash,

or **£29 10s. 6d.** down and 12 monthly instalments of **£7 15s. 0d.,** including insurance.

The 8hp Standard Combination Model 180. The catalogue illustration suggests the machine retained dummy rim brakes rather than the internal expanding drum brakes being fitted to the De Luxe and Sports models.

The 8hp De Luxe Combination Model 180. This factory photograph shows the machine with the earlier style front forks but fitted with Michelin Comfort balloon tyres and drum brakes. The gear change lever is located under the lower top tube rather than on the saddle tube.

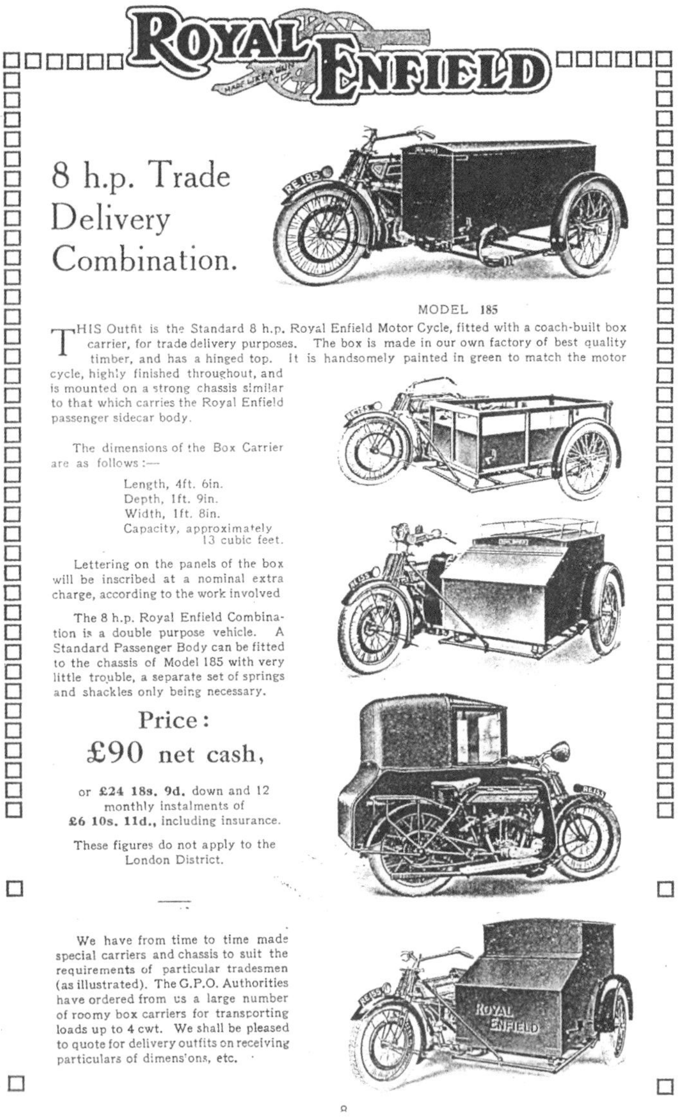

ROYAL ENFIELD

8 h.p. Trade Delivery Combination.

MODEL 185

THIS Outfit is the Standard 8 h.p. Royal Enfield Motor Cycle, fitted with a coach-built box carrier, for trade delivery purposes. The box is made in our own factory of best quality timber, and has a hinged top. It is handsomely painted in green to match the motor cycle, highly finished throughout, and is mounted on a strong chassis similar to that which carries the Royal Enfield passenger sidecar body.

The dimensions of the Box Carrier are as follows:—

Length, 4ft. 6in.
Depth, 1ft. 9in.
Width, 1ft. 8in.
Capacity, approximately 13 cubic feet.

Lettering on the panels of the box will be inscribed at a nominal extra charge, according to the work involved

The 8 h.p. Royal Enfield Combination is a double purpose vehicle. A Standard Passenger Body can be fitted to the chassis of Model 185 with very little trouble, a separate set of springs and shackles only being necessary.

Price:

£90 net cash,

or **£24 18s. 9d.** down and 12 monthly instalments of **£6 10s. 11d.,** including insurance.

These figures do not apply to the London District.

We have from time to time made special carriers and chassis to suit the requirements of particular tradesmen (as illustrated). The G.P.O. Authorities have ordered from us a large number of roomy box carriers for transporting loads up to 4 cwt. We shall be pleased to quote for delivery outfits on receiving particulars of dimens'ons, etc.

In addition to the standard passenger bodywork, the 8hp Combination could also be purchased with bodywork to meet customers' requirements. This illustration, which is taken from the sales brochure, gives examples of the typical commercial body styles and includes a de luxe passenger body. The illustrations appears to date back to 1918.

ACCESSORIES

The following Accessories only can be included above:—

ANY MODEL.—Cowey Speedometer, **£4 5 0**; Electric Lighting, Legshields, Hood, Windscreen and Side Curtains at Catalogue prices.

MODELS 200, 201, 201a, 202.—	
Lucas No. 331 Head Light and Generator	**£1 5 6**
Lucas No. 344 Rear Light and Generator	**13 0**
Lucas No. 63 E.B. Bulb Horn	**10 6**
MODELS 350, 351, 352.—	
Lucas No. 342 Head Light and Generator	**1 12 6**
Lucas No. 344 Rear Light and Generator	**13 0**
Lucas No. 63 E.B. Bulb Horn	**10 6**
MODELS 180, 182, 185.—	
Lucas No. 462 Head Lamp and Generator	**3 3 0**
Lucas No. 354 Sidecar Lamp and Generator	**1 2 6**
Lucas No. 344 Rear Light and Generator	**13 0**
Lucas No. 60 E.B. Bulb Horn	**15 6**

INSURANCE

A London Assurance Comprehensive Tariff Policy costs:—

£3 15 0 for Models 200, 201, 201a, 202, 350.
£3 16 0 for Model 351.
£3 18 0 for Model 350 with Sidecar.
£4 0 0 for Model 352.
£7 15 0 for Model 182 Solo.
£7 18 0 for Model 180 Combination.
£7 19 0 for Model 182 Combination.
£8 1 0 for Model 190 Combination.
£8 2 0 for Model 192 Combination.
£9 0 0 approximately for Country, and **£10 7 0** for London, for Model 185.

New 346cc Side-valve Model 350

Factory photographs, dated May 1925, of the new 346cc side-valve Model 350 fitted with the new 346cc side-valve engine

At the end of April 1925, less than nine months following the announcement of the new Enfield designed 8hp v-twin, the company released details of a new 350cc side-valve machine. The engine had been designed by E.O.Pardoe and was of Enfield's own manufacture. It was of orthodox design and owed much to the JAP engine it replaced. The bore and stroke remained 70 x 90 mm, The crankshaft was built-up with two heavy cast iron flywheels, the peripheries of which were polished to reduce oil drag. Double-row caged roller bearings supported both ends of the crankshaft and the big end. Lightness of the reciprocating parts was obtained through use of a flat topped piston and a Duralumin connecting rod. The gudgeon pin, which floated in both the small end and piston bosses, was fitted with aluminium end pads to avoid scoring the cylinder. The exhaust port was considered to be of a particularly large diameter and the valves were operated by rockers controlled by cams on a single camshaft.

Lubrication relied on the Enfield mechanical oil pump delivering oil directly to the timing gear case where it formed a pool below the timing gear. From here it was sucked through a non return valve into the crankcase, with surplus oil being returned as a mist through three release valves to the timing gears. An auxiliary hand-pump was fitted into the oil tank to supply extra oil when the engine was worked hard.

A slightly modified frame with the top tube sloping downwards from the steering head and duplex tank support tubes, which were bolted in place. Other modifications including the twin saddle post tubes, Terry saddle, valanced mudguards and shock absorbers were made along the lines of the ohv 2¾hp models.

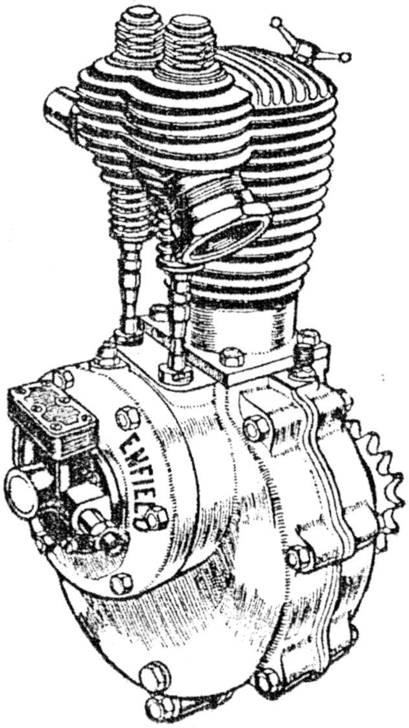

The new Ted Pardoe designed 346 cc side-valve engine.. The oil pump was mounted on the front of the timing chest and was not adjustable.

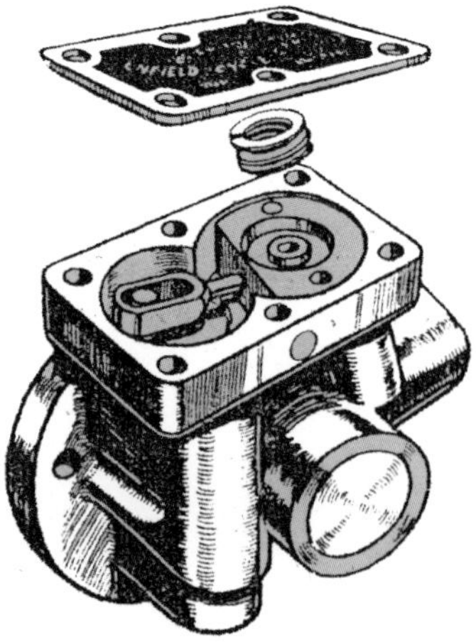

The Enfield mechanical oil pump was of the low speed oscillating plunger type.

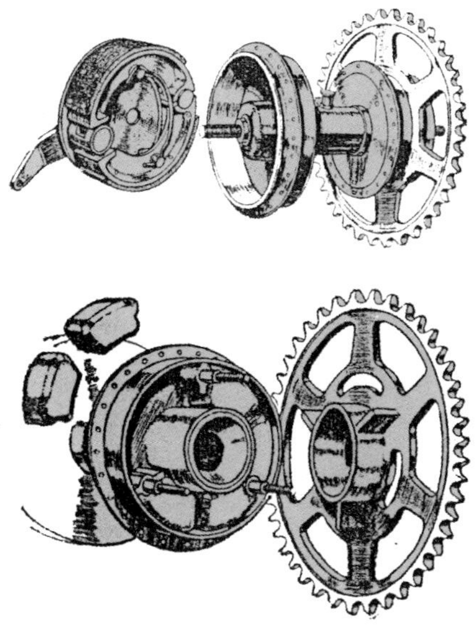

The components of the rear hub of the new side-valve Model 350. UPPER: The internal expanding brake. LOWER: The cush drive assembly.

The first 344cc JAP-powered TT machine made available to the South African rider, Charlie Young.

Testing

At this time all Enfield production engines were connected to the factory power shafting and turned for several hours to ensure they were fully run-in with all parts running smoothly and of the correct clearances. Upon completion each motorcycle was ridden round the company's test track behind the works. The track measured approximately one seventh of a mile and, typically, forty laps would be covered before the machine was returned for examination and rectification of identified faults. The machine was then taken on a lengthy road test before being passed as acceptable for delivery and forwarded first to the cleansing department and then to the despatch shop.

Tony Wilson-Jones

Tony Wilson-Jones joined Enfields in August 1925 to head the development work and would take responsibility for machine testing and the management of the Enfield TT machines and racing team. He was a graduate engineer and had been a premium apprentice at Daimler in Coventry, before working at Rover and BSA.

The Sporting Year

The Enfield Cycle Company did not enter its own race team for the 1925 Tourist Trophy Races in the Isle of Man, but did produce 344cc race machines for private riders. The first machine was produced for the successful South African rider Charlie Young and was intended for subsequent use in South Africa. This machine used a Sturmey-Archer gearbox with operating lever mounted directly on the box. The fuel tank was enlarged and a separate oil tank mounted in front of the twin saddle tubes. The front forks were fitted with shock absorbers on both the bottom and top links. The standard five inch internal expanding brakes were fitted to both wheels.

The four riders who entered the 1925 Junior TT on twin-port ohv JAP powered Royal Enfield machines, were South African Charlie Young, Yorkshireman George Reynard and Irishmen Stanley Woods and Gordon Burney. Hopes for the race would have been high as Reynard had recently won a 50 mile event at Southport and Woods, who had first ridden the course in 1922 when he finished in fifth position, had won the event in 1923 riding his Cotton at an average speed of 55.73mph. The race proved a disappointment for the Royal Enfield riders. Young achieved a creditable seventh position, but was over 24 minutes after race leader Wal Handley on his Rex-Acme. Reynard was sixty five minutes behind the leader and had to settle for seventeenth position whilst Burney and Woods were both forced to retire.

The lubricators fitted to the valve gear of the ohv Enfields.

Gordon Burney was more successful in the Ulster Grand Prix. The Junior Race was initially led by H.Chambers on his AJS, who by the fourth lap had opened out a five minute lead on Burney and the Sunbeam rider J.Forsythe. Burney managed to reduce this to 2½ minutes by the sixth lap and to just 45 seconds on the eighth and final lap before Chambers was forced to retire. He went on to achieve a comfortable victory at an average of 65.60mph.

ROYAL ENFIELD
MADE LIKE A GUN

1925 SUCCESSES

A.C.U. 1,000 MILES STOCK MOTOR CYCLE TRIAL
4 machines entered—gaining
4 GOLD MEDALS AND TEAM PRIZE

INTERNATIONAL SIX DAYS' TRIAL
3 machines entered—gaining
3 GOLD MEDALS

COLMORE CUP TRIAL.—2 Gold Medals.

YORKSHIRE EVENING NEWS TRIAL.—SILVER CUP.

DUBLIN AND DISTRICT M.C.C. TRIAL.—FREEMAN CUP. GOLD MEDAL.

RACES AT SOUTHPORT SANDS.—First in 50 Miles Race.

TAVISTOCK M.C.C. HILL CLIMB.—First in 350 c.c. Class.

WEST OF ENGLAND SPEED TRIALS AT BRENTOR.—First in 350 c.c. Class. Third in Unlimited Class.

CORNELSCOURT SPEED TRIALS, IRELAND.—2 Firsts. 1 Second.

NEWTON ABBOT M.C. SPEED TRIAL.—Fastest Time of Day.

BELFAST AND DISTRICT M.C. HILL CLIMB.—First in 350 c.c. Class.

SPEED TRIALS AT SOUTHPORT SANDS.—3 FIRSTS IN 50 MILES RACES.

REDDITCH M.C. AND C.C. TRIAL.—TERRY CUP. GOLD MEDAL.

LUNE VALLEY M.C. RACES AT HEYSHAM.—1 First. 2 Seconds.

TRACK RACE, ST. POLTEN, OESTERREICH.—FIRST.

RAMBLERS M.C.C. SPEED TRIALS.—Fastest Time in 350 c.c. Class.

BELFAST M.C.C. SPEED TRIALS.—Fastest Time in 250 c.c. Class. Second in 50 Miles Championship.

KNOCK M.C.C. TWELVE HOURS TRIAL.—Silver Vase.

DUMFRIES A.C. HILL CLIMB.—HOPE BELL CUP.

RAVENSCOURT PARK M.C.C. SPORTING TRIAL.—Gold Medal.

BATH M.C.C. NOVICES TRIAL.—First.

PONTEFRACT M.C. FREAK HILL CLIMB.—First in 350 c.c. Class.

ULSTER M.C.C. RACE.—1 First. 1 Second.

CARDIFF M.C.C. RACE AT PORTHCAWL.—FIRST IN 10 MILES SIDECAR RACE.

DUBLIN UNIVERSITY M.C. RACES.—2 Firsts. 2 Seconds.

M.C.C. LONDON TO EDINBURGH RUN.—3 GOLD MEDALS.

BELFAST AND DISTRICT M.C.C. TWELVE HOURS TRIAL.—Gold Medal.

TEPLITZ - EICHWALD - ZINNWALD MOUNTAIN RACE.—FIRST.

ROTHERHAM AND DISTRICT M.C.C. TEAM TRIAL.—Silver Cup.

DUBLIN AND DISTRICT M.C.C. 24 HOURS TRIAL.—FREEMAN CUP. GOLD MEDAL. GOLD-CENTRE MEDAL. SILVER MEDAL.

BRADFORD M.C. AND L.C.C. BRADFORD-LONDON-BRADFORD RUN.—SILVER CUP.

LEINSTER M.C. & L.C.C. SPEED TRIALS. — FIRST IN TWO 25 MILES CHAMPIONSHIPS

CARDIFF M.C.C. RELIABILITY TRIAL.—First Class Award.

TADCASTER SPEED TRIALS.—11 FIRSTS. 1 SECOND.

HONITON M.C. & L.C.C. TRIAL.—CLUB CUP. TRADERS' BOWL.

RUGGA-RUGGA SPEED TRIALS.—Fastest Time in 2 Classes.

LE GRANDE PRIX SUISSE D'ENDURANCE.—2 GOLD MEDALS.

FRASERBURGH AND DISTRICT M.C. SPEED TRIAL.—First in Sidecar Class.

BELFAST AND DISTRICT M.C. TRIAL.—McDOWELL CUP.

RACE GRABUNKA BIELANERWALD.—Gold Medal.

ULSTER GRAND PRIX
FIRST IN 350 c.c. CLASS

A page from the 1926 sales brochure recording some of the sporting successes achieved by Royal Enfield machines during 1925.

CHAPTER SIXTEEN

Model Year 1926

The major innovation for the 1926 Season was the introduction into the range of the redesigned two-stroke models. These new machines were fitted with a two-speed gearbox and brought a final end to the Enfield patent two-speed and free engine gear, which had served the company so well over many years. Also incorporated into the range was the redesigned side-valve 3.46 hp Standard Model 350, which had been announced in April 1925 and has been described in the previous chapter. In other respects the range contained few surprises.

With the exception of the Model 352 OHV Twin-port model, modest price reductions were applied across the range,

1926 Models

Model 200	2.25 hp	Sports Two-stroke	225cc	£33 10s
Model 201	2.25 hp	Standard Two-stroke	225cc	£33 10s
Model 201A	2.25 hp	Open Frame Two-stroke	225cc	£35
Model 350	3.46 hp	Standard Side-valve	346cc	£45
Model 350	3.46 hp	Standard Side-valve Sports Comb.	346cc	£55
Model 350	3.46 hp	Standard Side-valve Touring Comb.	346cc	£57
Model 351	3.46 hp	OHV Sports Solo	346cc	£50
Model 351	3.46 hp	OHV Sports Combination	346cc	£60
Model 351	3.46 hp	OHV Touring Combination	346cc	£62
Model 352	3.44 hp	OHV Twin-port Model	344cc	£75
Model 182	9.76 hp	Sports Solo	976cc	£70
Model 182	9.76 hp	Sports Combination	976cc	£84
Model 180	9.76 hp	Standard Combination	976cc	£84
Model 180	9.76 hp	Two-seater Combination	976cc	£88
Model 190	9.76 hp	De-Luxe Combination	976cc	£95
Model 190	9.76 hp	De-Luxe Two-seater Combination	976cc	£99
Model 185 or 175	9.76 hp	Deliver Combination or Small Milk Float	976cc	£84
Model 165	9.76hp	Delivery Combination with Large Milk Float	976cc	£90
Model 155	9.76 hp	Delivery Combination	976cc	£109

The 4.88 hp De-Luxe Side-valve, Model 502, was equipped with a Lucas 'Magdyno' lighting set and speedometer. Electric lighting sets were available for £5 5s on the other models.

Redesigned Two-stroke Models

The Royal Enfield 225cc two-stroke model, which had been under development for two years before making its first public appearance in August 1914, was first offered for sale in 1915. It had proved a popular machine, but by the mid-twenties the styling had become very dated and the Royal Enfield Patent Two-speed Gear was losing favour. A redesigned range of 225 cc two-stroke models was announced in September 1925. The engine retained measurements of 64 x 70 mm, giving a capacity of 225 cc, and was little changed. However the cylinder barrel was redesigned for improved efficiency and now had two exhaust ports and flange-fitted pipes The inlet stub was bolted on the left of the cylinder and had a right-angle bend allowing the Brown and Barlow carburettor to face forward.

A two-speed constant-mesh gearbox was fitted Gear selection was by a sliding dog controlled by a hand operated lever mounted on the lower tank rail; the gear change quadrant being screwed to the side of the tank.

The frame was designed with the upper tank rail sloping downwards from the headstock and with an increased incline at the rear. This afforded a reduced saddle height of 26½ inches. The styling following that set by the 350cc models. The front forks were now of the duplex girder design with parallel links and a coil spring on each side. The redesigned frame allowed the use of a more stylish fuel tank of 1 7/8 gallons capacity. The dated stirrup front brake and dummy rim rear brake of the earlier models were replaced by Enfield drum brakes and the rear hub fitted with an Enfield cush drive,

A three model range of two-stroke machines was offered. The Model 201 was the standard model and fitted with aluminium footboards and legshields. The Model 200 was a sports version fitted with long twin exhaust pipes, footrests and TT handlebars. The open frame Model 201A retained its distinctive drum-shaped fuel tank and was little changed in appearance although it was also fitted with the new engine, transmission and front forks.

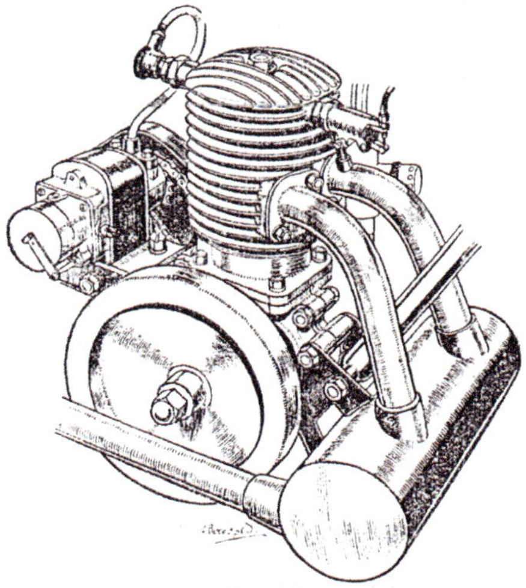

The new 225cc two-stroke engine employed twin exhaust ports. On the standard model, illustrated, these fed into a large cylindrical silencer placed across the frame in front of the engine.

2¼hp Two-strokes Lightweights

General Specification

Engine: 225cc single cylinder two-stroke.
Bore and stroke: 64 x 70mm
Lubrication: Oil mixed with petrol
Ignition: Magneto
Gearbox: Enfield two speed gear, with kick starter
Gear ratios: 8.1 to 1 and 5 to 1
Transmission: Chain, with cush drive in rear hub
Front forks: Royal Enfield
Brakes: Enfield internal expanding to both wheels
Carrier: Detachable with rear mudguard section
Handlebar: Adjustable Touring or Sporting type (Model 201A: Touring or Flat type)
Tyres: Dunlop Cord 24 x 2¼ in
Finish: Black enamel
Weight: 160 lbs (Model 200: 155 lbs)
Length: 6 ft 4 in
Width: 2 ft 8½ in
Saddle height: 26½ in (Model 201A: 27 in)
Price: £33 10s (Model 201A: £35)

ROYAL ENFIELD

2·25 h.p. Standard Model 201 (*Tax 30/-*)

SPECIFICATION:

ENGINE.—Royal Enfield Two-Stroke; bore and stroke 64 × 70 m/m.; capacity 225 c.c. Two exhaust ports. Roller bearings throughout.

LUBRICATION.—Oil mixed with petrol.

IGNITION.—High tension Magneto placed behind cylinder.

CARBURETTER.—Special two-lever type.

GEAR.—Royal Enfield two-speed. Handlebar-controlled clutch and kickstarter. Standard ratios 8.1 to 1 and 5 to 1.

TRANSMISSION.—Roller Chains. Patent Cush Drive in rear hub.

FRAME.—Low and comfortable riding position. Aluminium Footboards and protective Legshields. Royal Enfield Front Forks.

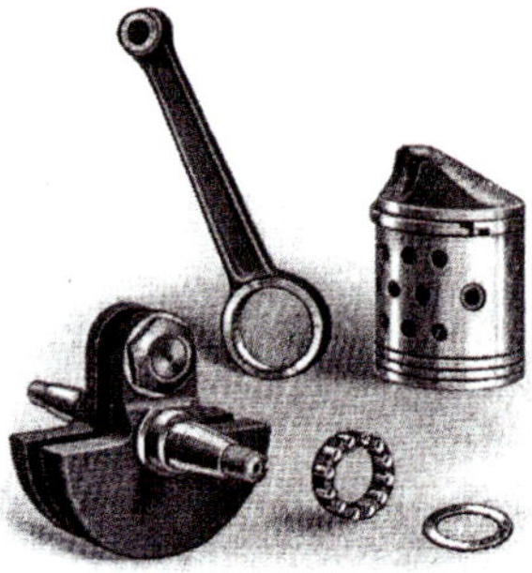

Piston, connecting rod, crank shaft, and roller bearing details of Royal Enfield two-stroke engine.

TANK.—New shape Tank of pleasing design. Combined Filler Cap and Oil Measure.

HANDLEBAR.—Touring or Sporting type; both adjustable.

BRAKES.—Internal expanding, front and rear.

SADDLE.—Best quality; long suspension springs.

TYRES.—Dunlop Cord 24 × 2¼ in.

MUDGUARDS.—Strong and efficient.

CARRIER.—Detachable carrier. By unfastening two nuts, the carrier and rear mudguard are lifted clear of the back wheel and tyre.

TOOL BAG.— Strong leather, containing a complete set of tools.

FINISH.—Enamelled in Best Black Enamel. Tank artistically panelled and lined. Bright parts heavily plated.

WEIGHT.—Approx. 160 lbs.

LENGTH.—6 ft. 4 in.

WIDTH.—2 ft. 8½ in.

SADDLE HEIGHT.—26½ in.

Price **£33 10 0** *Net Cash*

or **£9 6 3** down and twelve monthly instalments of **£2 10 4,** including insurance

This Model can be supplied fitted with Electric Lighting Set at an extra charge of **£7 10 0**

2.25 hp Standard Model 201

2.25 hp Open-frame Model 201A Model

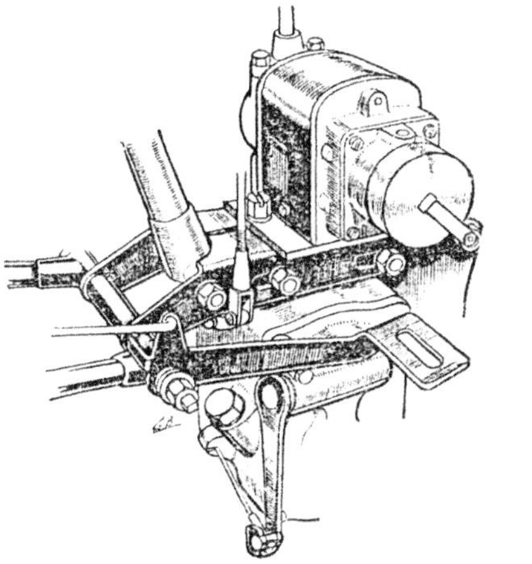

Details of the brackets for mounting the gearbox and magneto behind the engine.

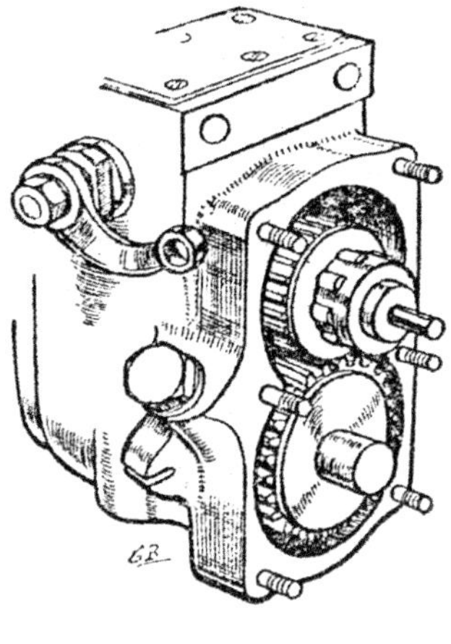

The new two-speed constant mesh gearbox. Standard ratios are 5 and 8.75 to 1.

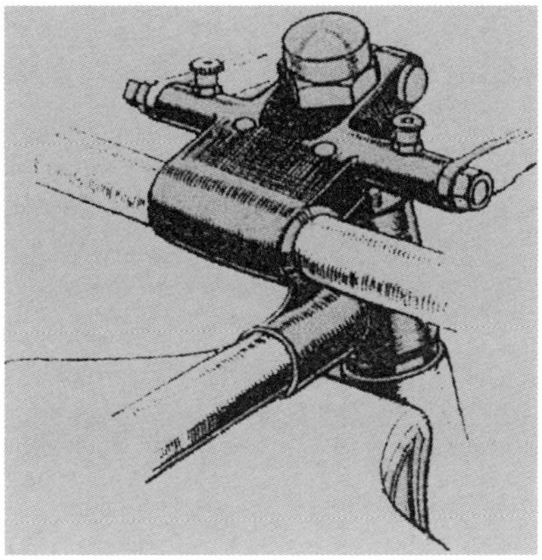

Steering head and handlebar fitting.

ABOVE and CENTRE: Works photographs of the new 225cc Two-stroke Model 200 Sports announced in September 1925 for the 1926 Season. The footboards of the standard models were replaced by footrests and forward reaching handlebars were fitted. The stylish twin exhaust pipes were swept round to connect with a single silencer on the right hand side of the machine.

350cc Model Range

The 350cc model range had continued to sell well and remained largely unchanged for the 1926 season. All models now used the lowered frame previously fitted to the overhead-valve models. The side-valve model also gained the larger diameter brakes of the OHV machines and larger tyres. The lightweight Sports and Touring sidecars were no longer listed separately for attachment to the 350cc models, but as a combination with the Model 350.

3.46hp Standard Model 350

Engine: 346cc side-valve single
Bore & stroke: 70 x 90 mm
Lubrication: Mechanical pump and auxiliary hand pump
Ignition: High tension magneto
Carburettor: Two-lever type
Gears: Sturmey-Archer three-speed
Standard ratios 5.9, 8.9 and 17.3 to 1
Transmission: Chain, with cush drive in rear hub
Frame: Adjustable footrests. RE front forks
Saddle: Terry or Lycett Spring Seat
Brakes: Internal expanding front and rear (5in drums)

Handlebar: Touring or Sports
Tyres: Dunlop Cord 26 x 2½ in
Finish: Best quality black enamel
Tank: Panelled green and lined
Optional: Footboards and leg shields £1 5s
Weight: 210 lbs
Length: 6ft 9in, Width: 2ft 8½in
Saddle height: 29in
Price: £45
Price with Electric Lighting Set: £53

3.46hp Model 350 Sidecars

Two models available for either sporting or touring use. The motorcycle specification was as Model 350, but with gear ratios 6.7, 10 and 19.5 to 1.
Chassis: Enfield design with four point attachment
C springs at rear and coil springs in front
Tyre: Dunlop Cord 26 x 2½ in
Bodywork: Sturdily built, polished aluminium panels
Upholstery: Black leathercloth
Overall dimensions: Width 4ft 9in, Length 6ft 9in

Sporting Sidecar: Fitted with hammock-type seat and tool locker behind seat
Length of sidecar: 6ft 4in
Weight of sidecar: 103 lb
Price: £55

Touring Sidecar: Fitted with spring seat
Length of sidecar: 4ft 9½in
Weight of sidecar: 198 lbs
Price: £57, Price with curved windscreen: £58

Model 350 Sporting Sidecar

Model 350 Touring Sidecar

3.46hp OHV Sports Model 351

Specification as Model 350
Gear ratios: 5.3, 6.7 and 10.1 To 1
Handlebars: Sports
Weight: 224 lbs
Length: 6ft 9in, Width: 2ft 9 ½in
Price: £50
Price with Electric Lighting Set: £58

Model 351

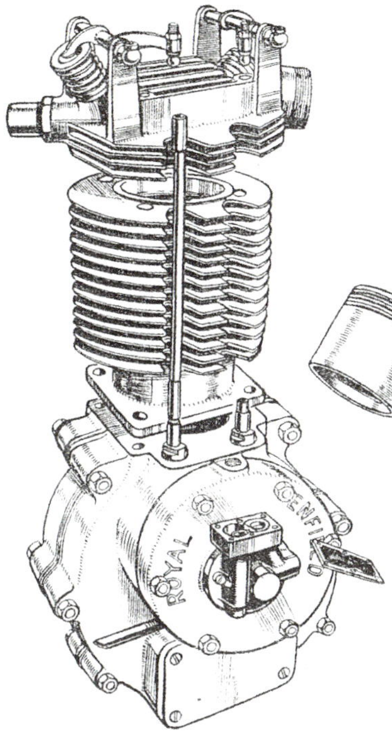

The 3.46hp OHV Sports engine fitted to the Model 351

3.44hp OHV Super Sports Model 352

Engine: JAP 344cc ohv Super Sports
Bore & stroke: 74 x 80 mm
Lubrication: Mechanical pump and auxiliary hand pump
Ignition: High tension magneto
Carburettor: Racing type
Gears: Sturmey-Archer three-speed (close ratio) 5.3, 6.7 and 10.1 to 1
Transmission: Chain, with cush drive in rear hub
Frame: Adjustable footrests,. RE front forks
Saddle: Terry or Lycett Spring Seat
Brakes: 5 inch internal expanding drums front and rear
Handlebar: TT type, adjustable
Tyres: Dunlop Cord 26 x 2½ in
Mudguards: Lightweight racing type
Finish: Best quality black enamel
Tank: Panelled green and lined
Weight: 224 lbs
Length: 6ft 9in. Width: 2ft 9½in
Saddle height: 29in
Price: £75.

The optional footboards and legshields for Models 350 and 351.

The 3.44hp Super Sports Model 352

The 9.76hp Models

Seven examples of the Royal Enfield big twin in solo or sidecar form were exhibited at the Motorcycle Show at Olympia in September 1925. These included the Sports Model 182 either as a solo or combination, three touring combinations, with Standard, De Luxe or Two-seater bodywork, and three examples of the commercial bodies. The commercial bodies available at this time were the box carrier, the Dairyman's open box and the GPO style Trade Delivery box.

9.76hp Sports Solo Model 182

Engine: Side-valve V-twin
Capacity: 976cc Bore & stroke: 85.5 x 85mm
Lubrication: Automatic mechanical pump and auxiliary hand pump
Carburettor: Two-lever
Gearbox: Sturmey-Archer three-speed, Standard ratios 4.2, 6.3 and 12.1 to 1
Transmission: Roller chains. Cush Drive Hub in rear wheel
Brakes: Internal expanding 8in drum brakes front and rear. Foot operated
Handlebar: Semi-TT, adjustable
Tyres: Dunlop Cord 700mm x 80mm
Saddle: Terry or Lycett Spring Saddle
Weight: 308 lbs
Dimensions: Length 7ft 2in, Width 2ft 8in
Saddle height: 30in
Finish: Black enamel, tank panelled in green
Price: £70. Lucas Magdyno Electric Lighting Set and Bulb Horn, £10 extra

9.76hp Sidecar Chassis Specification

Chassis: Enfield design with three point attachment. Quickly attached or removed
Springing: Leaf springs at front, C springs and shackles at rear
Dunlop Cord Balloon tyres available at an extra cost of £3 10s

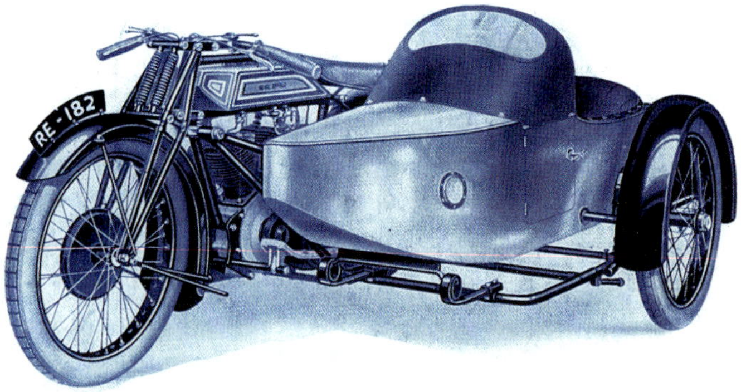

9.76hp Sports Combination Model 182

Specification as Sports Solo Model 182.
Body: Polished aluminium finish with aluminium beading. Upholstered in black leathercloth. Hammock type seat with tool locker behind. Celluloid panelled windscreen swings back to open door
Weight of sidecar, chassis, etc: 126 lbs
Total weight of combination: 434 lbs
Dimensions: Length 7ft 2in, Width 5ft 3in
Saddle height: 30in
Price: £84. Lucas Magdyno Electric Lighting Set and Bulb Horn, £11 extra

9.76hp Standard Combination Model 180

Specification as Sports Solo Model 182.
Body: Coach built body with locker at rear. Black leathercloth upholstery. Hammock type seat with tool locker behind. Curved celluloid panelled windscreen
Weight of sidecar, chassis, etc: 154 lbs
Total weight of combination: 462 lbs
Dimensions: Length 7ft 2in, Width 5ft 3in
Saddle height: 30in
Price: £84. Lucas Magdyno Electric Lighting Set and Bulb Horn, £11 extra

9.76hp De Luxe Combination Model 190
Specification as Standard Model 180.
Electrical equipment: Lucas Magdyno, battery (in sidecar), headlamp (incorporating switch), sidecar lamp, rear lamp and black bulb horn
Weight of sidecar, chassis, etc: 175 lbs
Total weight of combination: 490 lbs
Dimensions: Length 7ft 2in, Width 5ft 3in
Saddle height: 30in
Price: £95

9.76hp Two-seater Sidecars
The above sidecar was available with either the Model 180 or Model 190 and was designed to carry an adult and a child seated in tandem.. It cost an extra £4.
Dimensions: Length 7ft 11in, Width 5ft 6in

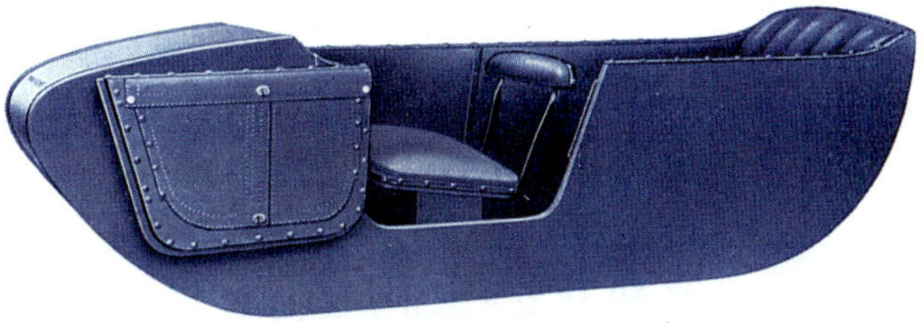

The special longer body to carry two adults, illustrated above, was available for an extra £5 5s.
Dimensions: Length 8ft 2in, Width 5ft 6in

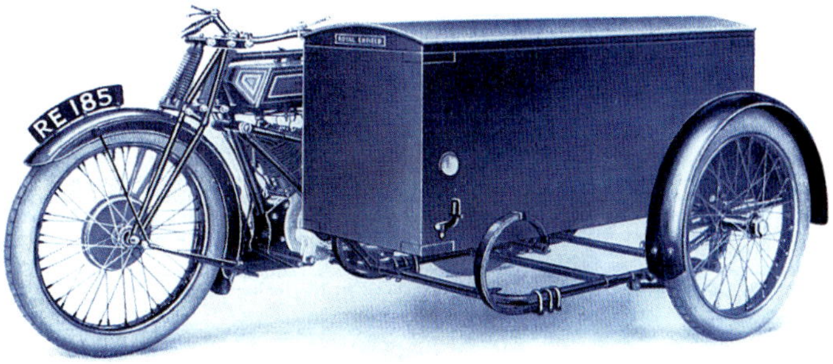

9.76hp Trade Delivery Combination Model 185
Specification as Standard Model 180.
Box dimensions: Length 4ft 6in, Depth 1ft 9in, Width 1ft 8in
Weight of sidecar, chassis, etc: 168 lbs
Dimensions: Length 7ft 2in, Width 5ft 3in
Price: £84

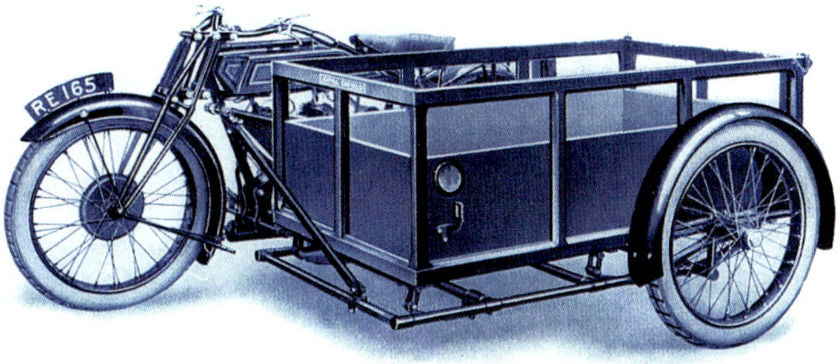

9.76hp Dairyman's Combination Model 165
Specification as Standard Model 180.
Body dimensions: Length 4ft 6in, Width 2ft 10in, Depth 1ft 7in
Weight of sidecar, chassis, etc: 196 lbs
Total weight of combination: 490 lbs
Dimensions: Length 7ft 2in, Width 6ft
Price: £95
Model 175, similar to Model 165, but with body dimensions and sidecar chassis as Model 185, available for £84.

9.76hp Trade Delivery Combination Model 155
Specification as Standard Model 180.
Tyres: Dunlop Cord 26 x 3 in Light car tyres
Parcel capacity: 31 cu ft, load 3-4 cwt
Weight of sidecar, chassis, etc: 245 lbs
Total weight of combination: 490 lbs
Dimensions: Length 7ft 2in, Width 6ft
Price: £109

The 1926 TT Machines

The Royal Enfield TT Models were little changed in design for 1926. The machines continued to use the two-port JAP OHV engines, with modifications to use the Enfield mechanical pump and lubrication system, assisted by a foot controlled auxiliary pump. Enfield brakes were fitted; the front 5in diameter drum was retained whilst a larger 8in diameter brake with aluminium cooling fins was fitted to the rear. The larger 2½ gallon capacity fuel tank was used with a separate saddle tube oil tank. In other respects the machine used all standard components. A close-ratio Sturmey-Archer gearbox was employed with gate change

Motor Cycling commented 'It is not often that the public have the opportunity of purchasing a sports model which is an exact replica of the T.T. racers, yet the 1926 TT Royal Enfields have already been introduced as a standard sports model which may be purchased by any who wish to acquire a really fast but tractable mount. Motor Sport, in its preview of the TT machines, commented that about half a dozen trade entries in the races were also using Enfield brakes.

Two factory views of the 1926 TT Machines fitted with the JAP 3.44 hp Racing Engine. The additional oil tank behind the seat tubes carried a supply of lubricant for the chains.

The Three-Fifty Racing (DOUBLE PORT)

LIKE the 250 c c Racing, this Engine has won SEVERAL CHAMPIONSHIPS, WORLD'S RECORDS and THOUSANDS OF MEDALS during the season, at HOME and ABROAD.

Amongst other Records won was the most gruelling of all THE TWENTY - FOUR HOUR WORLD'S RECORD.

A.C.U. Rating 3.44 H.P.
BORE - - 74 m/m.
STROKE - 80 m/m.
CAPACITY - 343.7 c.c.
WEIGHT - 66 lb.

SPECIFICATION.—Roller Bearing Big end, and Roller Bearing Pulley side, Special Valves Aluminium Piston, Magneto Chain Case, Magneto Chain, Magneto Sprockets, Magneto, Platform, Inlet Pipe, &c.

J.A.P. RACING ENGINES ARE GUARANTEED THE SAME IN EVERY DETAIL AS THOSE WHICH COMPETE ON THE TRACKS AND IN ALL RACES.

An illustration from the 1927 JAP Catalogue showing the Three-Fifty Racing engine of the type fitted to the Model 352.

Racing Successes

South African Charlie Young provided Enfields with an early success in 1926 at the South African Tourist Trophy Meeting held at the Kragga Kamma Circuit six miles outside Port Elizabeth on New Year's Day. He entered his 344cc Royal Enfield in the 350cc Class Race and led from the start to win at an average speed of 59.15mph. The 600cc Class Race was held on the following day and Charlie again entered his 344cc machine. This time there was a duel with Flook riding a 500cc Norton who led for the first two laps before being forced to retire. Young won at 60.43mph.

George Reynard on his 246cc JAP powered Enfield at the 1926 Lightweight TT. He was forced to retire on the fifth lap.

Charlie Young with his 344cc JAP-engined machine which he used to win both the 350cc and 600cc South African Tourist Trophy Races.

The Royal Enfield works riders for the Isle of Man TT Races were under the team management of Tony Wilson-Jones. The team consisted of Gordon Burney, George Reynard, Charlie Young, Cecil Barrow and Fred Bicknell. The first four were entered in the Junior Race for 350cc machines with Bicknell as reserve, whilst only Burney, Reynard and Barrow were entered in the Lightweight Race for 250cc machines with Young and Bicknell as reserves. The Junior and Lightweight machines were identical apart for the different engines. The Junior machines used the 344cc ohv (74 x 80mm) 350 Racing JAP and the Lightweight machines used the 246cc ohv (62.5 x 80mm) 250 Racing JAP. Both engines weighed an identical 66 lb.

Cecil Barrow was unfortunate to come off at Sulby Bridge during practice for the Lightweight Race and suffered a broken collar bone. His Junior Race machine was passed to reserve-man Bicknell. Young took over Barrow's 250 machine for the Lightweight Race and was able to make the fastest practice time in class. All four riders made it to the start of the Junior Race. Reynard suffered engine problems on the fifth lap, which forced his retirement. Burney, who was in sixth position at this time, maintained this position to the end of the race and achieved an average speed of 60.85mph, finishing almost 23 minutes behind the winner, Alec Bennett on his Velocette. Young struggled to maintain tenth position and was riding flat-out through Baaregarroo on the approach to the 13th Milestone when his forks broke. In the inevitable accident, he suffered extensive injuries to his face, hands, knees and back. Bicknell finished the race well down the field in twenty-sixth spot.

With both Cecil Barrow and first-reserve Charlie Young out of action, the Enfield riders for the Lightweight TT were Gordon Burney, George Reynard and Fred Bicknell. The first retirement in the Enfield team occurred when Burney dropped out on the second lap. Reynard suffered a broken valve spring on his fifth lap and despite efforts to continue was forced to retire. Bicknell finished the race with an average speed of 48.47mph, well down on the winning speed of 60.20mph set by Paddy Johnston on his Cotton but sufficient to gain him sixth place.

Other Royal Enfield racing successes during 1926 included a win for George Reynolds in the 50-Mile Handicap Race at Saltburn in July, a second place for T.Stewart in the 350cc Class at the Ulster Grand Prix and a win for Cecil Barrow in the Hutchinson Hundred, a 100-mile handicap race for machines up to 1,000cc. Barrow entered his Isle of Man Lightweight machine running on alcohol fuel and won at an average of 74.39mph.

Royal Enfield enjoyed sporting success of a different kind when a trade delivery outfit took part in the ACU Six Days' Trial for Commercial Sidecars. It completed the arduous course without losing points and received a special certificate of merit.

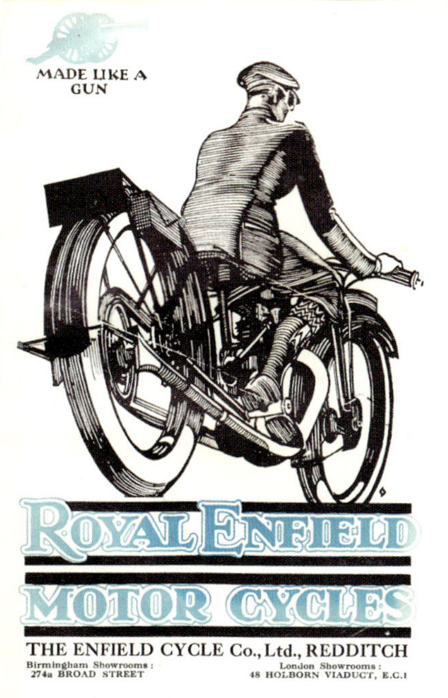

ROYAL ENFIELD

MADE LIKE A GUN

A few successes during 1926.

SOUTH AFRICAN T.T. RACES.
FIRST IN 350 C.C. RACE. FIRST IN 600 C.C. RACE

SCOTTISH SIX DAYS' TRIAL.
1 SILVER CUP. 1 GOLD MEDAL.

INTERNATIONAL SIX DAYS' TRIAL.
TWO GOLD MEDALS.

"HUTCHINSON" HUNDRED MILES RACE.
FIRST.

ULSTER GRAND PRIX.
Handicap Cup and Replica of Trophy.

Czecho-Slovakian T.T. Races.—First in 350 c.c. Race.
Colmore Cup Trial.—1 Gold Medal; 1 Silver Medal.
Ramblers' M.C.C. Trial.—Silver Vase.
Yorkshire News Trial.—Silver Watch.
Victory Cup Trial.—2 Gold Medals.
Dublin University M.C.C. Open Trial.—First Class Award.
Dublin and District M.C.C. Open Trial.—First Class Award.
Tunbridge Wells M.C.C. Trial.—Best Solo Performance; Best Novice Performance.
Redditch M.C. and C.C. Open Trial.—Nife Cup; 2 Gold Medals.
Luton and District M.C. Trial.—Dandy Cup.
M.C. and A.C. 24 Hours' Trial.—1 Gold Medal; 1 Silver Medal.
Athy M.C. and C.C. Speed Trial.—1 First Prize; 1 Second Prize.
Southport M.C. Speed Trial.—First in 350 c.c. Flying Kilo.
Travers Trophy Trial.—Gold Medal.
Natal M.C.C. Hill Climb.—3 First Prizes.
Cheshire Centre, A.C.U. Trial.—Gold Medal.
Carshalton Pillion Trial.—Silver Cup.
London-Land's End Trial.—2 Gold **Medals**; 7 Silver Medals.
Tadcaster M.C. Hill Climb.—4 First Prizes.
Scarboro' Two Days' Trial.—Best Solo Performance; Silver Rose Bowl; Silver Vase; Silver Cup.
Leinster M.C. and C.C. Novice Trial.—3 Gold Medals.
S.E. Durham M.C.C. Speed Trial.—3 First Prizes.
Kickham Memorial Open Trial.—John Bull Cup; 2 Silver Cups.
Races at Budapest.—First in 350 c.c. Class.
Knowle and District M.C. Trial.—Whitmore Trophy.
Hendon Athos M.C. Steeplechase.—Silver Cup and Medal.
Race Round Denmark, Three Days.—First Prize.
Knock M.C.C. All-night Run.—Silver Vase.
Bradford-London-Bradford Run.—3 Silver Cups.
Bearwood Amateur M.C.C. Trial.—Frank Hallam Cup.
Knock M.C.C. Five Miles' Championships.—First in 350 c.c. Championship.
Irish Championships at Magilligan.—2 Championships (50 miles); First in 50 Miles Handicap.
Dublin University M.C.C. Hill Climb.—1 First Prize; 2 Second Prizes.
Dordrecht Races, Holland.—2 First Prizes.
Cheshire Centre A.C.U. Races at Southport.—First in 25 Miles Race, 350 c.c.; 2 Second Prizes.
Athy M.C. and C.C. 75 Miles Race.—Fastest Time in 350 c.c. Class.
Tadcaster M.C. Speed Trials.—2 First Prizes; Fastest time of the day.
Southport M.C. 50 Miles Races.—First in 350 c.c. Race.
Irish 25 Miles' Championships.—First in 350 c.c. Sidecar Championship.
Dungannon M.C.C. 100 Miles Race.—Fastest time of the day.
Yorkshire Centre, A.C.U. Speed Trial.—1 Championship; Gold Cup; First in 4 Miles Race, 250 c.c.
N. Liverpool M.C.C. Races.—2 First Prizes.
Hartlepools M.C.C. Speed Trials.—3 First Prizes; 1 Second Prize.
Bradford M.C.C. Freak Hill Climb.—4 First Prizes.
York and District M.C. Speed Trial. First. Fastest time of the day, 250 c.c.
Irish 100 Miles Championships.—First in 350 c.c. Championship.
Rand M.C.C. (S. Africa) Races.—First in 5 Miles Scratch Race (350 c.c.).
Dublin and District M.C.C. Speed Trials.—First in 350 c.c. Class.
Presteigne A.C. Trial.—Best Sidecar performance.
Tadcaster and D.M.C. Speed Trials.—5 First Prizes; 5 Second Prizes.
Wye Valley A.C. Trial.—H.A.C. Bowl.
Irish 10 Miles Championships.—2 Championships First in 25 Miles Race.
Leinster M.C. and C.C. Trial.—2 First Class Awards.
M.C.C. High-Speed Test, Brooklands.—Gold Medal.
Ashford M.C. Trial.—President's Cup.
Western Centre Trial.—2 First Class Awards. Share of Dunlop Team Trophy.
Redditch M.C. and C.C. Trial.—Huins Cup; Garages Cup; 2 Gold Medals.
N. Wilts. M.C. Trial.—Goodyear Cup.
Camberley and District M.C. Trial.—Silver Cup.
Levis A.C. Trial.—First Class Award.
Southampton M.C. and L.C.C. Trial.—Angus Cup.
London-Gloucester-London Trial.—Silver Cup; 2 Silver Spoons.
Dublin University Trial.—Gold Medal.
Drogheda and District M.C.C. Trial.—Tredagh Cup and Replica.
London-Exeter-London Trial.—1 Gold Medal; 2 Silver Medals.

CHAPTER SEVENTEEN

Model Year 1927

The Scottish Six Days' Trial at the end of June 1926 was the first public appearance of the Enfield Cycle Company's new 346cc overhead-valve engine. Fred Bicknell and T.H.Garner both rode 346cc ohv-powered solos in the event, Garner was forced to retire on the second day when his crankcase was damaged by a rock, but Bicknell went on to secure a silver cup. A third rider on an Enfield big twin sidecar outfit won a gold medal. The new engine was fitted to the 3.46hp ohv models for the 1927 season, which were revealed at the Motorcycle Show at Olympia in October1926.

Also displayed at Olympia was a new mid-range machine fitted with a new Enfield-designed 488cc side-valve engine, similar in design to the 346cc side-valve engine introduced the previous year, and four-speed gearbox.

The 1927 Sales Brochure.

1927 Model Range

Model 201	2.25 hp	Standard Two-speed Two-stroke	225cc	£32 10s
Model 201a	2.25hp	Open-frame Two-stroke	225cc	£34
Model 200	2.25hp	Sports Two-speed Two-stroke	225cc	£32 10s
Model 350	3.46hp	Standard Side-valve	346cc	£45
Model 351	3.46hp	Overhead-valve Sports	346cc	£50
Model 352	3.44hp	OHV (twin-port) Sports	344cc	£75
Model 500	4.88hp	Standard Side-valve 4-speed	488cc	£52 10s
Model 180	9.76hp	Side-valve V-twin Standard	976cc	£66 10s
Model 182	9.76hp	Side-valve V-twin Sports	976cc	£68 10s
Model 190	9.76hp	De-Luxe V-twinCombination	976cc	£95

Prices are for machines without lighting sets, except Model 190

With the new model, the Enfield Cycle Company could boast a range comprising nine solo machines together with a range of sidecars to cater for all tastes. Arguably the only remaining gap in the range was for a sporting 500 and that would not be filled before the 1929 season. With the exception of the JAP 344cc twin-port racing engine fitted to the Sports Model 352, all the machines were fitted with modern engines of Enfield design and manufacture.

Royal Enfield 346cc ohv Engine

The new overhead-valve engine retained the 70mm bore and 90mm stroke of the side-valve engine, which was first announced in April 1925. It followed the same lines as the side-valve in respect of the crankcase design. The cylinder was fitted with a detachable hemispherical head held in place by four studs whose extensions provided the supports for the rocker gear. The light tubular steel push-rods, relaying tappet movement to the overhead rockers, were contained in telescopic covers. These allowed oil mist from the timing chest to be used to lubricate rocker gear. The oil passed through the hollow rockers before escaping through special passages to a breather pipe venting in the primary chaincase and lubricating the primary

An early version of the new 3.46hp Overhead-valve Model 351. Photographs of this machine were used for magazine promotions of the Olympia Show. The only discernible difference to the production models appears to be the screw fixing for the exhaust pipe.

chain. An ingenious arrangement of scraper ribs along the rocker shaft prevented dust from settling on the rocker sleeve housing. The inclined valves were of tulip design and closed by a pair of concentric springs, the outer of which tapered towards the top. The compression ratio was 5 to 1.

The new ohv 350 engine.

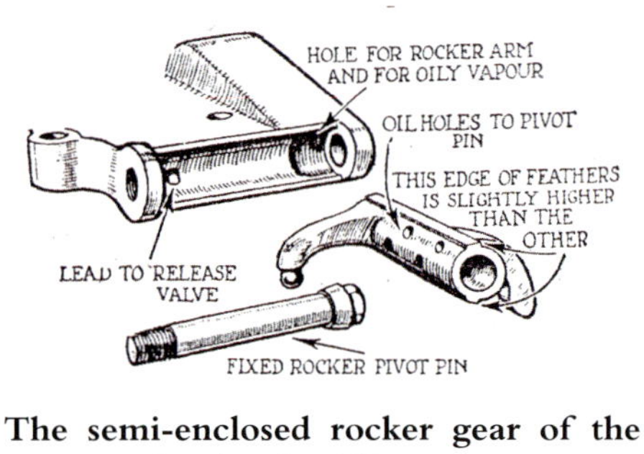

The semi-enclosed rocker gear of the new overhead-valve 350.

Two-stroke Lightweight Singles

The revised range of the popular Enfield 225cc two-stroke models had been introduced the previous year and, other than a reduction in price of £1 on each model, no change was considered necessary.

2.25 hp Standard Two-stroke Model 201

Capacity: 225cc.
Bore & stroke: 64 x 70mm.
Carburettor: Two-lever.
Gearbox: Royal Enfield 2-speed
Gear ratios: Standard: 5 & 8.1 to 1
Forks: Royal Enfield
Handlebar: Tourer or sporting
Brakes: 4 in drums front and rear.
Tyres: Dunlop Cord 24 x 2¼ in
Weight: 160lbs
Finish: Black enamel, tank finished in two shades of green enamel
Price: Without lights £32 10s.

Dimensions
Length: 6ft 4in
Width: 2ft 8½in
Saddle height: 26½in

Capacities
Fuel: 1 galls

2.25 hp Open-frame Two-stroke Model 201A

Specification as Model 201, but designed with open frame and circular tank.
Fitted with a dress-guard extending rearward from the tank.
Price; £34

2.25 hp Sports Two-stroke Model 200

Specification as Model 201.
Fitted with sports-type clip-on handlebar and footrests instead of footboards
Fitted with twin exhaust pipes to a cylindrical exhaust on the off-side of the machine.
Prices as Model 201

Mid-range Singles

The 350 model range remained largely unchanged for the 1927 Season, Both the Standard side-valve, Model 350, and Sports overhead-valve Model 351, were fitted with engines of Enfield design and manufacture. The Model 351 was designed as a fast quiet touring machine. The Sports Model 352, which was designed for competition use, continued to be available with the JAP 344cc ohv (twin-port) Racing engine. The frame design retained the duplex seat stays and under-tank rails. The Druid-style front forks were of Enfield design and fitted with shock absorbers. Dunlop Cord 27 x 2¾ in tyres were now fitted. The brake drums were redesigned with a flared flange designed to deflect mud from the working surfaces. The long exhaust expansion box was improved both in respect of appearance and effectiveness.

3.46 hp Standard Side-valve Model 350

Type: Four-stroke Side-valve
Capacity: 346cc.
Bore & stroke: 70 x 90mm.
Carburettor: B & B Two-lever.
Gearbox: Sturmey-Archer 3-speed
Gear ratios: Standard: 5.9, 8.9 & 17.3 to 1
Sidecar: 6.7, 10.0 &19.5 to 1
Handlebar: Touring or Sports
Brakes: 5 in drums front and rear.
Tyres: Dunlop Cord 27 x 2¾ in
Length: 6ft 11½ in Width: 2ft 8½in
Saddle height: 29in Weight: 210 lbs
Finish: Black enamel, tank finished in two shades of green.
Price: Without lights £45. Footboards and legshields extra 25s.
With M.L. lighting set £52

The footboards and legshields available with the Model 350.

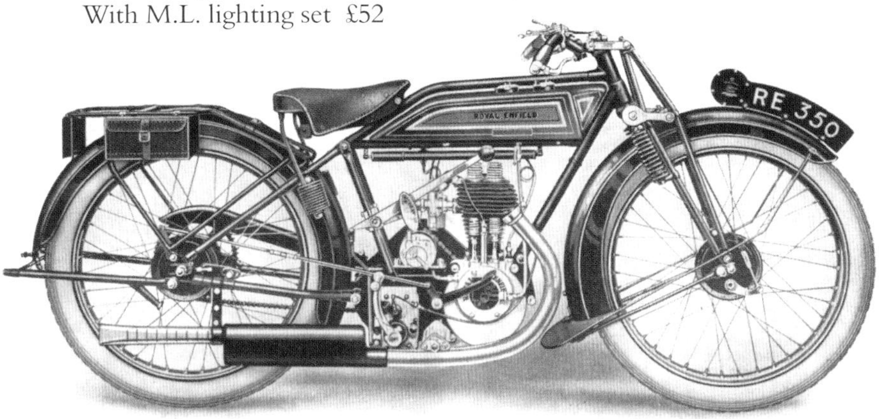

3.46 hp Standard OHV Model 351

Type: Four-stroke Overhead-valve
Capacity: 346cc.
Bore & stroke: 70 x 90mm.
Carburettor: B & B Two-lever.
Gearbox: Sturmey-Archer 3-speed
Gear ratios: Standard: 5.3, 7.7 & 11.5 to 1
Sidecar: 5.9, 7.8 & 12.8 to 1
Handlebar: Sports type, adjustable
Brakes: 5 in drums front and rear.
Tyres: Dunlop Cord 27 x 2¾ in
Length: 6ft 9in
Width: 2ft 9½in
Saddle height: 29in
Weight: 224 lbs
Finish: Black enamel, tank finished in two shades of green cellulose.
Price: Without lights £50. With M.L. lighting set £55 5s.

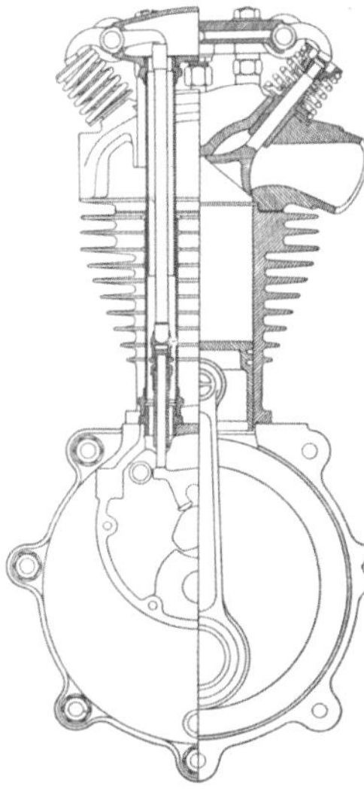

Sectional diagram of 3·46 h.p. O.H.V. engine.

3.44 hp OHV (twin-port) Model 352

Type: Four-stroke Overhead-valve Super-sports
Capacity: 344cc.
Bore & stroke: 74 x 80mm.
Carburettor: Racing type.
Gearbox: Sturmey-Archer 3-speed (close ratio)
Gear ratios: 5.3, 7.7 & 11.5 to 1
Handlebar: Adjustable
Brakes: 5 in drums front and rear.
Tyres: Dunlop Cord 27 x 2¾ in
Mudguards: Lightweight racing type, no carrier.
Length: 6ft 9in
Width: 2ft 9½in
Saddle height: 29in
Weight: 242lbs
Finish: Black enamel, tank finished in two shades of green cellulose.
Price: Without lights £75.

4.88hp Four-speed Gear Model 500

The Model 500 was a new model in the Enfield range for 1927 designed to fill the gap between the 350 models and the large capacity twins. Its 488cc side-valve engine, a larger version of the 346cc side-valve, was of Enfield design and manufacture. The four-speed gearbox was also a new Enfield designed unit. The new frame dispensed with the duplex saddle and lower tank tubes of the 350 models but incorporated duplex chain stays. The gear change lever moved within a gate attached to the lower tank tube. Drum brakes of 6½in diameter were used front and rear.

The new machine created considerable interest when exhibited at the 1926 Olympia Show and The Motor Cycle lost little time in assessing the road performance with a passenger sidecar attached. The machine was found to perform 'its duties with a pleasing absence of fuss' and a smoothness 'which is thoroughly delightful to experience'. Top gear 'pulled down to 15mph without snatch, and on bottom gear it was found possible to travel appreciably slower than walking pace without slipping the clutch'. The fuel consumption was 65mpg.

4.88hp Four-speed Gear Model 500

SPECIFICATION:

ENGINE.—488 c.c. Single Cylinder; bore and stroke 85·5 × 85 m/m. Aluminium alloy piston. Roller bearing big end.

LUBRICATION.—Automatic mechanical pump and auxiliary hand pump.

IGNITION.—High tension Magneto.

CARBURETTER.—Two-lever type.

GEAR.—**Enfield four-speed.** Handlebar-controlled clutch and kick-starter. Standard ratios: SOLO, 5·1, 6·6, 8·7, and 15·8; SIDECAR, 5·8, 7·5, 9·8, and 17·9.

TRANSMISSION.—Roller chains. Patent Cush Drive in rear hub.

FRAME.—Low saddle position. Royal Enfield Front Forks with shock absorbers. Footrests. Sidecar lugs incorporated.

SADDLE.—Terry or Lycett Spring Seat; long suspension springs.

BRAKES.—Internal expanding, front and rear (6½ in. drums).

HANDLEBAR.—Touring or Sports; adjustable.

TYRES.—Dunlop Cord 27 × 2·75 in.

MUDGUARDS.—Strong and efficient.

CARRIER.—Detachable carrier. By unfastening two nuts, the carrier and rear mudguard are lifted clear of the back wheel and tyre.

STANDS.—Front and rear.

TOOLS.—Complete set in two pannier bags on carrier.

FINISH.—Best quality Black Enamel. Tank panelled and lined. Bright parts heavily plated.

WEIGHT.—Approx. 245 lbs.

LENGTH.—6 ft. 9½ in.

WIDTH.—2 ft. 8½ in.

SADDLE HEIGHT.—29 in.

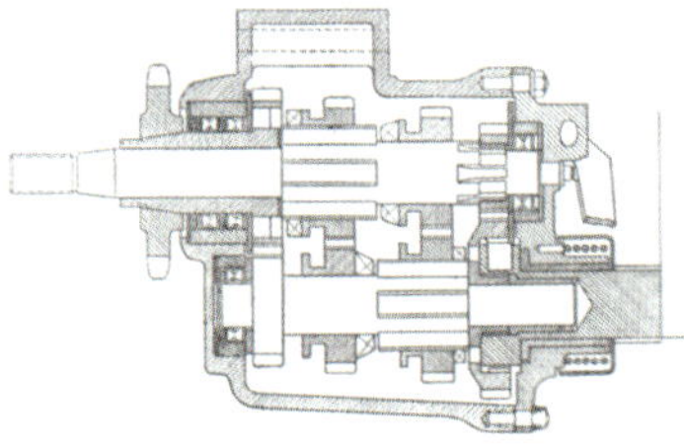

Diagram of Four-speed Gearbox.

Price £52 10 0 net cash

or **£14 12s. 6d.** down, and twelve monthly instalments of **£3 19s. 1d.,** including Insurance.

Electric Lighting Set, **£7** *extra.*

Price with Sidecar or Box Carrier (illustrated on next page) **£68**

or (with Sidecar) **£18 11s. 0d.** down, and twelve monthly instalments of **£5 2s. 0d.,** including Insurance.

4.88hp Model 500 fitted with the Lucas Maglita lighting Set, which was an optional extra costing £7 for the Model 500 and smaller capacity models.

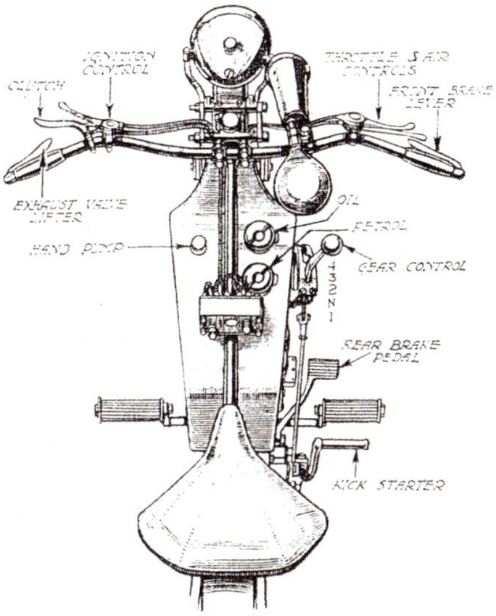

Controls of the 4.88hp Four-speed Gear Model 500.

Sidecars for 350 and 500 Models

Both the Sports, Model 14, and the Touring, Model 12, sidecars remained available for use with the 3.46hp road models. The specifications were unchanged.

For the 4.88hp Model 500 an interchangeable sidecar and delivery box was made available. The strong chassis was made easily detachable and the sidecar body mounted on springs; coil springs at the front and C springs at the rear. The passenger body was upholstered in leathercloth and a celluloid windscreen was supplied as standard. By unfastening four bolts the passenger body could be replaced by the commercial box. Both bodies were finished in dark green to match the motorcycle.

3.46hp Model 350 Sidecars

Chassis: Enfield design with four point attachment.
C springs at rear and coil springs in front.
Tyre: Dunlop Cord 26 x 2.75in
Overall dimension of combination: Width 4ft 9in. Length 6ft 9in.

Touring Sidecar Model 12

Length of sidecar: 4ft 9½in.
Weight of sidecar: 98lb
Price: £12.
Price with curved windscreen: £13

Model 12

Sports Sidecar Model 14

Length of sidecar: 6ft 4in.
Weight of sidecar: 100lb
Price: £12.

Model 14

4.88p Model 500 Sidecars

Chassis: Enfield design with four point attachment.
Cee springs at rear and coil springs in front.
Tyre: Dunlop Cord 26 x 2.75in
Overall dimension of combination: Width 4ft 10½in.

Price: £15 10s.
Additional price for interchangeable box: £9

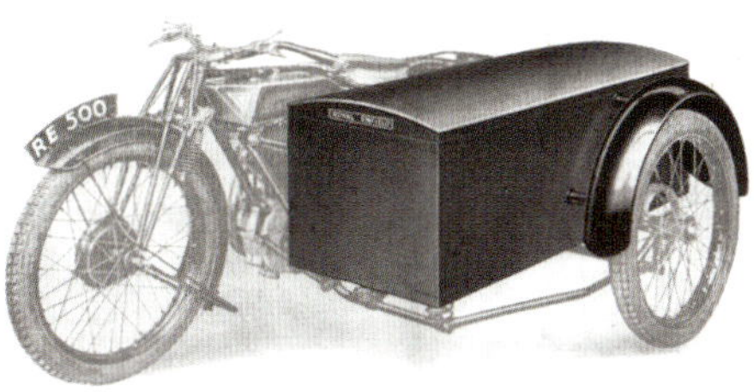

Box dimensions:
Length: 4ft 4in
Width: 1ft 8in
Depth: 1ft 6½in

9.76hp Models

The cylinder castings on the 9.76hp twins exhibited at the Olympia Show in October 1926 had been made more massive. They were given more substantial cooling fins around the exhaust ports, whilst the ports themselves were also increased in diameter to improve gas flow. These required correspondingly larger diameter exhaust pipes and modified silencers. On the Sports model these took the form of a twinned exhaust system on the off-side. The sparking plugs were relocated to the side of the combustion chambers and the valve caps were fitted with aluminium cooling cones. Forked conrods were now fitted, rather than the side by side type, allowing the cylinders to be fitted in line.

The new cylinder design for the 9.76hp twins featured larger exhaust ports and cooling fins.

Other changes included transferring the magneto drive from the offside to the nearside of the engine, which did away with the timing chest extension and loss of the Royal Enfield inscribed cover plate. A pie-crust solder joint used to attach the base of the fuel tanks on the 9.76hp models was a handsome feature of the 1927 models. This would be the last year that the slab-sided design of fuel tank would be used across the the full range of machines.

9.76hp Sports Model 182

Type: Side-valve V-twin
Capacity: 976cc.
Bore & stroke: 85.5 x 85mm.
Carburettor: Two-lever.
Twin exhaust and silencers
Gearbox: Sturmey-Archer 3-speed
Gear ratios: Solo: 4, 5.8 & 11.7 to 1
Sidecar: 4.7, 6.9 & 13.8 to 1
Brakes: 8in front & rear drums.
Handlebars: Semi TT, adjustable
Tyres: Dunlop Cord 26 x 3.25 in
Length: 7ft 2in
Width: 2ft 8in
Saddle: Terry or Lycett.
Saddle height: 30in
Weight: 308 lbs without lights.
Finish: Black enamel, tank finished in two shades of green cellulose.
Prices: Without lights £68 10s. With Lucas Magdyno lighting set £78 10s

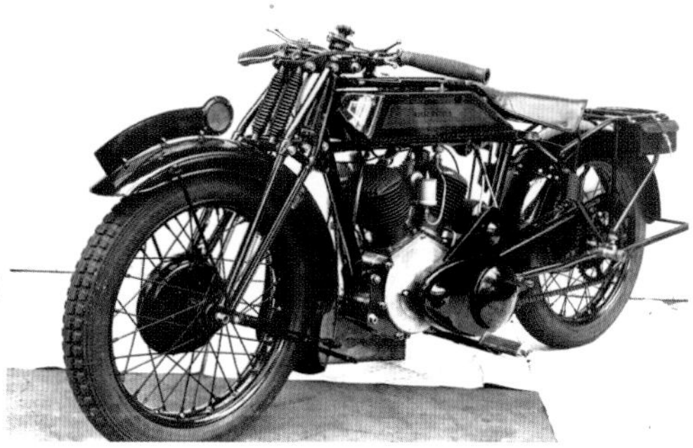

Nearside view of the 9.76hp Sports Model 182. When used solo, the footboards were replaced by footpegs.

Sports Sidecar for Model 182

Chassis: Enfield design, easily detachable.
Body: Polished aluminium panels, Upholstery in black leathercloth, Hammock seat.
Weight: 126 lb
Price: £15 10s

9.76hp Standard Model 180

Specification as Model 182
with following amendments.
Gear ratios: Solo: 4, 6.9 & 13.8 to 1
Handlebars: Adjustable
Length: 7ft 2in
Width: 2ft 6in
Weight: 308 lbs without lights.
Finish: Black enamel,
Tank finished in two shades
of green cellulose.
Prices: Without lights £66 10s.
With Lucas Magdyno lighting set £78 10s

Sidecars for 9.76hp Standard Model 180

The 1926 range of sidecar models was carried through to 1927 without significant change. The Standard Combination Model 180 comprising the Model 180 and Touring Sidecar was still available, but the Model 180 and Touring Sidecar were listed separately. The illustrations of the Two-seater sidecars were unchanged but the quoted overall lengths of the combinations were reduced; the child/adult model by 3in and the double adult by 5in. The style of bodywork fitted to the Dairyman's Combination Models 165 and 175 was revised, although the dimensions were unchanged. The prices of the combinations were unchanged from the previous year with the exception of the Model 155 box carrier, which was reduced substantially from £109 to £95.

Touring Sidecar for Model 180

Touring Sidecar Model 180 with hood, windscreen and side curtains fitted. These items cost an additional £5 15s.

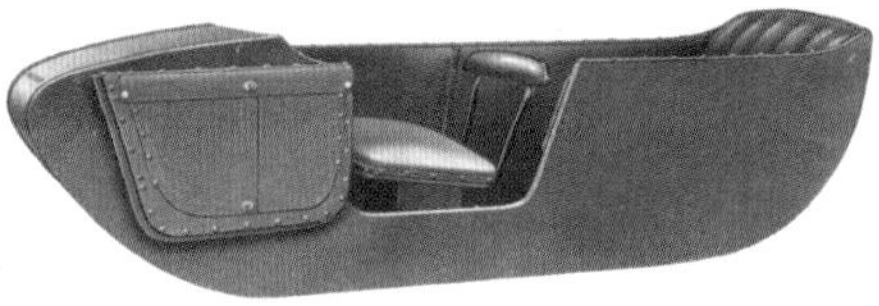

Two-seater sidecar for an adult and child seated in tandem (TOP) or for double adult (BOTTOM).

Trade Delivery Combination Model

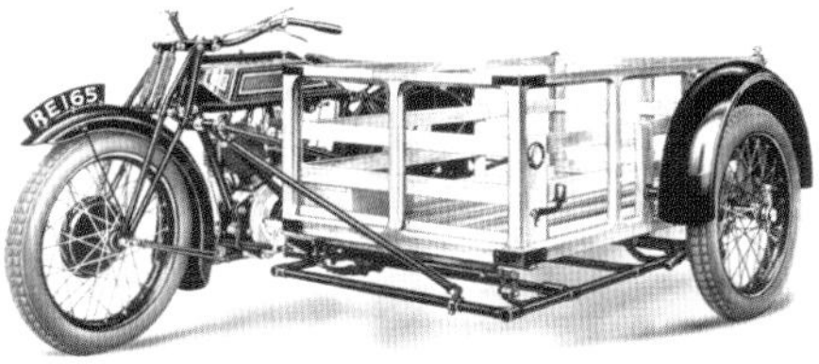

Dairyman's Combination Model 165 & Model 175

9.76hp De Luxe Combination Model 190

Large Box Carrier Model 155

The large flat surfaces of the box carriers provided ample space for business promotion. Lettering on the box was provided at a small extra charge, according to the work done.

LEFT: 'Simplicimus' of Motor Cycling with the 1926 Royal Enfield Two-stroke that he subjected to 2,000 miles of hard use with a minimum of maintenance.

In Praise of the Two-stroke

Motor Cycling magazine ran a regular feature in the mid-twenties titled 'Two-Stroke Topics'. The feature sub-titled 'Gossip About the Most Simple Type of Motorcycle' was written by 'Simplicimus'.

'Simplicimus' collected a two-stroke Enfield from the works in January 1927 with the intention of 'running it as far as it would go without touching it with a spanner'. He was forced to clean a plug after 1,000 miles, which should have brought the test to an end but he was clearly enjoying the machine to such an extent that he dismissed the trouble as trivial and continued. He handed the machine back after 2,000 miles and reported that in addition to the plug incident he had blown up each tyre once and oiled the chains on a single occasion. He claimed to have treated the machine brutally. His travels took him to Nottingham, where he called in on George Brough. Together they travelled to a hostelry 16 miles distant with 'Simplicimus' on George's SS100 Brough Superior and George on the Enfield. Naturally, 'Simplicimus' was the first to arrive but George did his best giving the Enfield 'a real education' and was flat out nearly all the way. 'Simplicimus' reported that with the throttle wide he could travel at 45mph and occasionally exceed the 50-mark; good performance for a 225cc engine.

An old speedometer, which 'Simplicimus' fitted for the test, ceased working near half distance. Nevertheless he was able to estimate the fuel consumption at just over 100mpg. He cut down the oil to ½ pint in a gallon, or 16 to 1.

New London Service Depot

Despite a turn down in the British economy following the General Strike in May 1926, the Enfield Cycle Company had maintained growth of both the bicycle and the motorcycle manufacturing sides of the business and was able to return good profits. It continued to invest in order to improve its products and provide an improved service to customers.

In November 1927 the company open a spacious new service depot and spares department at Hatton Wall off Hatton Garden, London. The building was set up with a well-equipped workshop, staffed by Enfield trained mechanics, to carry out all manner of servicing and repairs. The extensive spares department carried every single part for all Royal Enfields dating back to 1915.

The company's London show rooms and offices at Holborn Viaduct remained in use and displayed the complete range of Enfield cycles and motorcycles.

Works Extension

The popularity of the Enfield sidecars was such that the company decided to build a large new extension at Redditch specifically for the manufacture of sidecars. The new shop was completed towards the end of 1927. Taking into account the sports grounds for the employees' recreation, the total area occupied by the company amounted to a total of 24 acres.

Isle of Man TT Successes

The Royal Enfield machines for the Junior and Lightweight TT races were little changed from the the previous year. Indeed they remained little changed from the twin-port Model

1927 Junior TT machine 344cc OHV JAP 2-port.

352 available to the public. The 8in rear brake used in 1926 was replaced by a 6½in brake and both hubs were fitted with journal bearings instead of rollers. The specification included Enfield forks incorporating shock absorbers and a steering damper, an M.L. Magneto, new racing-type Brown & Barlow carburettor, Terry saddle and Dunlop tyres The Junior machine was on 27 x 2.75 wired-on tyres, whilst the smaller machine used 26 x 2.375. The engines of both machines were of JAP twin-port racing type.

The Enfield Works riders, under the team management of Tony Wilson-Jones, in both the Junior and Lightweight races were G.L.Reynard, J.G.Burney, T.Stewart and C.S.Barrow. In the Junior race the team achieved their best post-war racing success. George Reynard achieved fourth place, coming home twelve minutes after race leader Freddie Dixon on his HRD, Cecil Barrow was seventh, Gordon Burney eleventh and T.Stewart fourteenth. The team was the first works team to finish and was awarded the Manufacturers' Team Prize.

In the Lightweight Race, Reynard was forced to retire, but Barrow was able to finish in seventh position and Burney in eleventh. Stewart improved on his performance in the Junior Race to finish fourteenth.

SUCCESSES IN 1927

Royal Enfield machines have again been consistently successful in the principal competitions during the past year. Space is not available in this List for a complete record of these successes, but a few of the more outstanding performances are mentioned below.

Tourist Trophy Races:
JUNIOR T.T. TEAM PRIZE AND 5 REPLICAS.

Cookstown 70 Miles Road Race:
FIRST, SECOND AND FASTEST LAP.

London-Land's End Trial:
9 GOLD MEDALS AND SILENCE PRIZE.

London-Gloucester Trial:
TEAM PRIZE AND 3 SILVER CUPS.

Scottish Six Days' Trial:
3 SILVER CUPS, 2 GOLD MEDALS.

International Six Days' Trial:
3 GOLD MEDALS, 1 SILVER MEDAL.

A.C.U. One-Day Sporting Trial:
SECOND BEST TRADE PERFORMANCE.

Victory Cup Open Trial:
FRANK HALLAM CUP.

Matchless Cup Open Trial:
LADIES' TROPHY.
PRIVATE OWNERS' SIDECAR CUP.

N.W. Centre Sidecar Championship:
PRESIDENT'S ROSE BOWL.
MANUFACTURERS' VASE.

Dublin and District Open Trial:
DUNLOP CUP.

Leinster M.C.C. Open Novices Trial:
McDOWELL CUP.

Cardiff-Leicester-Cardiff Open Trial:
CUP FOR BEST 600 C.C. SOLO.

Southern Open Trial, S.E. Centre:
BEST PERFORMANCE OF CENTRE MEMBER.

Reliance Cup Open Trial:
BURNS CHALLENGE CUP.

Crystal Palace Races:
LEICESTER CUP.

Irish End-to-End Two Days' Trial:
BEST SIDECAR PERFORMANCE.

Ulster M.C.C. Reliability Trial:
BEST SIDECAR PERFORMANCE.

Redditch M.C.C. Autumn Trial:
HUINS CHALLENGE CUP.

Ashford M.C.C. Senior Trophy Trial:
SENIOR TROPHY.

Wood Green and District M.C. Championship Trial:
PRESIDENT'S CUP.

Also

FIRST IN THREE IRISH CHAMPIONSHIPS,
in addition to many First and Second Prizes in Hill Climbs, Sand Races, etc., and Silver Cups, Gold Medals, Souvenirs, and other Awards in Reliability Trials.

Enfield testers on the company's test track at the rear of works . The photograph has been reproduced from the company's photograph album and is believed to date from the period from late 1927 to early 1928.

The photograph, reproduced from the company's photograph album, is believed to date from late 1927 to early 1928 and was taken in the Works Test Shop.

CHAPTER EIGHTEEN

Model Year 1928

With its 1928 model range of motorcycles announced in time for the Motorcycle Show at Olympia in November 1927, Royal Enfield could justifiably claim to offer machines to appeal to all types of rider. With fourteen different models, the range extended from small capacity two and four-stroke machines to the large capacity V-twins. The mid-range of 350cc and 500cc models offered touring and sports singles and included competition machines suitable for trials or racing. The sidecar rider was also well catered for with lightweight designs suitable for use with 350cc machines to luxury touring or trade models for the 1,000cc V-twins.

1928 Model Range

Model 201	2.25 hp	Standard Two-speed Two-stroke	225cc	£32 10s
Model 201a	2.25hp	Open-frame Two-stroke	225cc	£32 10s
Model 200	2.25hp	Sports Two-speed Two-stroke	225cc	£32 10s
Model 202	2.25hp	Standard Three-speed Two-stroke	225cc	£35
Model 203	2.25hp	Sports Three-speed Two-stroke	225cc	£35
Model 250	2.25hp	Four-stroke	225cc	£36
Model 350	3.46hp	Standard Side-valve	346cc	£44 10s
Model 351	3.46hp	Standard Overhead-valve	346cc	£50
Model 355	3.46hp	Sports Overhead-valve	346cc	£52 10s
Model 352	3.44hp	Double-port Overhead-valve	344cc	£78 15s
Model 500	4.88hp	Standard Side-valve	488cc	£48 5s
Model 180	9.76hp	Three-speed Standard	976cc	£67 10s
Model 182	9.76hp	Four-speed Sports	976cc	£73 10s
Model 190	9.76hp	De-Luxe Combination	976cc	£93 15s

Prices are for machines without lighting sets, except Model 190

The 225cc four-stroke Model 250 was a totally new model which filled the gap in the model range for a utilitarian four-stroke. This was an attractive machine fitted with the new style of fuel tank. In previous years Enfield's had employed a flat tank design, which by the late 1920's was becoming very dated. The new saddle tank design, with pie-crust seams, greatly enhanced the appearance of the machine and would also feature on approximately half of the other models in the range. Amongst the machines with the new tank design were two new three-speed versions of the 225cc two-strokes. The 9.76hp Model 182 was also redesigned to appeal to the demands of the sporting rider and now incorporated a new Enfield designed four-speed gearbox.

The Druid-pattern front forks were being gradually phased out in favour of a design using a single central spring in compression

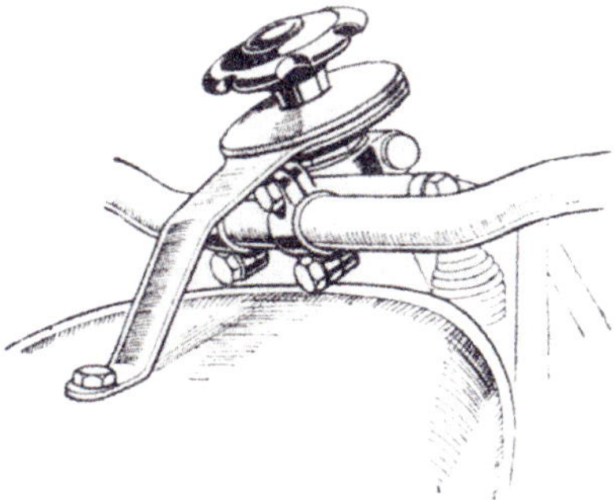

The design of steering damper for use with machines fitted with the new saddle tank.

New Lightweight Four-stroke

Royal Enfield had previously met the requirements of the utilitarian end of the motorcycle market with its range of 225cc two-stroke machines. These had proved successful and were amongst the best available, being capable of long service with minimum attention; they provided reliable and economic day to day transport. Many riders, however, disliked two-strokes and would have preferred a similarly lightweight and economical four-stroke. Enfields recognised the gap in the Royal Enfield range and launched the new Model 250 for the 1928 season. The model featured on the Enfield stand at the Olympia Show in November 1927, but does not appear to feature in the pre-show advertising.

ROYAL ENFIELD

MADE LIKE A GUN

SOLO MOTOR CYCLES FROM - - £31
SIDECAR COMBINATIONS FROM £54.10s.

ALL Royal Enfield Motor Cycles have internal expanding brakes to both wheels, patent cush drive rear hub, detachable mudguard and carrier, etc. All the four-stroke models have aluminium pistons, mechanical lubrication, spring-top saddles, and four of them have saddle tanks (cellulose finish). Enfield steering dampers are a standard fitment on all machines except 2·25 h.p.

Several of our 1928 models are illustrated on this and the opposite page. Our preliminary List describes in detail these and several others. Post your request for a copy to-day. Ask for Folder "E" and particulars of our

Gradual Payments

NOV 11th WEAR A FLANDERS POPPY

STAND No. 36 OLYMPIA OCT. 31 —NOV. 5

2·25 H.P. Two-stroke Model 201 with Enfield Two-speed Gear **£31**
Deposit .. £8-13-9.

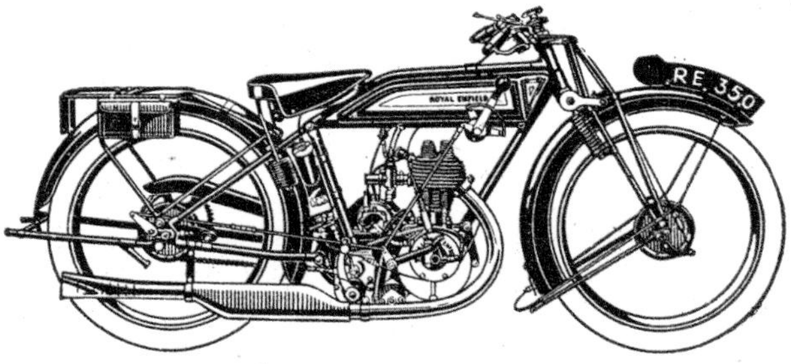

3·46 H.P. Side-valve Model 350 with Sturmey-Archer Three-speed **£42.10s.**
Deposit .. £11-11-3.

In answering this advertisement it is desirable to mention "The Motor Cycle." C12

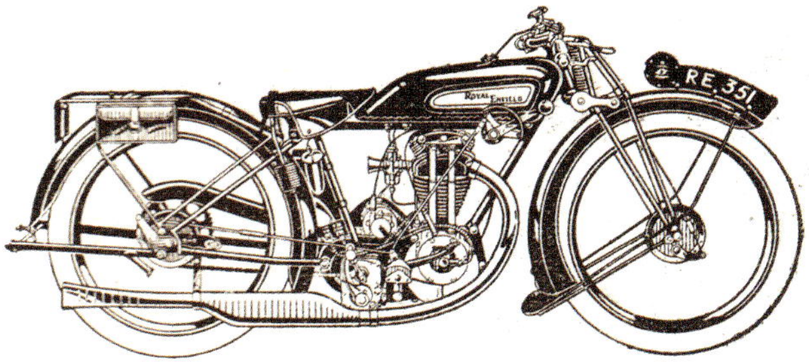

3·46 H.P. Standard O.H.V. Model 351 with Sturmey-Archer 3-speed **£47.10s.**
Deposit .. £12–16–3.

3·46 H.P. Sports O.H.V. Model 355 with Sturmey-Archer 3-speed **£50**
Deposit .. £13–8–9.

4·88 H.P. Standard Model 500 with Enfield 4-speed Gear **£46**
Deposit .. £13.

9·76 H.P. Sports Solo Model 182 with Enfield 4-speed Gear **£70**
Deposit .. £19–8–6.

9·76 H.P. Sports Combination (Polished aluminium Sidecar) **£86**
Deposit .. £23–9–6.

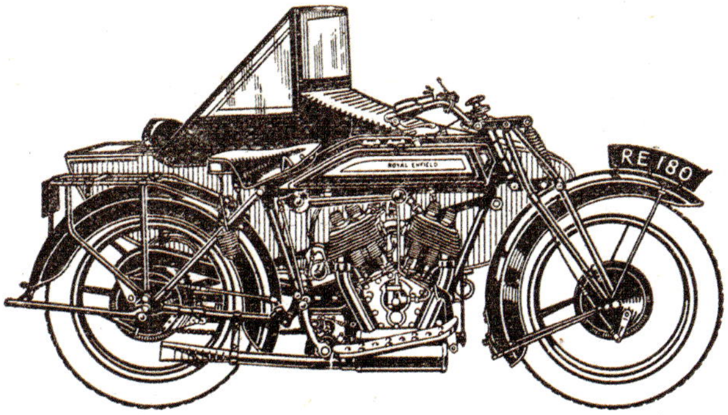

9·76 H.P. Standard Combination (Coach-built Sidecar) **£82.10s.**
Deposit .. £22–11–9.

OTHER ROYAL ENFIELDS FROM £33.10s. to £90.

ROYAL ENFIELD

MADE LIKE A GUN

THE ENFIELD CYCLE CO., LTD., Head Office and Works **REDDITCH**

Birmingham Office & Showrooms : 274a, Broad St. London Service Depot : 5, 7, 9, Hatton Wall, E.C.1. London Office & Showrooms : 48, Holborn Viaduct. E.C.1.

In answering this advertisement it is desirable to mention "The Motor Cycle."

The engine used the same 64mm bore and 70mm stroke as the two-strokes. The Motor Cycle tester, writing in April 1928 found much to praise about the new machine. He found it quite capable of maintaining 36-40mph, at which speed the fuel consumption was estimated at 95mpg. Oil lubrication was set by the factory and oil consumption estimated at 1,000mpg. He noted that 'Long periods of full throttle work on any gear failed to make the little engine show any signs of overheating.'

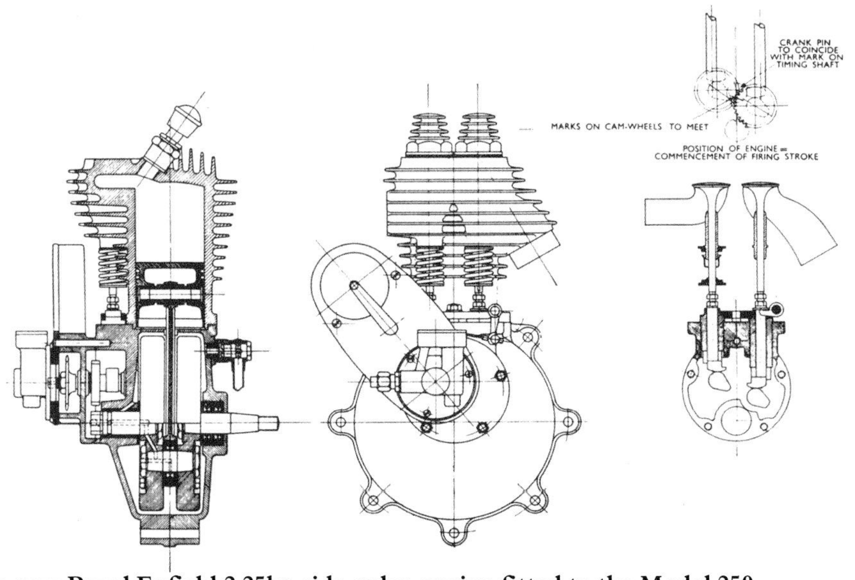

The new Royal Enfield 2.25hp side-valve engine fitted to the Model 250.

2.25 hp Side-valve Model 250

Type: Four-stroke
Capacity: 225cc
Bore & stroke: 64 x 70mm
Carburettor: Amal Two-lever
Gearbox: Royal Enfield 3-speed
Gear ratios: 7.2, 10.6 & 20.8 to 1
Brakes: 5 in drums front and rear
Tyres: Dunlop Cord 25 x 3 in
Fuel tank: Approximately 2 gallons
Length: 6ft 5in Width: 2ft 8½in
Saddle height: 25in
Weight: 190 lbs
Finish: Black enamel,
tank finished in two shades of green cellulose
Price: Without lights £36. With Lucas lighting set £41 5s

The prices quoted for the machines, with the exception of the De-Luxe Combination, Model 190, were exclusive of a lighting set. It was usual to offer the ML Maglita on the lower priced models at an extra cost of £5 5s. However a dynamo lighting set could be provided on request. The above photograph taken at the works shows a Model 250 fitted with a dynamo lighting set.

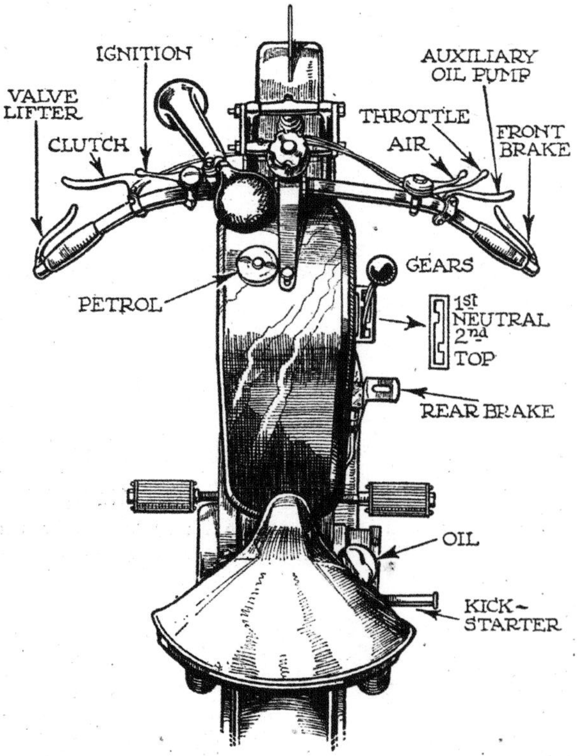

The riding controls of the Model 250.

225cc Two-stroke Models

For 1928 the range of Royal Enfield two-stroke machines was expanded from three to five different models. The earlier range of two-speed machines, comprising the Standard Model 201, Open-Frame Model 201A and Sports Model 200, was continued and joined by two new three-speed machines, the Standard Model 202 and the Sports Model 203. In addition to the three-speed gearbox of Enfield manufacture the new machines also boasted the new handsome saddle tank. Further, the engines of the new models also benefited from aluminium pistons and a detachable aluminium cylinder head giving an improved combustion chamber.

2.25 hp Standard Two-stroke Model 201

Type: Two-stroke
Capacity: 225cc
Bore & stroke: 64 x 70mm
Carburettor: Two-lever
Gearbox: Royal Enfield 2-speed
Gear ratios: Standard: 5 & 8.1 to 1
Forks: Royal Enfield
Handlebar: Tourer or sporting
Brakes: 4 in drums front and rear
Tyres: Dunlop Cord 24 x 2½ in
Weight: 160 lbs Saddle height: 26½in
Finish: Black enamel, tank finished in two shades of green enamel
Price: Without lights £32 10s, With Maglita lighting set £37 15s

Dimensions
Length: 6ft 6in
Width: 2ft 8½in

Capacities
Fuel: 1 galls

2.25 hp Open-frame Two-stroke Model 201A

Specification as Model 201, but designed with open frame and circular tank.
Fitted with a dressguard extending rearward from the tank.
Prices as Model 201

2.25 hp Sports Two-speed Two-stroke Model 200

Specification as Model 201.
Fitted with sports-type handlebar and footrests instead of footboards.
Fitted with twin exhaust pipes to a cylindrical exhaust on off-side of machine.
Prices as Model 201

2.25 hp Standard Three-speed Two-stroke Model 202

The specification is as Model 201 with following changes:
New engine with aluminium pistons and detachable cylinder head.
Saddle tank finished in cellulose
Spring seat saddle
Price without lights £35.
Price with Maglita lighting £40 5s

2.25 hp Sports Three-speed Two-stroke Model 203

The specification is as Model 202 with following changes:
Sports silencing system
Footrests
Sports handlebar

Middle Range 350 and 500 Machines

All the 350 models were now fitted with a separate oil tank located behind the saddle down tube. This required moving the clamp for the hand change lever on the Model 350. The exhaust systems were modified and of similar style to that used on the Model 500 of the previous year. The bottom of the range side-valve 350 and 500 models retained the old flat-sided fuel tank whilst the remaining models were fitted with the new saddle tank, which required minor frame modifications.

A factory photograph of a 350 Side-valve model. It appears to have been built using a Model 351 frame with saddle, possibly as a development model or to special order.

3.46 hp Standard SV Model 350

Type: Four-stroke Side-valve
Capacity: 346cc
Bore & stroke: 70 x 90mm
Carburettor: Two-lever
Gearbox: Sturmey-Archer 3-speed
Gear ratios: Standard: 5.9, 8.9 & 17.3 to 1
Sidecar: 6.7, 10.0 & 19.5 to 1
Handlebar: Touring or Sports
Brakes: 5 in drums front and rear
Tyres: Dunlop Cord 27 x 2¾ in
Length: 6ft 11½ in
Width: 2ft 8½in
Saddle height: 29in
Weight: 230 lbs
Finish: Black enamel, tank finished in two shades of green.
Price: Without lights £44 10s, With M.L. lighting set £49 15s

3.46 hp Standard OHV Model 351

Type: Four-stroke OHV
Capacity: 346cc
Bore & stroke: 70 x 90mm
Carburettor: Two-lever
Gearbox: Sturmey-Archer 3-speed
Gear ratios: Standard: 5.3, 7 & 11.5 to 1
Sidecar: 5.9, 7.8 &12.8 to 1
Handlebar: Sports type, adjustable
Brakes: 5 in drums front and rear
Tyres: Dunlop Cord 27 x $2\frac{3}{4}$ in
Length: 6ft $1\frac{1}{2}$1in
Width: 2ft $9\frac{1}{2}$in
Saddle height: 27in
Weight: 242 lbs
Finish: Black enamel, tank finished in two shades of green cellulose

3.46 hp Sports OHV Model 355

Similar specification as Model 351, but with different frame, giving lower saddle height, and a specially tuned engine with high compression piston. High level exhaust and no carrier.
Saddle height: 26in
Weight: 242 lbs
Price: £52 10s.

3.44 hp Double-port OHV Model 352

Specification as Model 355 but with a special engine tuned for road racing and no kick start.
Price £78 15s

LEFT: The factory photograph is of an early version of the Model 352 for the 1928 season. The machine retains a flat fuel tank, but a separate oil tank is attached to the seat down tube. The company was in the process of updating the Royal Enfield range with saddle fuel tanks and the saddle tank model was displayed at Olympia and illustrated in the catalogue.

4.88hp Four-speed Model 500

Type: Side-valve
Capacity: 488cc
Bore & stroke: 85.5 x 85mm
Carburettor: Amal Two-lever
Gearbox: Enfield 4-speed
Gear ratios: Standard: 5.1, 6.6, 8.7 & 15.8 to 1
Sidecar: 5.8, 7.5, 9.8 & 17.9 to 1
Brakes: 6½ in drums front and rear
Tyres: Dunlop Cord 27 x 2.75 in
Length: 6ft 11½in
Width: 2ft 8½in
Saddle height: 29in
Weight: 245 lbs
Finish: Black enamel, tank finished in beige cellulose with red panel
Prices: Without lights £48 5s. With ML Maglita lighting set £53 10s

9.76hp Twins

At the Olympia Show in November 1927, the big twin Enfields were displayed for the first time as machines definitely planned for sports and solo use. The Sports Model 182 had hitherto been an adaptation of the touring Model 180. For 1928 it featured a shorter dropped frame providing a very low saddle height, a new four-speed gearbox of Enfield design and manufacture, redesigned exhaust system and improved forks with steering damper

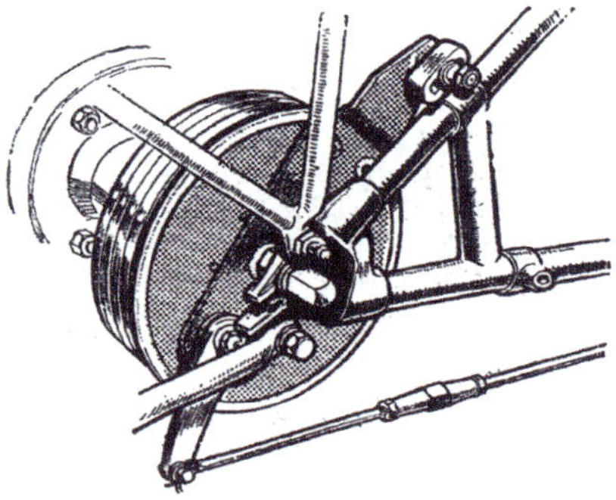

The 9.76hp Sports Twins were fitted with a strut between the rear chain stays and forks to spread the brake torque reaction.

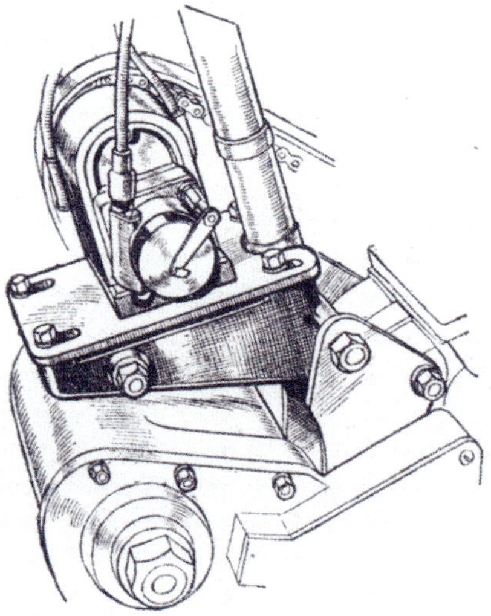

The adjustable magneto mounting platform on the 9.76hp Sports Twin.

9.76 hp Three-speed Standard Model 180

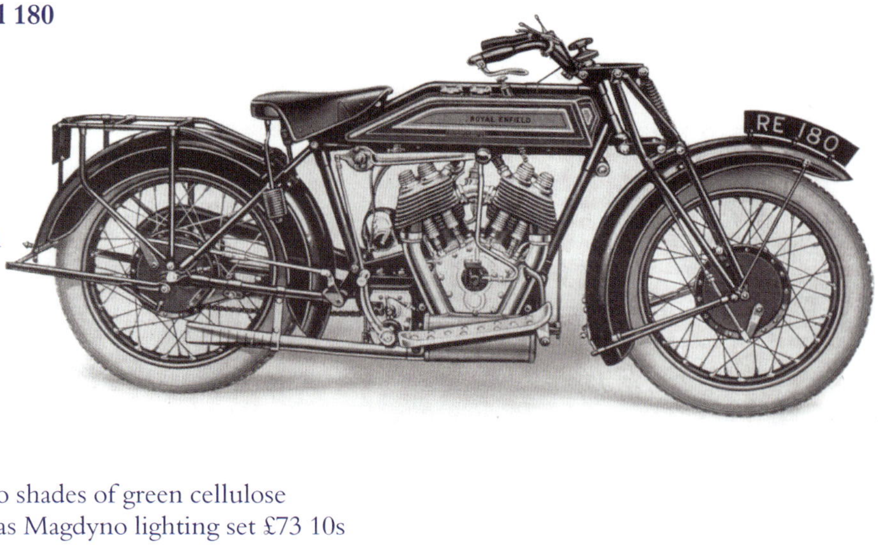

Type: Side-valve V-twin
Capacity: 976cc
Bore & stroke: 85.5 x 85mm
Carburettor: Two-lever
Gearbox: Sturmey-Archer 3-speed
Gear ratios: Standard: 4.7, 6.9 & 13.8 to 1
Brakes: 8in front & rear drums
Handlebars: Sports or touring adjustable
Tyres: Dunlop Cord 26 x 3.25 in
Length: 7ft 2in
Width: 2ft 8in
Saddle height: 30in
Weight: 345 lbs without lights
Finish: Black enamel, tank finished in two shades of green cellulose
Prices: Without lights £67 10s. With Lucas Magdyno lighting set £73 10s

9.76 hp Four-speed Sports Model 182

Specification as Model 180
with following amendments:
Twin exhaust and silencers
Gearbox: Enfield 4-speed
Gear ratios:
Solo: 4.2, 5.4, 7.1 & 13 to 1
Sidecar: 4.7, 6.1, 7.9 & 14.5 to 1
Brakes: 8in front & rear drums
Handlebars: Semi-TT
Saddle height: 29in
Weight: 336 lbs without lights
Finish: Black enamel, tank finished in two shades of green.
Prices: Without lights £73 10s, With Lucas Magdyno lighting set £79 10s

9.76hp De-Luxe Combination Model 190

Specification as Model 180 with following amendments:
Fitted Lucas Magdyno lighting set
Weight: 133lb
Overall length of combination: 7ft 2in
Overall width of combination: 5ft 3in
Price: £93 15s

Also available with Model 22
Two-seater Sidecar.
Price: £98

Balloon tyres 10s per wheel extra

Sidecars and Combinations

The sidecar range for 1928 was largely unchanged from the previous year. A new Sports Sidecar, Model 8, was made available to complement the Touring Sidecar, Model 10, previously available with the 4.88hp Model 500. Both chassis could take the optional Model 8 Delivery Box. A new Occasional Two-seater sidecar, Model 22, was made available for use with the 9.76hp machines. The Touring Model 180 and Sports Model 182 sidecarsremained in production, but were now identified as Model 2 and 4, respectively.

Price adjustments were made across the range and presumably were made based upon the previous year's experience of production costs and to allow for the current market conditions.

1928 Sidecars

Sidecars suitable for 3.46hp machines					
Model 14	Sports Sidecar	Weight	100 lbs	Price	£12 10s
Model 12	Touring Sidecar	Weight	98 lbs	Price	£12 10s
Interchangeable Sidecars and Delivery Box for Model 500					
Model 10	Touring Sidecar	Weight	133 lbs	Price	£16 15s
Model 8	Sports Sidecar	Weight	118 lbs	Price	£16 15s
Model 6	Delivery Box (box & mudguard only)		Price	£9 10s	
Sidecars suitable for 9.76hp machines					
Model 2	Touring Sidecar	Weight	154 lbs	Price	£18 15s
Model 4	Sports Sidecar	Weight	126 lbs	Price	£16 15s
Model 22	Occasional Two-seater	Weight	199 lbs	Price	£23
Outfits complete with 9.76hp Three-speed Standard Model 180					
Model 190	De-Luxe Combination			Price	£93 15s
Model 190	De-Luxe Combination with Model 22 Sidecar			Price	£98
Model 185	Trade Delivery Outfit	Box weight	168 lbs	Price	£86 5s
Model 165	Dairyman's Outfit	S/c weight	196 lbs	Price	£88 10s
Model 175	Dairyman's Outfit based upon Model 185 chassis			Price	£86 5s
Model 155	GPO Type Combination			Price	£94 10s

Lightweight Sidecars for 3.46 hp Models 350 & 351

Sports Sidecar Model 14

Weight: 100lb
Overall length of combination: 6ft 11½im
Overall width of combination: 4ft 9in
Price: £12 10s

Touring Sidecar Model 12

Weight: 98lb
Overall length of combination: 6ft 11½im
Overall width of combination: 4ft 9in
Price: £12 10s Windscreen: £1 extra

Lightweight Sidecars for 4.88 hp Model 500

Touring Sidecar Model 10

Weight: 133lb
Overall length of combination: 6ft 11½in
Overall width of combination: 4ft 10½in
Price: £16 10s

Sports Sidecar Model 8

Weight: 118lb
Overall length of combination: 6ft 11½im
Overall width of combination: 4ft 10½in
Price: £16 10s

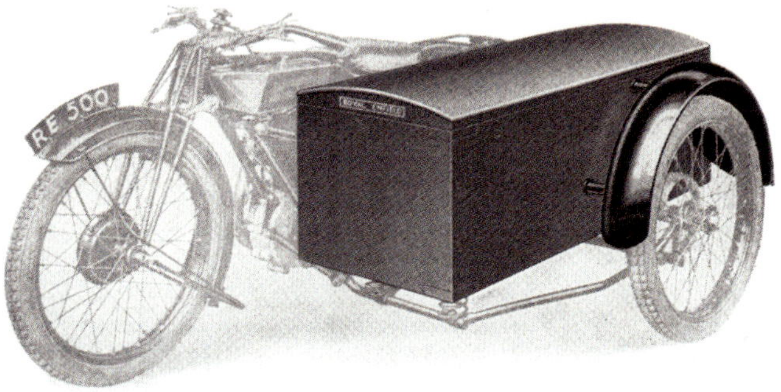

Model 6 Delivery Box

The Delivery Box Model was interchangeable with the Sports or Touring bodies by unfastening only four nuts.
Length: 4ft 4in Depth: 1ft 6½in With: 1ft 8in
Price of Delivery Box with mudguard: £9 10s

A collapsible hood was available with Models 180 and 190 for an additional charge of £2. Made of black waterproof material on a metal frame, it could be raised or lowered in a few seconds.

Sidecars for 9.76hp Machines

Touring Sidecar Model 2

Weight: 133lb
Overall length of combination: 6ft 11½im
Overall width of combination: 4ft 10½in
Price: £16 10s

Touring Sidecar Model 4

Weight: 133lb
Overall length of combination: 6ft 11½im
Overall width of combination: 4ft 10½in
Price: £16 10s

Occasional Two-seater Sidecar Model 22

Weight: 133lb
Overall length of combination: 6ft 11½im
Overall width of combination: 4ft 10½in
Price: £16 10s

Delivery Outfit Model 185

Weight: 133lb
Overall length of combination: 6ft 11½im
Overall width of combination: 4ft 10½in
Price: £16 10s

Dairyman's Outfit Model 165

Weight: 133lb
Overall length of combination: 6ft 11½im
Overall width of combination: 4ft 10½in
Price: £16 10s

Large Box Carrier Model 155

Weight: 245lb
Box dimensions: Length: 4ft 9in Depth 3ft 3in
Width 2ft 9½ in Capacity 31 cu ft
Overall length and width as Model 165
Price: £95

The lightweight sidecars for the 3.46hp machines were increased in price by 10s. The Model 10 Touring Sidecar for the 4.88hp Model 500 was increased in price by £1 5 to £16 15s and was the same price as the new Model 8 Sports Sidecar. The Model 180 De Luxe Combination was reduced in price by £1 5s with the single-seater body but by only £1 if supplied as a two-seater. The trade combinations suffered the greatest increase in price. The Model 165 now cost an additional £2 5s, whereas the Dairyman's Outfits increased in price by £3 10s for the wide body Model 165 version and £2 5s for the narrow body Model 175 version. By contrast the large GPO type box carrier was reduced in price by 10s.

1928 TT Races

The Enfield works team included George Reynard, Cecil Barrow and Gordon Burney, who had formed part of the team that won the team prize for Enfields in the 1927 Junior race. The newcomer to the team was the Swedish rider Edward Magner who had a good reputation in his native Sweden but was participating for the first time in the island.

Works photographs of the Royal Enfield TT machines to be used in the 1928 TT Races. The upper photograph is of the 350cc machine for the Junior Race, which is fitted with the JAP 3.44hp OHV 2-port racing engine. The carburettor is a Binks Rat-trap. The lower photograph is of a similar machine, but fitted with the JAP 2.45hp OHV 2-port racing engine, entered in the Lightweight Race

Edward Magner, the Swedish agent, was already a very successful racer in his native Sweden before joining the Enfield race team and travelling to Britain for the Isle of Man TT.

The specification of the machines was practically standard Model 352, but in the Lightweight Race the 250cc two-port JAP engine was used. Specifications included Amal carburettors, M.L. Magnetos, Sturmey-Archer three-speed gears and Terry saddles. Dunlop tyres were used; the Lightweight machines on 27 x 2.75 ins and the Junior machines on the new size of 27 x 3 ins.

Whereas in 1927 the JAP engines had the edge on all the rival makes in the Junior Race, it was to be the Velocette engine, together with the new Velocette innovation of the 'positive-stop' gear change, that would hold sway in 1928. Dodson, Arcangeli and Franconi set the pace from the start with their Sunbeams, with Alec Bennett enjoying a relaxed ride in fourth place on his Velocette. By the second lap Bennett had taken the lead and held it through to the finish. He completed the fifth lap at a record speed of 70.28mph and completed the race at an average of 68.65mph to gain his fifth TT win, He was followed home by his team mate, Harold Willis, at 67.16mph. George Reynard was the fastest Royal Enfield rider and achieved an average speed of 62.9mph which gave him sixth place. Despite troubles the other Enfield riders completed the course with Barrow gaining a fourteenth place, at an

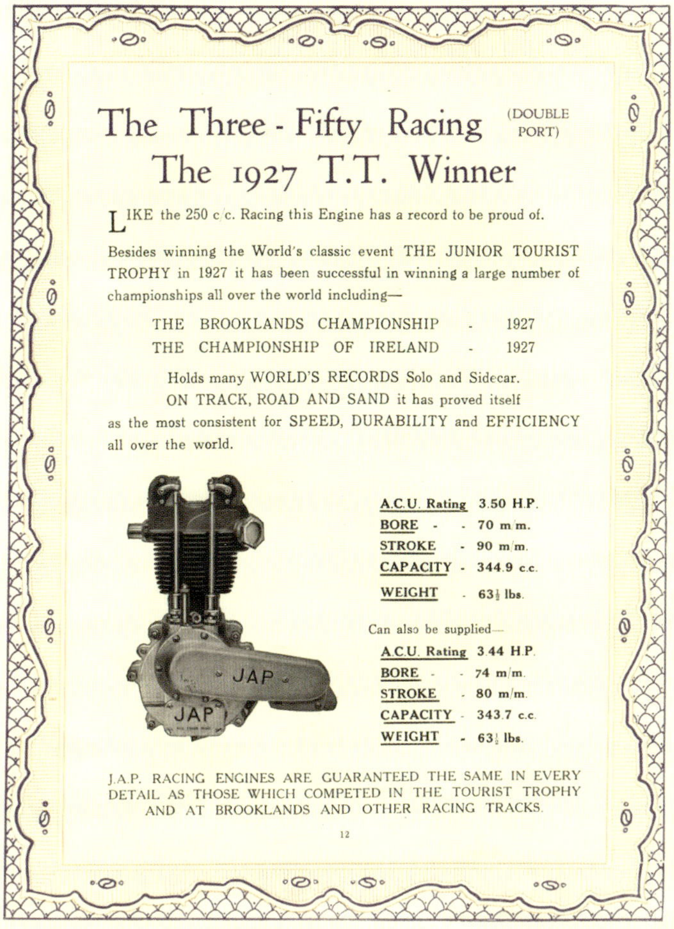

The Three - Fifty Racing (DOUBLE PORT)
The 1927 T.T. Winner

LIKE the 250 c/c. Racing this Engine has a record to be proud of.

Besides winning the World's classic event THE JUNIOR TOURIST TROPHY in 1927 it has been successful in winning a large number of championships all over the world including—

THE BROOKLANDS CHAMPIONSHIP - 1927
THE CHAMPIONSHIP OF IRELAND - 1927

Holds many WORLD'S RECORDS Solo and Sidecar.
ON TRACK, ROAD AND SAND it has proved itself
as the most consistent for SPEED, DURABILITY and EFFICIENCY all over the world.

A.C.U. Rating	3.50 H.P.
BORE	70 m/m.
STROKE	90 m/m.
CAPACITY	344.9 c.c.
WEIGHT	63½ lbs.

Can also be supplied—

A.C.U. Rating	3.44 H.P.
BORE	74 m/m.
STROKE	80 m/m.
CAPACITY	343.7 c.c.
WEIGHT	63½ lbs.

J.A.P. RACING ENGINES ARE GUARANTEED THE SAME IN EVERY DETAIL AS THOSE WHICH COMPETED IN THE TOURIST TROPHY AND AT BROOKLANDS AND OTHER RACING TRACKS.

12

The Two-Fifty Racing (DOUBLE PORT)

THIS Engine holds nearly EVERY WORLD'S RECORD in the 250 c/c class from the Flying Kilometre to the 24 Hours. Such has never been the position of any Proprietary or other Engine before. In addition it has to its credit the following:—

LIGHTWEIGHT T.T.	1924	GRAND PRIX DE L'EUROPE	1927
LIGHTWEIGHT T.T.	1925	SWEDISH T.T.	1927
FRENCH GRAND PRIX	1926/1927	FASTEST LAP. Also THIRD,	
LEINSTER HUNDRED	1927	FOURTH, SEVENTH, EIGHTH, TENTH	
BROOKLANDS CHAMPIONSHIP	1927	and ELEVENTH PLACES in the	
BELGIAN GRAND PRIX	1927	LIGHTWEIGHT T.T.	1927

A TRULY WONDERFUL RECORD!

A.C.U. Rating	245.2 H.P.	BORE	62.5 m/m.
STROKE	80 m/m.	CAPACITY	245.2 c.c.
WEIGHT	58 lbs.		

SPECIFICATION.—Roller Bearing Big end, and Roller Bearing Pulley side, Special Valves, Aluminium Piston, Magneto Chain Case, Magneto Chain, Magneto Sprockets, Magneto Platform, Inlet Pipe, &c.

J.A.P. RACING ENGINES ARE GUARANTEED THE SAME IN EVERY DETAIL AS THOSE WHICH COMPETE ON THE TRACKS AND IN ALL RACES.

11

average speed of 57.3mph, Magner a sixteenth place with an average speed of 56.5mph and Burney in twentieth place with an average speed of 53.1mph.

In the Lightweight Race Barrow improved on his Junior Race speed and completed the course at an average of 58.92mph, which was sufficient to give him second place to race winner Frank Longman who won at 62.9mph riding his OK Supreme. Burney also improved on his Junior performance to lap at an average of 55.6mph which was adequate to give him tenth place. He was followed home in eleventh place by C Lord, also riding a Royal Enfield, at an average of 51.9mph. Both Magner and Reynard failed to finish.

G.L.Reynard (Number 8) at the start of the 1928 Lightweight TT.

Enjoy the
Pleasure
of—
Effortless
Speed!
ROYAL ENFIELD
MADE LIKE A GUN

CHAPTER NINETEEN

Model Year 1929

The 1929 model range, comprising thirteen different motorcycle models and twelve different sidecars, was designed to meet all customer requirements. Emphasis was moving away from the big twins and towards the mid-range 350 and 500 models. The 4.88hp Standard side-valve model was now joined by a De-Luxe version, equipped with electric lighting and speedometer, and a two-port overhead-valve sports model. The 350 range, comprising side-valve, standard overhead-valve and twin-port overhead-valve Sports and Super Sports models had proved popular and was continued. The popularity of the big twin machines was reducing and only a single version of the 976cc big twin was retained. This was the Sports Model 182 of the previous year, but now offered with either a three- or four-speed gearbox. The range of twelve sidecar models was unchanged.

1929 Models

Model 201	2.25 hp	Standard Two-stroke	225cc	£28 17s 6d
Model 201A	2.25 hp	Open Frame Two-stroke	225cc	£30
Model 202	2.25 hp	Standard Three-speed Two-stroke	225cc	£32 10s
Model 203	2.25 hp	Sports Three-speed Two-stroke	225cc	£32 10s
Model 250	2.25 hp	Side-valve Four-stroke	225cc	£36
Model 350	3.46 hp	Standard Side-valve	346cc	£43
Model 351	3.46 hp	Standard OHV	346cc	£45
Model 355	3.46 hp	Sports Two-port OHV	346cc	£50
Model 352	3.46 hp	Super-Sports Two-port OHV	346cc	£75
Model 501	4.88 hp	Standard Side-valve	488cc	£45 10s
Model 502	4.88 hp	De-Luxe Side-valve	488cc	£53 15s
Model 505	4.88 hp	Two-port OHV	488cc	£52 10s
Model 182	9.76 hp	Side-valve Twin	976cc	£62 10s

Royal Enfield motorcycles had for many years been easily identified by the fuel tanks being artistically finished in two shades of green. When the new 1929 machines were exhibited at the Olympia Motor Cycle Show in November 1928, The Motor Cycle in its show review commented 'Something of a sensation was caused by the appearance of the new 488cc Royal Enfields, with their beige and red saddle tanks'. The new colour schemes were perhaps aimed to draw attention to the machines during a period of tough competition and market uncertainty. The prices of most Enfield models were also reduced by as much as £3 in order to retain a competitive edge.

225cc Two- and Four-stroke Models

The Two-speed Sports Model 200 was withdrawn from the Royal Enfield range for 1929 reducing to four the number of

The 1929 Brochure.

Works photograph of a 2.25hp Two-stroke. The machine appears to be a Model 201 with the leg shields and foot boards replaced by foot pegs. The photograph shows the chain cover and kick-start mechanism.

two-stroke machines. The two-speed models, Standard Model 201 and Open-frame Model 201A, were both retained and offered at a reduced price. These two models were the last Enfield machines to be offered with the early vintage style flat tanks, all the remaining models featuring more modern styled saddle tanks. Neither model would be made available for 1930

The Two-stroke Three-speed Models 202 and 203 together with the Four-stroke Model 250 were continued unchanged.

2.25 hp Standard Two-stroke Model 201

Type: Two-stroke
Capacity: 225cc
Bore & stroke: 64 x 70mm
Carburettor: Amal Two-lever
Gearbox: Royal Enfield 2-speed
Gear ratios: 5 & 8.1 to 1
Brakes: 4 in drums front and rear
Tyres: Dunlop Cord 24 x 2½ in
Length: 6ft 6in Width: 2ft 8½in
Saddle height: 26½in
Weight: 160 lbs
Capacities: Fuel: 1 gallon
Finish: Black enamel, tank finished in two shades of green enamel
Price: Without lights £28 17s 6d, With ML Maglita lighting set £34 2s 6d

2.25 hp Open-frame Two-stroke Model 201A

Specification as Model 201 but with open frame and circular tank.
Weight: 164 lbs
Saddle height: 27in
Finish: Black enamel, tank enamelled black with green centre panels.
Price: Without lights £30
With Maglita lighting set: £35 5s

2.25 hp Standard Three-speed Two-stroke Model 202

Type: Two-stroke with aluminium piston and detachable cylinder head.
Capacity: 225cc
Bore & stroke: 64 x 70mm
Carburettor: Amal Two-lever
Gearbox: Royal Enfield 3-speed
Gear ratios: Standard: 5, 7.35 & 14.5 to 1
Brakes: 4 in drums front and rear
Tyres: Dunlop Cord 24 x 2½ in
Length: 6ft 5in Width: 2ft 9in
Spring seat saddle Saddle Height: 25in
Fuel: 1 gallon Weight: 160 lbs
Finish: Black enamel, tank finished in two shades of green cellulose
Price: Without lights £32 10s, With ML Maglita lighting set £37 15s

2.25 hp Sports Three-speed Two-stroke Model 203

The Model 203 specification is similar to Model 202 but with two long exhaust pipes and two silencers, footrests, sports type handlebars and without legshields.

2.25 hp Side-valve Four-stroke Model 250

Type: Four-stroke
Capacity: 225cc
Bore & stroke: 64 x 70 mm
Carburettor: Amal Two-lever
Gearbox: Royal Enfield 3-speed
Gear ratios: Standard: 7.2,10.6 & 20.8 to 1
Brakes: 5 in drums front and rear
Tyres: Dunlop Cord 25 x 3 in
Length: 6ft 5in Width 2ft 8½in
Saddle height: 25in
Fuel capacity: 1 gallon
Weight: 190 lbs
Finish: Black enamel, tank finished in two shades of green cellulose
Price: Without lights £36
With Lucas lighting set £41 5s

The handsome saddle tanks fitted to all models, except the Models 201 and 201A, were of all welded steel construction, thus removing the risk of leakage at soldered seams. Knee grips were fitted as standard only to the 3.46hp and 4.88hp models.

RIGHT: A promotional advert published in the motorcycling magazines at the time of the 1928 Olympia Show. It is not clear why the 2.25hp Four-stroke was omitted from the list of available machines.

3.46hp Four-strokes

The range of 346cc four-stroke machines continued to offer a choice to meet all user requirements. The standard models, side-valve or overhead-valve, were available for general use whilst two-port overhead-valve machines were available for the more sporting rider and the racer.

The Standard Side-valve Model 350 was considerably redesigned to bring it into line with the other models in the range. The frame was based upon the previous year's sporting models with a straight top tube in-line from the steering head towards the back hub and the absence of a lower tank tube. The machine was fitted with a saddle tank and a separate oil tank was fastened to the seat tube. The engine also received attention in order to make it more robust. Changes included a strengthened crankcase with stiffening webs, a larger diameter mainshaft and a redesigned cylinder casting with a greater area of cooling fins. Similar improvements were made to the single-port overhead-valve engine of the Model 351, which also benefited from revised cam contours.

The frame design for the Two-port OHV Sports Model 355 and Super-sports Model 352 was based upon that used for the 1928 TT Race machines and incorporated a lower tank tube running parallel to the top tube forward from the seat lug before bending down towards the steering head.

3.46 hp Standard Side-valve Model 350

Type: Four-stroke Side-vale
Capacity: 346cc Bore & stroke: 70 x 90 mm
Carburettor: Amal Two-lever
Gearbox: Sturmey-Archer 3-speed
Gear ratios: Standard: 5.9, 8.6 & 15.7 to 1 Sidecar: 9.6, 7.8 &17.7 to 1
Brakes: 5 in drums front and rear
Tyres: Dunlop Cord 27 x 2¾ in
Length: 6ft 11in Width: 2ft 9½in Saddle height: 27in
Weight: 230 lbs Fuel: 1 gallons
Finish: Black enamel, tank finished in two shades of green cellulose
Price: Without lights £43, With Lucas or M.L. lighting set £48 5s

3.46 hp Standard OHV Model 351

Specification as Model 350 with following exceptions.
Type: Four-stroke Overhead-valve Compression ratio: 5¼ to 1
Gear ratios: Standard: 5.3, 7 & 12.7 to 1 Sidecar: 5.9, 7.8 &14.2 to 1

3.46 hp Sports Two-port OHV Model 355

Specification of Model 355 as Model 351 with following changes:
Compression ratio: 6 to 1, increased to 6½ to 1 by removal of compression plate
Brakes: 5in drum front, 6½in rear
Tyres: Dunlop Cord 26 x 3 in
Weight: 304 lbs Length: 7ft

3.46 hp Super-Sports Two-port OHV Model 352

Specification exactly similar to Model 355, but with a special engine tuned for road racing. The push rods are not enclosed and the overhead rockers run on roller bearings. No kick-starter fitted.

4.88hp Four-strokes

The 488cc Side-valve Standard Model 500 of 1928 was discontinued and replaced by the Model 501. The new model used a frame of similar design to the Sports Model 355 and was fitted with a saddle tank. The engine gained deeper fins and a stiffer crankcase. The magneto chain was moved to the off-side and both the magneto chain and the primary chain were fully enclosed. Whereas the previous year's model had benefited from the fitment of a four-speed gearbox, the new model was fitted as standard with a three-speed Sturmey-Archer box; the option of the Enfield four-speed box now cost an extra £2. A de luxe version of the Model 501, identified as the Model 502, was made available for an extra £8 5s. This used a primary chaincase made of cast aluminium, rather than sheet metal, a full Lucas 'Magdyno' lighting set, with control switch and ammeter set into the back of the lamp, and a Smiths speedometer, flush mounted into the top of the fuel tank.

The new overhead-valve 500 machine was fitted with a new two-port 488cc engine of Enfield design and manufacture. The engine was generally similar to the 346cc engine of the Model 355 although the design dispensed with cam followers and the tappets were operated directly from the cam wheels. The crankcase and timing chest were linked allowing oil mist from the crankcase to lubricate the timing gear and tappets. Further passages allowed the oil mist to pass along the push rod tubes and lubricate the overhead rocker box, From here surplus oil was drained through a pipe to the primary chain.

RIGHT: The new 488cc overhead-valve engine layout followed the best current practice with an aluminium alloy piston, deeply finned cylinder, stiffened crankcase and drive shaft. Roller bearings were fitted to big ends and main shaft bearings. The push rods and rockers were enclosed and automatically lubricated.

4.88 hp Standard Side-valve Model 501

Type: Four-stroke Side-valve
Capacity: 488cc Bore & stroke: 85.5 x 85mm.
Carburettor: Amal Two-lever
Gearbox: Sturmey-Archer 3-speed
Gear ratios: Standard: 5, 7.3 & 14.7 to 1
Sidecar: 5.6, 8.2 & 16.5 to 1
Brakes: 6½ in drums front and rear
Tyres: Dunlop Cord 26 x 3.25 in
Length: 7ft Width: 2ft 9½in Saddle height: 27in
Weight: 280 lbs Fuel: 1¾ gallons Oil: 2½ pints
Finish: Black enamel, tank finished in beige cellulose with red panel

4.88 hp De Luxe Side-valve Model 502

Specification as Model 501, but supplied with aluminium primary chaincase, Lucas Magdyno lighting set and Smith speedometer.
Weight: 308 lbs Price: £53 15s.

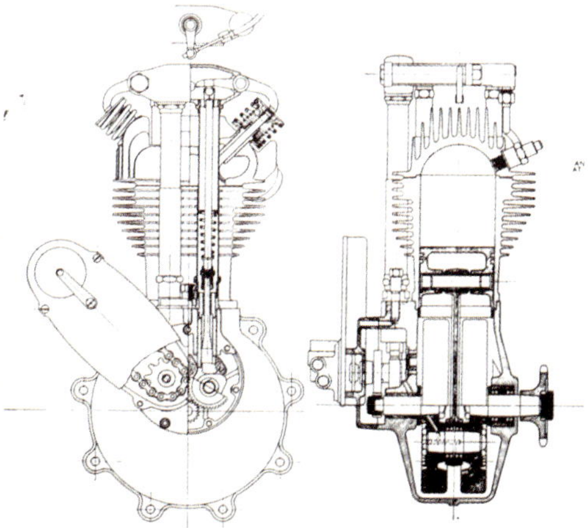

Both Models 501 and 502 were available with the Enfield four-speed box for an additional £2.
The standard ratios were:
Solo: 5, 6.5, 8.5 and 15.5 to 1
Sidecar: 5.6, 7.3, 9.6 and 17.4 to 1

4.88 hp Two-port OHV Models 505

Type: Overhead-valve Twin-port
Capacity: 488cc
Bore & stroke: 85.5 x 85 mm
Carburettor: Amal Two-lever
Gearbox: Sturmey-Archer 3-speed
Gear ratios: Standard: 4.7, 6.3 & 11.3 to 1
Sidecar: 5.3, 7 & 12.7 to 1
Brakes: 6½ in drums front and rear
Tyres: Dunlop Cord 26 x 3.25 in
Length: 7ft Width: 2ft 9½in Weight: 308 lbs Fuel: 1¾ gallons Oil: 2½ pints
Finish: Black enamel, tank finished in beige and red cellulose
Prices: Without lights £52 10s, With Lucas or ML lighting set £57 15s

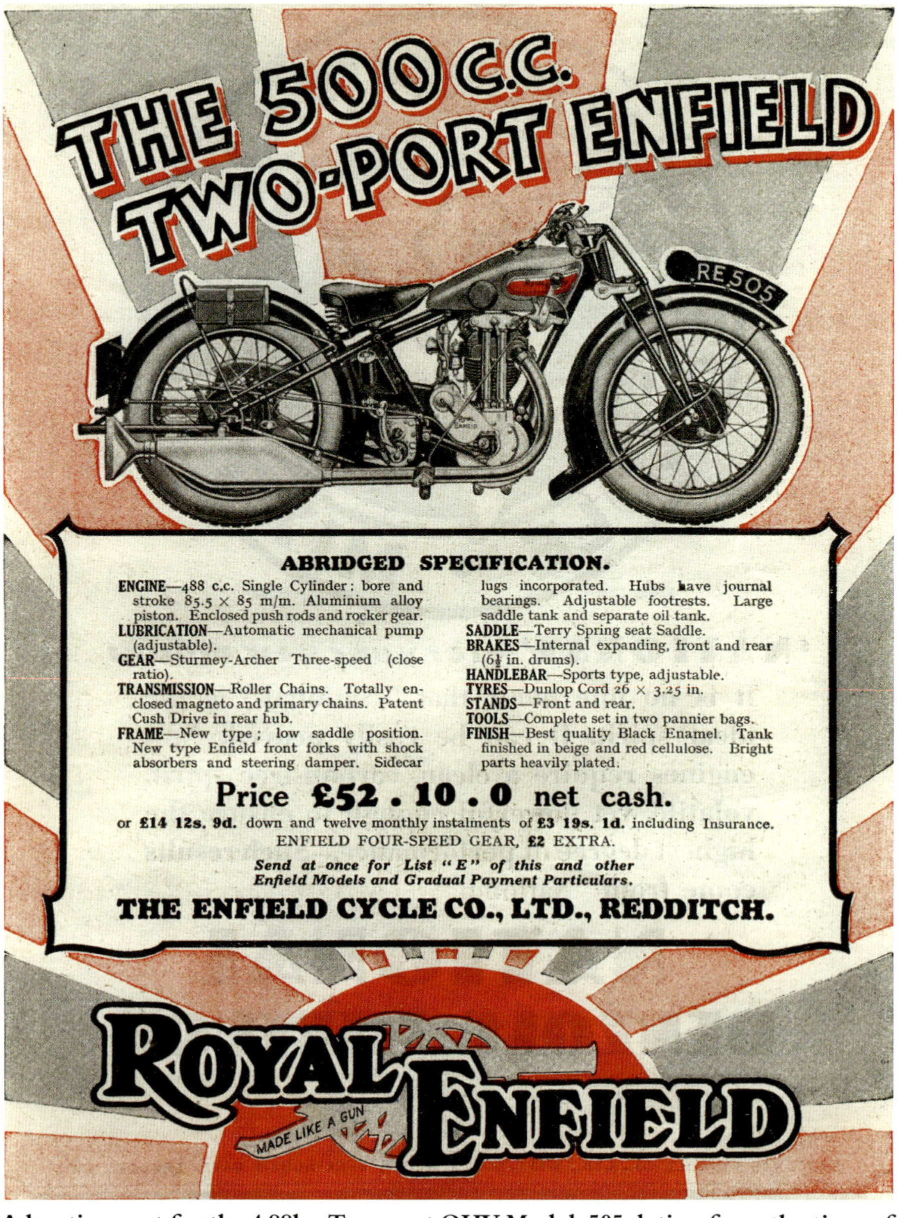

Advertisement for the 4.88hp Two-port OHV Model 505 dating from the time of the 1928 Motor Cycle Show.

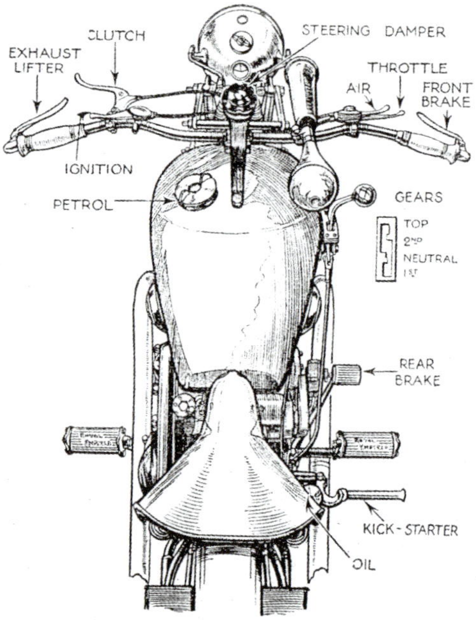

Controls of the 488cc Twin-port Model 505,

9.76hp Side-valve Twins

The Model 182 was the only 9.76hp Twin solo machine retained in the 1929 range. The specification was unchanged from the previous season's Four-speed Sports model with the exception that the Enfield 4-speed gearbox was replaced by a Sturmey-Archer 3-speed box, although the four-speed box remained an option. Perhaps reflecting the lessening popularity of the big twins, the price was reduced from £73 10s to £62 10s or £64 10s with the four-speed box.

A factory photograph showing the nearside view of the 488cc OHV 2-port Model 505. The machine illustrated was not fitted with a separate oil tank. Also unusual is the domed rear spindle nut with distance collar and the reverse method of chain adjustment.

Models 501 & 502
Solo: 5, 6.5, 8.5 & 15.5 to 1
Sidecar: 5.6, 7.3, 9.6 & 17.4 to 1
Model 505
Solo: 4.7, 6.1, 7.9 & 14.5 to 1
Sidecar: 5.3, 6.9, 8.9 & 16.4 to 1
Model 182
Solo: 4.2, 5.4, 7.1 & 13 to 1
Sidecar: 4.7, 6.1, 7.9 & 14.5 to 1

The 4.88hp and 9.76hp models were supplied with a three-speed gearbox as standard, but a four-speed box was available for an additional £2. The table above lists the standard gear ratios for the different models used solo or with a sidecar.

9.76 hp Side-valve Twin Model 182

Type: Side-valve V-twin
Capacity: 976cc
Bore & stroke: 85.5 x 85 mm
Carburettor: Amal Two-lever
Gearbox: Sturmey-Archer 3-speed
Gear ratios:
Standard: 4.2, 6.1 & 11.2 to 1
Sidecar: 4.7, 6.8 & 12.5 to 1
Brakes: 8in front & rear drums
Handlebars: Sports or touring adjustable
Tyres: Dunlop Cord 26 x 3.25 in
Length: 7ft 2in Width: 2ft 8 ½in Fuel: 2 gallons Oil: 2½ pints
Weight: 345lbs without lights.

9.76 hp Delivery Outfit Model 185

Coach-built box carrier with hinged lid, painted green to match the motorcycle. As supplied to bakers, butchers, billposters, boot repairers, drapers, fish mongers, grocers and others.
Box capacity: 13 cu ft
Inside dimensions: Length 4ft 5in, Width 1ft 7in, Depth 1ft 8in
Overall length: 7ft 2in, Width: 5ft 3in
Price: £77 10s

Trade Delivery Outfits

Although the Model 180 was discontinued as a solo for 1929, it was retained for use as a trade delivery outfit. The motorcycles retained the frame and slab sided fuel tanks used on 1928 models, but in other respects the specification was as the Model 182. Three different 9.76hp trade delivery outfits were offered as suitable for both the small tradesman and the large store.

A delivery box, Model 6, was also offered for use with the 4.88hp Standard Side-valve Model 501 and interchangeable with the Touring Sidecar Model 10.

9.76 hp Dairyman's Outfit Model 165

Suitable for dairymen, farmers and greengrocers. Fitted with a wide and exceptionally strong chassis. The milk float illustrated is capable of carrying three milk churns, besides cans, tools, spares, etc.
Box capacity: 13 cu ft
Inside dimensions: Length 4ft 6in, Width 2ft 7in, Depth 1ft 6in
Overall dimensions of combination: Length: 7ft 2in, Width: 6ft
Price: £79 10s

9.76 hp Dairyman's Outfit Model 175 (not illustrated)

A similar outfit to Model 165, but built on the standard chassis and of the same dimensions as Model 185. Designed to carry two milk churns instead of three.
Price: £77 10s

ABOVE and RIGHT: Photographs taken in 1929 showing the 9.76hp Three-speed Standard Model 180 with Dairyman's sidecar employed by the London Depot of the Enfield Cycle Co Ltd to advertise the Company's products and for the efficient delivery of bicycles to dealers in the London district.

9.76 hp GPO Type Outfit Model 155

As supplied to the GPO. Fitted to the wide chassis and capable of carrying loads up to 10cwt. The large box has a lid at the front and double doors at the rear.
Box capacity: 31 cu ft
Box dimensions: Length 4ft 9in, Width 2ft 9½in, Depth 3ft 3in
Overall length: 7ft 2in Width: 6ft
Price: £85

Delivery Box Model 6

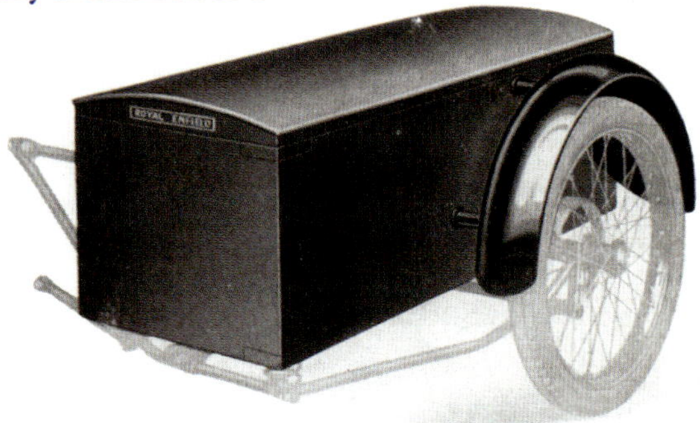

Suitable for 4.88 hp Standard Side-valve Model 501. The box carrier was strongly built of best quality timber and beautifully finished.
Inside dimensions: Length 4ft 1in, Width 1ft 6in, Depth 1ft 3in
Price: Box with mudguard only £9 10s

1929 Sidecar Range	
Suitable for 3.46hp and 4.88hp machines	
Model 14 Sports	£12 10s
Model 12 Touring	£12 10s
Model 8 Sports	£16 15s.
Model 10 Touring	£16 15s.
Model 6 Delivery Box	£16 15s
Suitable for 9.76hp Side-valve Model K	
Model 2 Touring	£18 15s
Model 4 Sports	£16 15s
Model 22 Occasional Two-seater	£23
Model 185 Delivery Box outfit	£77 10s
Model 165 Dairyman's outfit	£79 10s
Model 175 Dairyman's Narrow outfit	£77 10s.
Model 155 GPO Type outfit	£85.

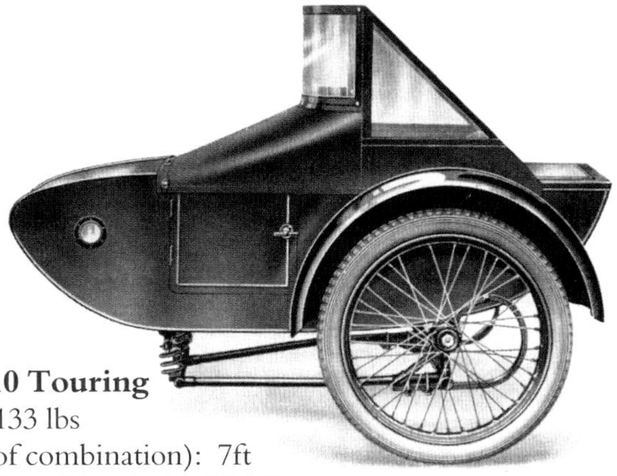

Model 10 Touring
Weight: 133 lbs
Length (of combination): 7ft
Width (of combination): 4ft 11in

Sidecar Range for 3.44hp and 4.88hp Models

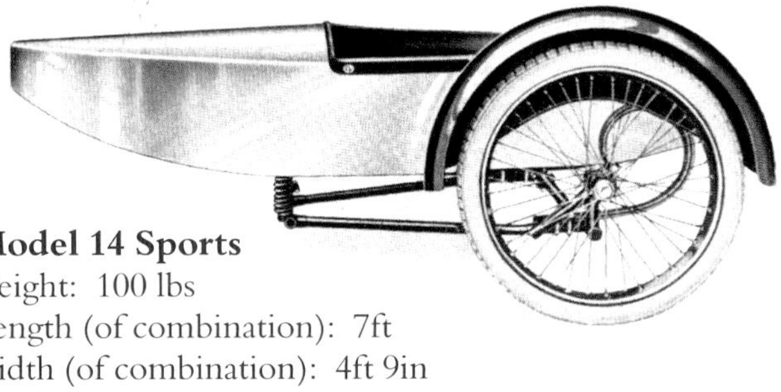

Model 14 Sports
Weight: 100 lbs
Length (of combination): 7ft
Width (of combination): 4ft 9in

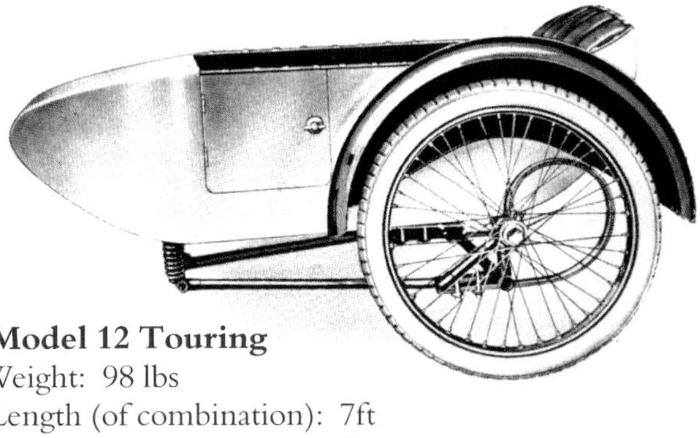

Model 12 Touring
Weight: 98 lbs
Length (of combination): 7ft
Width (of combination): 4ft 9in

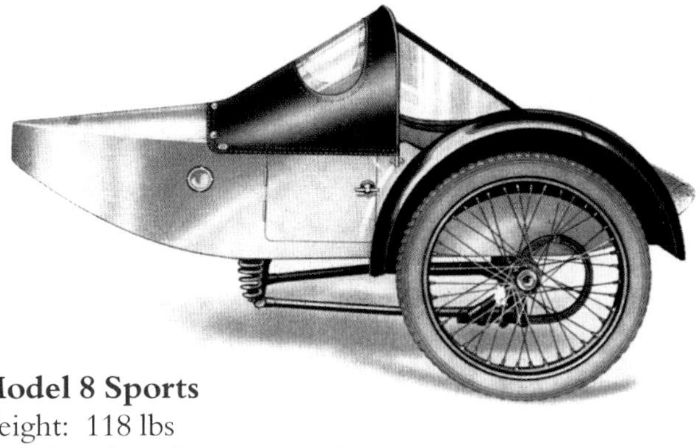

Model 8 Sports
Weight: 118 lbs
Length (of combination): 7ft
Width (of combination): 4ft 11in

Sidecar Range for 9.76hp Model

Model 2 Touring
Weight: 154lbs
Length (of combination): 7ft 2in
Width (of combination): 5ft 3in

Model 4 Sports
Weight: 126 lbs
Length (of combination): 7ft 2in
Width (of combination): 5ft 3in

Model 22
Occasional Two-seater Touring
Weight: 199 lbs
Length (of combination): 7ft 7in
Width (of combination): 5ft 3in

The Motor Cycle in its edition of 25th April 1929 reported on the arrival in London during the preceding week of two 976cc Royal Enfield sidecar outfits and their young Swedish riders. The machines had been ridden from Stockholm via Germany and Belgium and were about to continue across France and Spain before continuing on a long tour of Africa. The party consisted of Bertil Hult, a Swedish journalist, Ake Burstrom, a photographer, Oskar Gustavsson, mechanic, and Kai Thorenfelt, a wireless operator for the field wireless set carried on one of the machines.

Dirt Track Machines

Dirt track racing or speedway was introduced into Britain by visiting Australian riders. The first demonstration of the broadsiding style of riding had been given at Camberley Heath in May 1927. However, it was not until a proper dirt track meeting was held using a new dirt circuit at Crystal Palace in February 1928 that the sport gained popularity. Spectators revelled in the sounds of open exhausts, the aroma of the burnt Castrol R oil, the close racing and the spectacle of the broadsiding style of riding, which threw showers of dirt into the air at each turn. Circuits were quickly established across the country and crowds flocked to watch. The promotional benefits of success in this new form of racing was quickly appreciated and manufacturers lost little time in either adapting existing race machines or in developing new machines in order to compete.

Two views of the experimental dirt track model fitted with Twin-port Overhead-valve JAP engine, which dates from late 1928.

The Enfield Works photograph album includes photographs of two development machines and a fifth photograph showiing a production version which was offered for sale at £70. With Enfield's TT racing pedigree and the use of JAP engines the company might have anticipated doing well in the new sport, but failed to make its mark.

RIGHT: Two views of an experimental dirt track model with 500cc twin-port OHV JAP engine. At this stage the machine retains a front wheel brake. Although a gearbox is present it is a single gear machine; the fitment being used to mount a clutch and presumably provide the option to adjust the gear ratio.

This photograph dates from November 1929 and shows the Royal Enfield dirt track motorcycle in final form. The engine is the Enfield 500cc two-port. The primary chain enclosure has ben retained, as has the rear wheel brake, and a knee hook fitted. In this standard form the machine was priced at £70. It was also made available with a plated finish to the frame and forks and with pale green tanks and mudguards for an additional £3.

The 4.88hp De-Luxe Side-valve Model 502 was the only machine supplied as standard with a lighting set and speedometer. The other machines would be supplied with an electric lighting set for an additional £5 5s. On the two-stroke machines this would be a ML Maglita set, whilst the other machines would be offered with the choice of either a ML Maglita or Lucas electric lighting set. A sidecar lamp cost an additional 10s. An electric horn was available for £1 on those machines fitted with the Lucas equipment. A Cowey speedometer was priced at £3 5s.

At this period many riders still preferred acetylene lighting sets and would often transfer the lighting set from their old machine. The table (RIGHT) provides the prices of new acetylene lighting equipment.

ACCESSORIES

The following Accessories *only* can be included in the Specification :

Various Models.—Cowey Speedometer (to register 60 or 80 m.p.h.), £3 5s. 0d. ; Leg Shields, Electric Lighting, Windscreens, Balloon Tyres, Speedometer with instrument in Tank (on certain Models only), etc., at Catalogue Prices.

	£	s.	d.
Models 201, 201a, 202, 203, 250—			
Lucas No. 320 F.S. Head Light and Generator	1	2	6
Lucas No. 63 E.B. Bulb Horn	0	8	6
Models 350, 351, 355, 501 and 505—			
Lucas No. 420 F.S. Head Light and Generator	1	12	6
Lucas No. 350 Sidecar Light and Generator	0	16	6
Lucas No. 63 E.B. Bulb Horn	0	8	6
Model 182 and all Trade Delivery Combinations—			
Lucas No. 462 F.S. Head Light and Generator	2	10	0
Lucas No. 354 Sidecar Light and Generator	1	2	6
Lucas No. 60 E.B. Bulb Horn	0	13	6
All Models—			
Lucas No. 344 Rear Light and Generator	0	13	0

A photograph from the Enfield photo album to illustrate the ease with which the Model 6 delivery box could be removed to be replaced by the Touring Sidecar Model 10.

This factory photograph of a single-port overhead-valve solo machine provides a good illustration of the handlebar controls and the Lucas lighting set. The speedometer is flush fitted into the fuel tank.

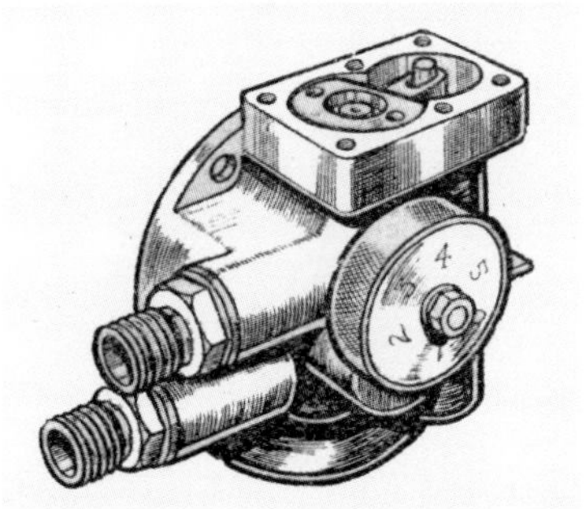

All four-stroke models were fitted with a mechanical oil pump of the oscillating plunger type driven by a spiral gear from the cam shaft. On selected models the flow could be adjusted; the minimum setting allowed for normal riding and this could be increased to three times the minimum for hard riding. The other models relied upon a hand pump to supply the extra oil as required.

Machines prepared for dispatch.

CHAPTER TWENTY

Model Year 1930

The 1930 season saw the introduction of new models and a rationalisation of the range together with the introduction of a revised model identification code. The alphabetic letters A to K replaced the numeric code of previous years. A single 225cc two-stroke machine, the Model A, replaced the four earlier two-stroke models including the Ladies Model 201A. The 225cc four-stroke Model 250 became the Model B. Model C identified a new 346cc lightweight model with side-valve engine inclined forward after the style popularised by the successful BSA Sloper models.

1930 Model Range

Model A	2.25 hp	Two-stroke	225cc	£32 10s
Model B	2.25 hp	Four-stroke	225cc	£36
Model C	3.46 hp	Lightweight Side-valve	346cc	£39 15s
Model CO	3.46 hp	Light Two-port OHV	346cc	£43 5s
Model D	4.88 hp	Standard Side-valve	488cc	£45 10s
Model E	4.88 hp	Standard Two-port OHV	488cc	£49
Model F	3.46 hp	De Luxe Side-valve	346cc	£45
Model G	3.46 hp	De Luxe Two-port OHV	346cc	£51 10s
Model H	4.88 hp	De Luxe Side-valve	488cc	£48 10s
Model HA	4.88 hp	Standard Side-valve	488cc	£45 10s
Model J	4.88 hp	De Luxe Two-port OHV	488cc	£55
Model JA	4.88 hp	Standard Two-port OHV	488cc	£49 17s 6d
Model K	9.76 hp	Side-valve Twin	976cc	£62 10s

Models A and C were also available as models AL and CL with a ML Maglita Electric Lighting Set for an additional £5 5s.

Models B and D to K were also available as models BL and DL to KL with a Lucas Dynamo Electric Lighting Set for an additional £5 10s.

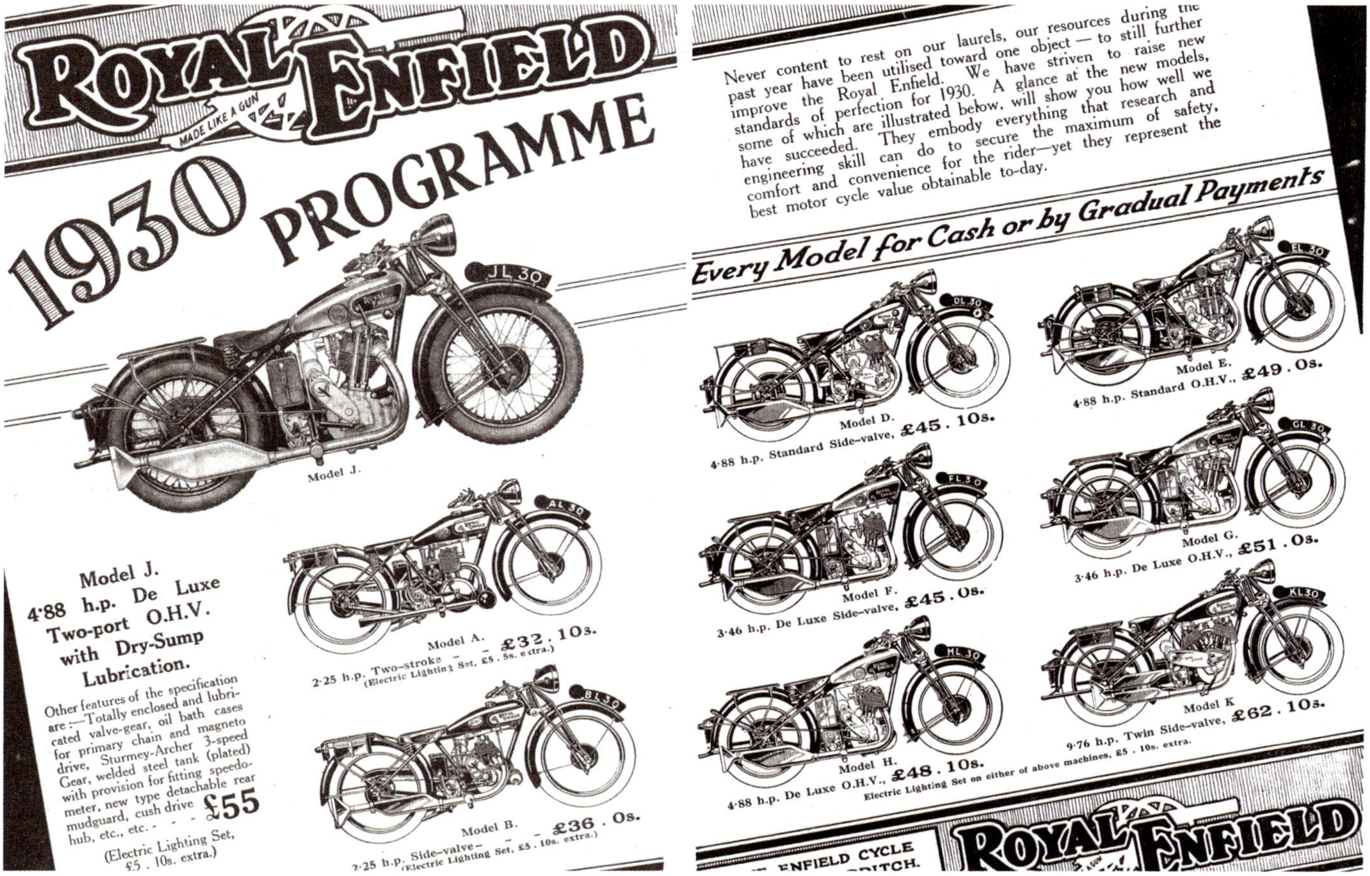

Advertisement published at the time of the 1929 Motor Cycle Show.

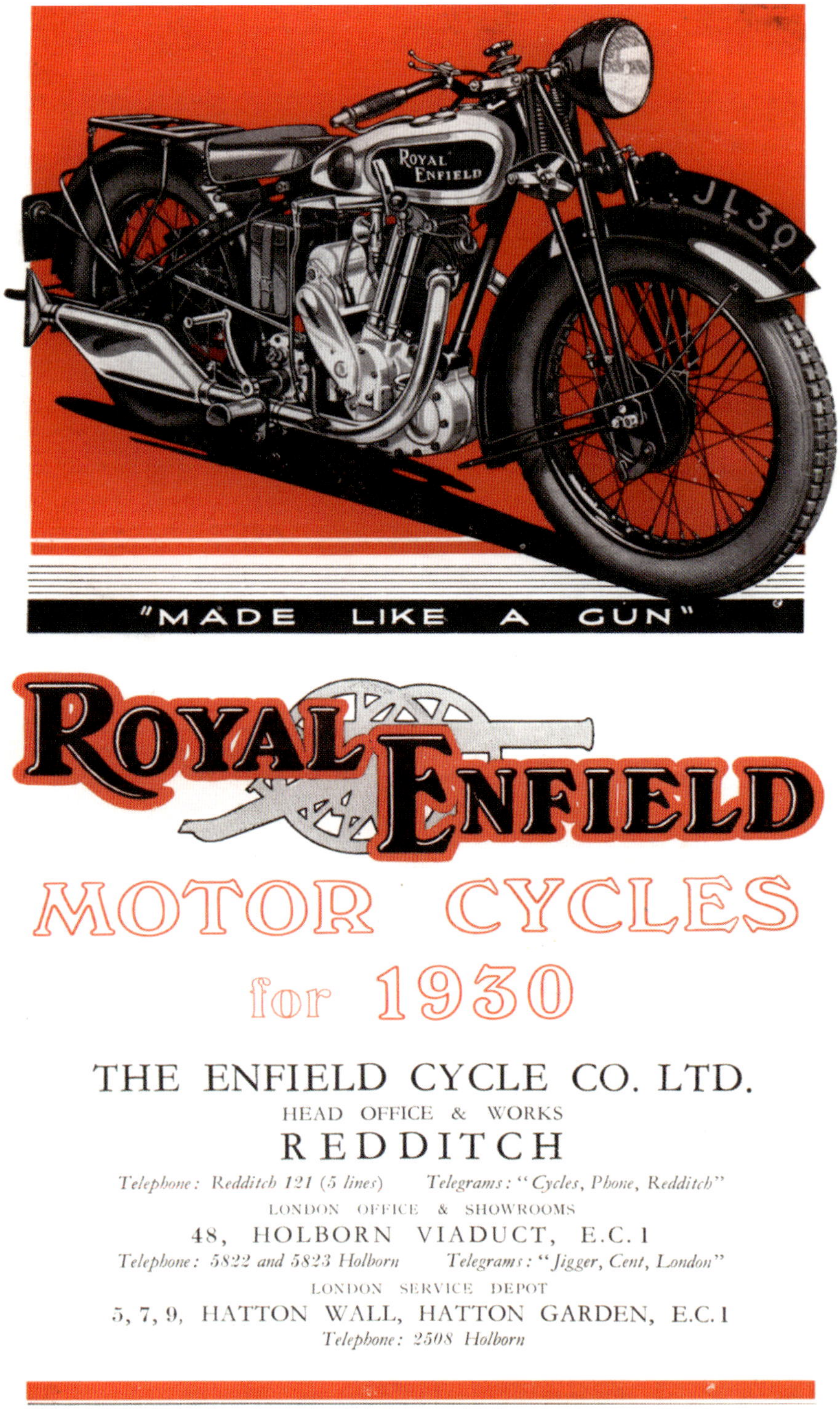

The 1930 Sales Brochure listed eleven motorcycle models and twelve sidecars.

The model letters D and E were used for the 488cc Standard side-valve and twin-port OHV models, respectively; both models retained the upright engine. The remaining single cylinder machines were fitted with inclined engines incorporating a dry-sump lubrication system. Model numbers F to J were used for the De Luxe models; letters F and H identified the side-valve models whilst letters G and J identified overhead-valve models of 346cc and 488cc respectively. A single side-valve 976cc vee-twin was offered and identified as the Model K.

The 488cc side-valve and two-port overhead-valve machines had proven popular during 1929 and it was at the agents' request that the machines were retained in the catalogue and modified only in detail. The other eight models in the Enfield range displayed at Olympia were either entirely new or modified to a considerable extent.

The simplicity of the new naming convention quickly broke down with the introduction in January of a lightweight 346cc twin-port OHV machine, identified as the Model CO. Models C, CO, F, G, H and J all featured inclined engines. Two additional models, standard versions of the De Luxe Models H and J were added in May and identified as Models HA and JA.

All models were available with or without electric lighting sets; the machines with lighting sets being identified by adding an additional letter L to the model identification letter.

2.25hp Two-stroke Model A

The new utility 225cc two-stroke was based upon the previous year's Model 202 and, like the other models in the range, featured the Royal Enfield welded-steel design of saddle tank, which also carried an improved design of hand-change gear lever. The same design was used for all the models of the range except Models D and E, which were unchanged from the previous year. It was fitted with the two-port Enfield engine, but the petroil lubrication previously employed was replaced by an automatic system with adjustable mechanical oil pump. The oil was carried in a separate compartment in the fuel tank. The magneto was situated behind the engine and replaced by a M.L. Maglita for the Model AL equipped with lighting. A central spring-up stand was fitted. The new frame was similar to that of the earlier Model 202, but with a lower saddle height and foot pegs rather than foot boards. The front fork was a lightweight version of the design used throughout the Royal Enfield range and employed two coil springs in tension.

2.25hp Side-valve Models B and BL

The Model B was largely unchanged from the Model 250 of the previous year. It was fitted with an improved silencer with large fish-tail outlet.The new Enfield design of front fork with twin springs was fitted, together with a revised mudguard. The new gear lever was also fitted, whilst the steering damper, fitted in 1929, was discontinued. There was insufficient space behind the engine to fit a Lucas Magdyno and the Model BL with electric lighting set was fitted with a separate Lucas dynamo mounted on the engine plates in front of the crankcase at the bottom of the front down tube.

2.25 hp Two-stroke Models A and AL

Type: 2-stroke. Capacity: 225cc. Bore & stroke: 64 x 70mm.
Carburettor: 2-lever Amal
Gearbox: Royal Enfield 3-speed Gear ratios: 5, 7.4 & 14.7 to 1
Brakes: 5in drums front and rear Tyres: Dunlop Cord 25 x 3
Length: 6ft 4in Width: 2ft 9in Fuel: $1\frac{5}{8}$ galls. Oil: 2½ pints
Weight: 180 lbs without lights, 197 lbs with lights
Finish: Black enamel, tank finished in two shades of green cellulose
Prices; Model A £32 10s
Model AL with Maglita lighting set £37 15s

2.25 hp Four-stroke Models B and BL

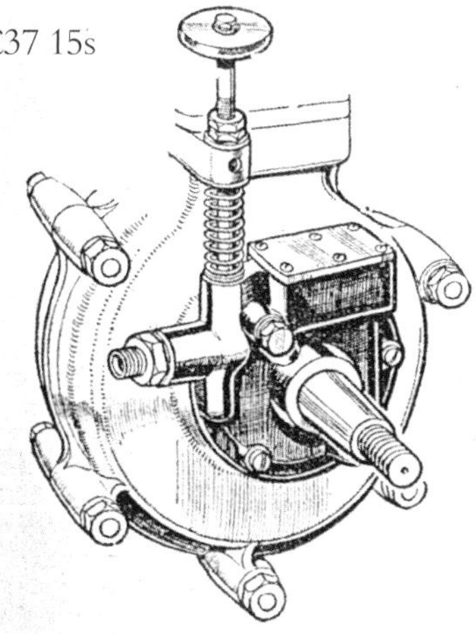

The new season's models were fitted with adjustable oil pumps. The illustration shows the pump fitted to the two-stroke models.

Type: 4-stroke side-valve
Capacity: 225cc.
Bore & stroke: 64 x 70mm
Carburettor: 2-lever Amal
Gearbox: Royal Enfield 3-speed
Gear ratios: 7.2, 10.6 & 21.2 to 1
Brakes: 5in drums front and rear
Tyres: Dunlop Cord 25 x 3
Length: 6ft 4in Width: 2ft 9in
Fuel: $1\frac{5}{8}$ galls. Oil: 2½ pints
Weight: 188 lbs without lights
208 lbs with lights
Finish: Black enamel, tank finished in two shades of green cellulose
Prices: Model B £36
Model BL with Lucas dynamo lighting set £41 15s

LEFT: Another works photograph, this time of the 2.25hp Side-valve Model BL. The headlamp would be set at a lower position on the production machines.

3.46 hp Lightweight Side-valve Models C and CL

Type: Side-valve   Capacity: 346cc   Bore & stroke: 70 x 90mm
Carburettor: 2-lever Amal
Gearbox: Royal Enfield 3-speed   Gear ratios: 5.8, 8.6 & 17.2 to 1
Brakes: 5in drums front and rear   Tyres: Dunlop Cord 25 x 3 in
Length: 6ft 10in   Width: 2ft 9in   Fuel: 1¾ galls   Oil: 2½ pints
Weight: 203 lbs without lights, 220 lbs with lights
Finish: Black enamel, tank finished in two shades of green cellulose
Prices: Model C £39 15s. Model CL with Maglita lighting set £45

Lightweight 'Sloper' Model C

The new 3.46 Lightweight Side-valve Model C was promoted as 'an ideal general purpose light solo machine'. It was a development of the previous years side-valve Model 350. The machine had been considerably lightened with the objective of bringing it within the 30/- (£1 10s) annual tax bracket. The engine was inclined in the frame. The three-speed Enfield gearbox was bolted to the rear engine plates and the frame slightly altered so that the chainstays ran below the gearbox. A centre stand, similar to the type fitted to Models A and B, rather than a spring-up rear stand, was fitted. A revised design of silencer was fitted. The front fork was the lightweight version of the new Enfield design with redesigned shock absorbers. The fuel tank was supported at the top of the front down tube and the rear of the tank rail; a method considered novel at the time.

Considerations of weight had prevented the use of the detachable carrier and mudguard. Compared with the Model 350 of the previous year, the Model C weighed 27 lb less at 203 lb and the saddle height was reduced by 1in to 26in.

The machine was supplied as standard with a three-speed Sturmey-Archer gearbox and the optional four-speed Enfield box was available for an additional £2.

Works photograph of the 346cc Lightweight Side-valve Model CL with Lucas Maglita electrical equipment. The fuel tank on all the 350cc and larger capacity machines of the range incorporated a blanked-off hole to mount a Smiths speedometer. The speedometer was a £2 optional extra. Also optional was the Lucas horn; bulb horns cost 8s 6d for the Model 63 or 13s 6d for the Model 60, whilst the electric horn was priced at 15s.

Light OHV 'Sloper' Model CO

The 3.46hp Light Two-port OHV, Model CO was announced in January 1930, almost four months after the other models of the range. It was intended as a lightweight sporting solo. The Motor Cycle considered the most outstanding features of the new model to be its light weight and low price, which was thought little short of remarkable given the specification. Indeed the 346cc ohv machine was the most inexpensive machine of its class. The engine was inclined forward in the

3.46 hp Light Two-port OHV Models CO and COL

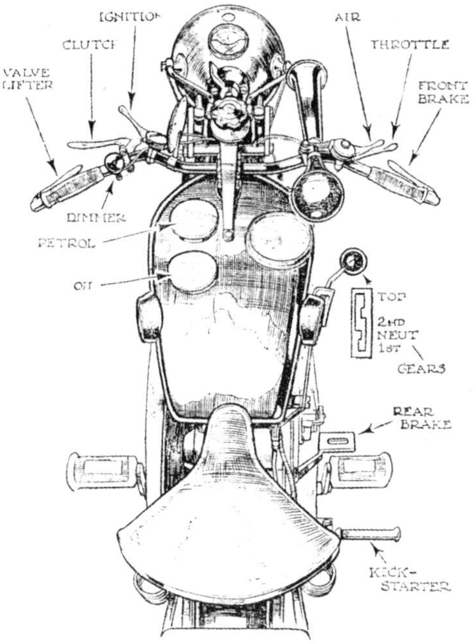

Type: OHV Twin-port Capacity: 346cc Bore & stroke: 70 x 90mm
Carburettor: Amal 2-lever
Gearbox: Royal Enfield 3-speed Gear ratios: 5.4, 7.1 & 14.2 to 1
Brakes: 5in drums front and rear Tyres: Dunlop Cord 25 x 3 in
Length: 6ft 10in Width: 2ft 9in Fuel: 1¾ galls Oil: 2½ pints
Weight: 230 lbs without lights, 249 lbs with lights
Finish: Black enamel, tank finished in two shades of green cellulose.
Prices: Model CO with ML Magneto £43 5s
Model COL with Lucas Magdyno lighting set £48 15s

The controls of the 3.46hp Twin-port OHV Model COL.

The Two-port OHV cylinder head with enclosed push rods, which was introduced in 1929.

Near-side view of the 3.46 hp Light Two-port OHV Model CO. The machine is thought to be a development version based upon the Model C and is fitted with the Model C's centre stand.

frame as the Model C and it shared the same 70 x 90mm dimensions as the Model C. It was of twin-port overhead-valve design with enclosed rockers and push rods. Lubrication was by means of an Enfield mechanical pump mounted on the magneto chain case. The Enfield three-speed gearbox was fitted with close ratio gears, although the standard three-speed box fitted to the Model C was an option. The same frame was used for both machines but, without the same requirement for weight saving to meet the taxation class, was equipped with the more satisfactory spring-up rear stand. The Enfield works had installed a chromium-plating plant and this was fully-operational by the start of the 1930 season. Chromium plating was made available as an alternative to nickel plate on the 1930 models at extra cost. The Model CO was the first Enfield machine to use chromium plate as standard on all the bright parts. The fuel tank was finished in two shades of green cellulose

The top speed was 65mph, with a 45mph maximum available in second gear. Fuel consumption, recorded in a road test performed by The Motor Cycle, was 95mpg.

488cc Standard Models D and E

The 488cc Standard Side-valve and Two-port OHV machines, Models D and E, had proved popular during 1929 and remained in production with very few changes. The two machines remained easily identifiable with beige and red tanks, which were now fitted with mountings for speedometers driven from the gearbox, although the speedometer remained a £2 extra. The machines were also fitted with the new Enfield design of front fork with redesigned shock absorbers and neat lugs for the headlamp brackets. The battery box on the DL and EL models was mounted on the rear mudguard below the saddle.

4.88 hp Standard Side-valve Models D and DL

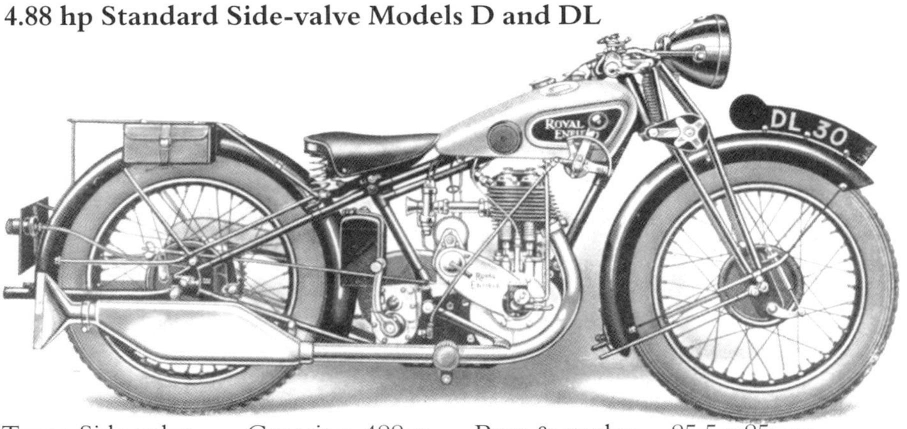

Type: Side-valve   Capacity: 488cc   Bore & stroke: 85.5 x 85mm
Carburettor: Amal Two-lever
Gearbox: Sturmey-Archer 3-speed
Gear ratios: Standard: 5, 7.3 & 14.7 to 1   Sidecar: 5.6, 8.2 & 16.5 to 1
Brakes: 6½ in drums front and rear   Tyres: Dunlop Cord 26 x 3.25 in
Length: 6ft 10in   Width: 2ft 9in   Fuel: 1¾ gallons   Oil: 2½ pints
Weight: 272 lbs without lights, 295 lbs with lights
Finish: Black enamel, tank finished in beige and red cellulose
Prices: Model D £45 10s, Model DL with Lucas Magdyno lighting set £51

4.88 hp Standard Two-port OHV Models E and EL

Type: OHV Twin-port
Capacity: 488cc.
Bore & stroke: 85.5 x 85mm
Carburettor: Amal two-lever
Gearbox: Sturmey-Archer 3-speed
Standard: 4.7, 6.3 & 11.3 to 1
Sidecar: 5.3, 7 & 12.7 to 1
Brakes: 6½ in drums front and rear
Tyres: Dunlop Cord 26 x 3¾ in
Length: 7ft   Width: 2ft 9in
Fuel: 1¾ gallons   Oil: 2½ pints
Weight: 309lbs without lights, 332lbs with lights
Finish: Black enamel, tank finished in beige and red cellulose
Prices: Model E £49,   Model EL with Lucas Magdyno lighting set £54 10s

Works promotional photographs of the 488cc OHV Standard Two-port Model EL. The implication is that it is the lady who is the rider and it is she who has taken her boyfriend for a ride, although hardly dressed for riding.

De Luxe 'Sloper' Models F, G, H and J

The De Luxe models, which were available with 346cc or 488cc side-valve or overhead-valve engines, were entirely new and were the first Royal Enfield production machines with dry sump lubrication. The engines were inclined forward and the same open diamond pattern of frame was used for all models. The 346cc models used the lighter version of the Enfield front fork and a smaller diameter front brake. The appearance of the machines was enhanced by an improved design of saddle tank. This was made longer, brought down further over the top tubes and given a flatter top. It was recessed at the rear to accept the nose of the saddle and finished in nickel plating with black side panels and gold lining.

3.46 hp De Luxe Side-valve Models F and FL

Type: Side-valve Capacity: 346cc Bore & stroke: 70 x 90mm
Carburettor: Amal Two-lever
Gearbox: Sturmey-Archer 3-speed
Gear ratios: Standard: 6.3, 9.2 & 16.7 to 1 Sidecar: 7.1, 10.4 & 19 to 1
Brakes: 5 in front, 6.5in rear drums. Tyres: Dunlop Cord 26 x 3in
Length: 7ft 1in Width: 2ft 9in Fuel: 2¼ gallons Oil: 2½ pints
Weight: 309lbs without lights, 332lbs with lights
Finish: Black enamel, tank plated with black top and side panels
Prices: Model F £45, Model FL with Lucas Magdyno lighting set £50 10s

3.46 hp De Luxe Two-port OHV Models G and GL

Type: OHV Twin-port Capacity: 346cc. Bore & stroke: 70 x 90mm.
Carburettor: Amal Two-lever
Gearbox: Sturmey-Archer 3-speed
Gear ratios: Standard: 5.4, 7.1 & 12.8 to 1, Sidecar: 6.2, 8.3 & 14.9 to 1
Brakes: 5in front, 6½ in rear drums Tyres: Dunlop Cord 26 x 3 in
Length: 7ft 1in Width: 2ft 9in Fuel: 2¼ gallons Oil: 2½ pints
Weight: 283 lbs without lights, 306 lbs with lights
Finish: Black enamel, tank plated with black top and side panels
Prices: Model G £51 10s, Model GL with Lucas Magdyno lighting set £57

Dry sump lubrication

Royal Enfield first introduced dry sump lubrication on the De Luxe 346cc and 488cc 'sloper' models F, G, H and J. A separate compartment with a capacity of 2 ½ pints was cast into the front of the crankcase. The circulation of the oil is maintained by two slow working pumps of the oscillating plunger type operated by a worm on the timing side end of the crankshaft. The feed pump, at the rear, has a bore of 5/16in and a stroke of 3/8in. It is used to draw oil from the container through a filter and force it through internal passageways into a nozzle in the main shaft. It is then forced along a duct through the flywheels to the double roller big end by centrifugal action. After lubricating the big end the oil is flung out into the crank case and drops into the sump. The return pump is used to draw the oil through a second filter and return it via a copper pipe into the oil compartment.

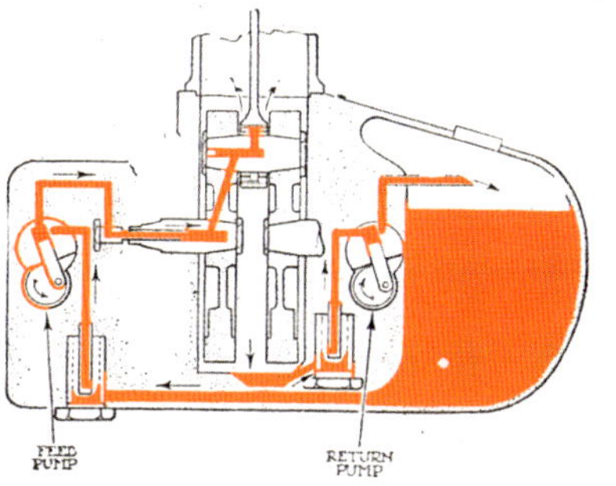

The diagram illustrates the flow of oil in the Royal Enfield dry-sump lubrication system. Oil is pumped through the crank pin and flywheel to the big end and thence by splash to the rest of the engine before collecting in a shallow sump and returned to the forward chamber.

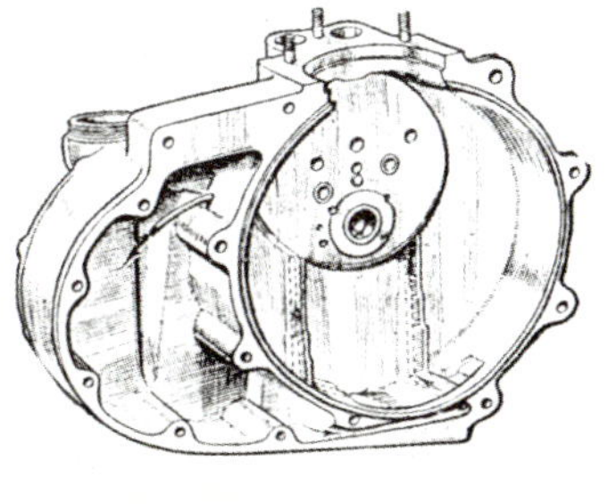

The new crankcase design. The oil flowing back to the chamber drips from the copper pipe visible through the filler opening.

4.88 hp De Luxe Side-valve Models H and HL

Type: Side-valve
Capacity: 488cc
Bore & stroke: 85.5 x 85mm
Carburettor: Amal Two-lever
Gearbox: Sturmey-Archer 3-speed
Gear ratios: Standard: 5, 7.3 & 13.3 to 1
Sidecar: 5.8, 8.5 & 15.5 to 1.
Brakes: 6½ in drums front and rear
Tyres: Dunlop Cord 26 x 3.25 in
Weight: 276 lbs without lights, 299 lbs with lights
Finish: Black enamel, tank plated with black top and side panels
Prices: Without lights (Model H) £48 10s, With Lucas Magdyno lighting set (Model HL) £54

Dimensions
Length: 7ft 1in
Width: 2ft 9in

Capacities
Fuel: 2¼ gallons
Oil: 2½ pints

4.88 hp De Luxe Two-port OHV Models J and JL

Type: OHV Twin-port
Capacity: 488cc
Bore & stroke: 85.5 x 85mm
Carburettor: Amal Two-lever
Gearbox: Sturmey-Archer 3-speed
Gear ratios: Standard: 4.8, 6.4 & 11.5 to 1
Sidecar: 5.3, 7 & 12.7 to 1
Brakes: 6½ in drums front and rear
Tyres: Dunlop Cord 26 x 3.25 in
Weight: 313 lbs without lights, 336 lbs with lights
Finish: Black enamel, tank plated with black top and side panels
Prices: Without lights (Model J) £55, With Lucas Magdyno lighting set (Model JL) £60 10s

Dimensions
Length: 7ft
Width: 2ft 9in

Capacities
Fuel: 1 ¾ gallons
Oil: 2½ pints

Standard 'Sloper' Models HA and JA

The Royal Enfield range was expanded to thirteen models in mid-May by the introduction of two additional 488cc 'Sloper' models, These were economy versions of the side-valve Model H and two-port OHV Model J. They were identified as Models HA and JA and described as Standard models. A three-speed Sturmey-Archer gearbox was fitted although the machines could be supplied with the four-speed Enfield gearbox for £2 extra. The front brakes were reduced to 5 inches. The Enfield quickly detachable rear mudguard was fitted and the stand was the spring-up type with retaining clip.

4.88hp De Luxe Two-port OHV Model JL. The machine is thought to be a prototype and is fitted with different head lamp brackets and luggage grid than production models.

Dry sump lubrication with crankcase oil compartment and an absence of external oil pipes gave an uncluttered appearance to the 'sloper' engines. This is the 488cc two-port OHV engine.

Factory illustration of what appears to be a development model of the 4.88hp De Luxe Two-port OHV. The gear change is that of the 488cc Standard model whilst the rear carrier is the type fitted to the 488cc Standard Side-valve machine. The speedometer drive gear is visible on the front wheel but the speedometer appears to be missing.

4.88 hp Standard Side-valve Models HA and HAL

Type: Side-valve
Capacity: 488cc
Bore & stroke: 85.5 x 85mm
Carburettor: Amal Two-lever
Gearbox: Sturmey-Archer 3-speed
Gear ratios: Standard: 5, 7.3 & 13.3 to 1
Sidecar: 5.8, 8.5 & 15.5 to 1
Brakes: 5in front, 6½in rear drums
Tyres: Dunlop Cord 26 x 3.25 in

Dimensions
Length: 7ft 2in
Width: 2ft 8in

Capacities
Fuel: 2¼ gallons
Oil: 2½ pints

Finish: Black enamel, tank finished in two shades of green. Bright parts chrome plate
Prices: Without lights (Model HA) £45 10s, With Lucas Magdyno lighting set (Model HAL) £51

4.88 hp Standard Two-port OHV Model JA and JAL

Type: OHV Twin-port
Capacity: 488cc
Bore & stroke: 85.5 x 85mm
Carburettor: Amal Two-lever
Gearbox: Sturmey-Archer 3-speed
Gear ratios: Standard: 4.8, 6.4 & 11.5 to 1
Sidecar: 5.5, 7.3 & 13.2 to 1
Brakes: 5in front, 6½in rear drums
Tyres: Dunlop Cord 26 x 3.25 in

Dimensions
Length: 7ft 2in
Width: 2ft 8in

Capacities
Fuel: 2¼ gallons
Oil: 2½ pints

Finish: Black enamel, tank finished in two shades of green. Bright parts chrome plate
Prices: Without lights (Model JA) £49 17s 6d, With Lucas Magdyno lighting (Model JAL) £55 7s 6d

The rear carrier was an optional extra costing 10s on the OHV model, but fitted as standard on the side-valve model. The bright parts were chromium plated, but petrol tanks were finished in two shades of green cellulose.

9.76hp Side-valve Twin Model K

The new frame fitted to the 9.76hp model had the effect of reducing the saddle height by 1½in to a more acceptable 27½in. The new front fork was fitted with shock absorbers and the front brake reduced in diameter to 7in. Foot operation of both front and rear brakes was retained. The magneto drive chain was enclosed in aluminium chain cases and the oil flow made adjustable. The saddle tank was redesigned to be in keeping with the rest of the model range and painted in two shades of Enfield green. The machine was fitted with the new detachable combined rear mudguard and carrier. The exhaust system was unusual in that a single common fishtail was employed for both silencers

9.76 hp Side-valve Twin Models K and KL

Type: Side-valve V-twin
Capacity: 976cc Bore & stroke: 85.5 x 85 mm
Carburettor: Amal Two-lever
Gearbox: Sturmey-Archer 3-speed
Gear ratios: Standard: 4.2, 6.1 & 11.2 to 1 Sidecar: 4.8, 7 & 12.7 to 1
Brakes: 7in front, 8in rear drums
Tyres: Dunlop Cord 26 x 3.25 in
Length: 7ft 3½in Width: 2ft 9in Fuel: 2 gallons Oil: 2½ pints
Weight: 326 lbs without lights, 347 lbs with lights
Finish: Black enamel, tank finished in two shades of green
Prices: Without lights (Model K) £62 10s
With Lucas Magdyno lighting set (Model KL) £68

Sidecars

The 1929 range of sidecar models was continued for the 1930 model year, although with a number of minor changes and upward price adjustment. The trade delivery bodies had previously only been offered as complete outfits with the 9.76 hp Side-valve twin as models 185, 165, 175 and 155, but could now be purchased separately as models 85, 65, 75 and 55.

The Sports Models 8 and 4 were described as of improved shape and to be slightly wider than previously. The body of the Trade Delivery sidecar, Model 65, was modified during the year and the open framework was replaced by solid panels up to two-thirds height; the capacity remained unaltered.

The factory photograph illustrates the revised bodywork, which was introduced during the year. Generally more practical than the previous design, the quoted capacity remained three milk churns, plus cans, tools, spares etc. A narrower version, Model 75, built on the standard chassis and with the same dimensions as the Model 85 box carrier, was able to carry two churns.

The Death of Lord Seymour

The death occurred on 23 January 1930, at age 79 years, of Lord Ernest James Seymour at his home in Kenilworth. Lord Seymour had been chairman of the company since its inception in 1898 and had proved to be a sound guiding hand through periods of depression as well as through periods of prosperity. His replacement, appointed at the end of March, was George Cartland who had also been a board member since the founding of the company. R.W.Smith and his son Major Frank Walker Smith continued as joint managing directors. R.W.Smith was by this time 75 years of age and his son was increasingly responsible for the day to day running of the company The death of Lord Seymour resulted in a vacancy on the board and this was filled by Major Smith's younger brother, G.H.Smith, who also became the company secretary.

1930 Sidecar Range	
Suitable motorcycle Models D, E, F, G, H & J	
Model 14 Sports	£13
Model 12 Touring	£13
Model 8 Sports	£17 5s.
Model 10 Touring	£17 10s.
Model 6 Delivery Box	£17
Suitable for 9.76 hp Side-valve Model K	
Model 2 Touring	£19 10s
Model 4 Sports	£18 10s
Model 22 Occasional Two-seater	£24
Model 85 Delivery Box	£17 10s
Model 65 Dairyman's Box	£20
Model 75 Dairyman's Narrow Box	£17 10s.
Model 55 GPO Type	£26.

Model 14 Sports
Weight: 100 lbs
Length (of combination): 7ft 1in
Width (of combination): 4ft 11in

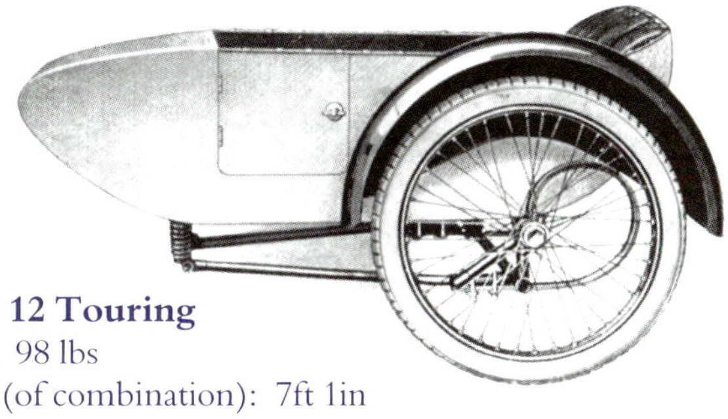

Model 12 Touring
Weight: 98 lbs
Length (of combination): 7ft 1in
Width (of combination): 4ft 11in

Model 8 Sports
Weight: 118 lbs
Length (of combination): 7ft 1in
Width (of combination): 5ft

Model 2 Touring
Weight: 154 lbs
Length (of combination): 7ft 3½in
Width (of combination): 5ft 3in

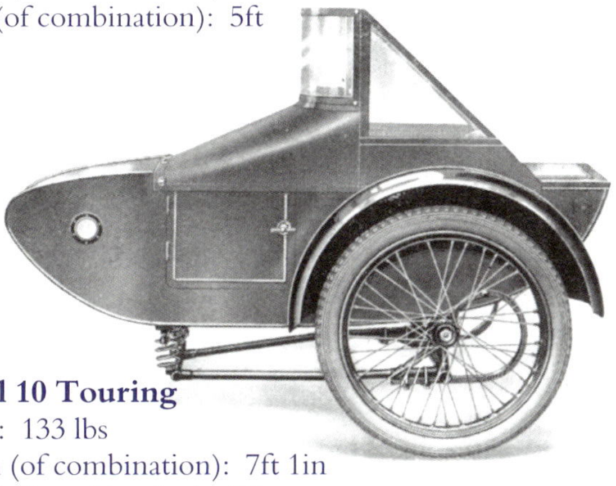

Model 10 Touring
Weight: 133 lbs
Length (of combination): 7ft 1in
Width (of combination): 5ft

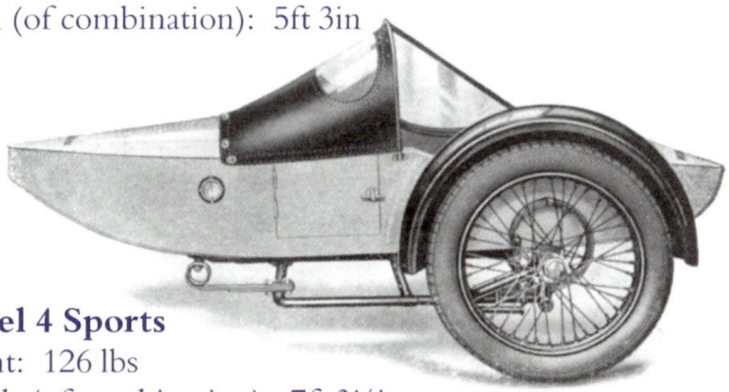

Model 4 Sports
Weight: 126 lbs
Length (of combination): 7ft 3½in
Width (of combination): 5ft 3in

Model 6 Delivery Box
Inside dimensions of box.
Length : 4ft 1in Width: 1ft 6in
Depth: 1ft 3in

Model 22 Occasional Two-seater Touring
Weight: 199 lbs
Length (of combination): 7ft 7in
Width (of combination): 5ft 3in

Model 85 Delivery Box
Inside dimensions of box:
Length: 4ft 5in Width: 1ft 7in
Depth: 1ft 8in
Length (of combination): 7ft 2in
Width (of combination): 5ft 3in

Model 65 Dairyman's Box
Inside dimensions of box:
Length: 4ft 6in Width: 2ft 7in
Depth: 1ft 6in
Length (of combination): 7ft 3½in
Width (of combination): 6ft

Model 55 GPO Type
Dimensions of box:
Length: 4ft 9in Width: 2ft 9½in
Depth: 3ft 3in Capacity: 31cu ft

The Model 22 Occasional Two-seater sidecar was designed to accommodate two adults in tandem. When not in use the rear compartment could be closed giving the appearance of a single seater. The sidecar was supplied with a single windscreen and an additional one was available for £1 10s.

Two interesting works competition machines. The absence of a front brake suggests both machines were intended for speedway use, The machine on the right has a knee hook confirming its intended use. Both the front forks and chain stays on this bike are of unusual design.

CHAPTER TWENTY ONE

Post 1930

The history of Royal Enfield post-1930 is outside the date limit set for this publication. The limit was not chosen for arbitrary reasons. Firstly, the end of of 1930 brings to a close the vintage period of motoring as recognised by both the Vintage Sports Car Club in respect of motorcars and the Vintage Motor Cycle Club in respect of motorcycles. Whilst I have an interest in the history of all early machinery, I have a particular interest in the cars and motorcycles of the Vintage and earlier periods. Secondly this book was written to complement the work by Gordon G. May, titled Royal Enfield - By Miles the Best, which provides a year by year model guide of all models from 1930 to 1970.

I have chosen to provide brief details of the 1931 model range as the 1931 models were announced in the motorcycling press in September 1930. They would have been in production at the time of the announcement and available for purchase by the public by the time of the Olympia Motor Cycle Show held in November; they thus scrape within the selected cut-off date.

I have also chosen to provide an overview of the subsequent history of the marque.

1931 Model Range

Model	C31	3.46 hp	Lightweight Side-valve	346cc	£38 10s
Model	CO31	3.46 hp	Light Two-port OHV	346cc	£42 10s
Model	F31	3.46 hp	De Luxe Side-valve	346cc	£42 10s
Model	G31	3.46 hp	De Luxe Two-port OHV	346cc	£45
Model	HA31	4.88 hp	Standard Side-valve	488cc	£45
Model	JA31	4.88 hp	Standard Two-port OHV	488cc	£48 10s
Model	H31	5.70 hp	De Luxe Side-valve	570cc	£48 10s
Model	J31	4.99 hp	De Luxe Two-port OHV	499cc	£52 10s
Model	K31	9.76 hp	Side-valve Twin	976cc	£60

Maglita Electric Lighting Set for Models C and CO: £5 5s.

Enfields and the Motor Cycle Industry in 1930

The late Twenties had proved a difficult period for the British Motor Cycle Industry with periods of industrial unrest, which saw many manufacturers go to the wall. British motorcycle production had fallen from a peak figure of 160,000 in 1927 to 126,000 in 1930, would drop to 74,700 in 1931 and to as low as 52,200 in 1933 before the turning point of the Great Depression. The Enfield Cycle Company had weathered the initial downturn in the market better than many other manufacturers. During the post-war period the works had been expanded year on year to meet demand for the company's products and was amongst the most modern and best equipped in the country. The company had always offered well engineered designs and by using the latest production methods the machines could be sold at competitive prices. The company was in a better position than most at the start of the Thirties.

1931 Model Range

'Nine Royal Enfields' announced the The Motor Cycle and Cycle Trader at the beginning of September 1930 in its review of the Royal Enfield 1931 Programme. This was followed by the succinct summary: 'Chromium Plating on All Models Without Extra Charge. - New Spring Fork, Detachable Rear Guard and Flange Fitting Carburetter Standard Throughout Range. - A Model for Every Public Need.'

A model for every public need was a surprising claim as the range did not include a two-stroke model; the first time in over sixteen years. No doubt the two-stroke models, which had been introduced in 1914, were less profitable owing to their lower price, but the absence is surprising at a period of increasing austerity. The omission would be rectified in 1932 with the re-introduction of the 225cc Model A at £31 and in 1933 with the introduction of the radically new 148cc (56 x 60mm) Cycar Model, which

Model C31 - 3.46hp Lightweight Side-valve

Model CO31 - 3.46hp Light Two-port OHV

Model F31 - 3.46hp De Luxe Side-valve

Model G31 - 3.46hp De Luxe Two-port OHV

Model HA31 - 4.88hp Standard Side-valve

Model JA31 - 4.88hp Standard Two-port OHV

Model H31 - 5.70hp De Luxe Side-valve

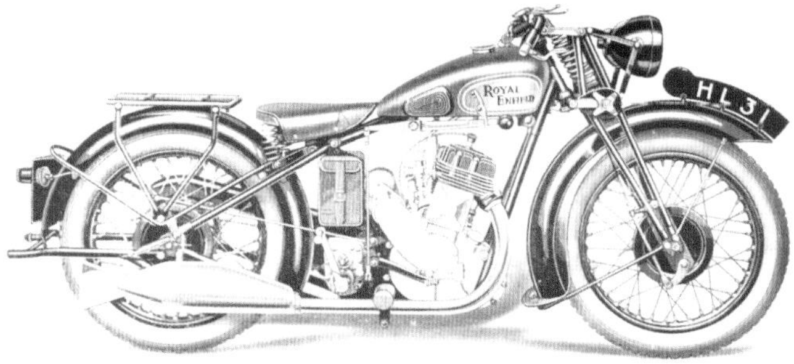

Type: Four-stroke Side-valve
Capacity: 570cc Bore & Stroke: 85.5 x 99.25 mm
Gear ratios: Standard: 5, 7.3 & 13.1 to 1
Sidecar: 5.8, 8.5 & 13.5 to 1
Brakes: 7 in drums front and rear
Length: 7ft 2in Width: 2ft 8in Saddle height: 27in
Petrol tank capacity: 2½ gallons
Finish: Black enamel and chrome. Tank in two shades of green cellulose
Price: £48 10s, With Lucas Magdyno lighting set £54
With Lucas Magdyno lighting set & instrument board £57

Model J31 - 4.99hp De Luxe Two-port OHV

Model K31 - 9.76hp Side-valve Twin

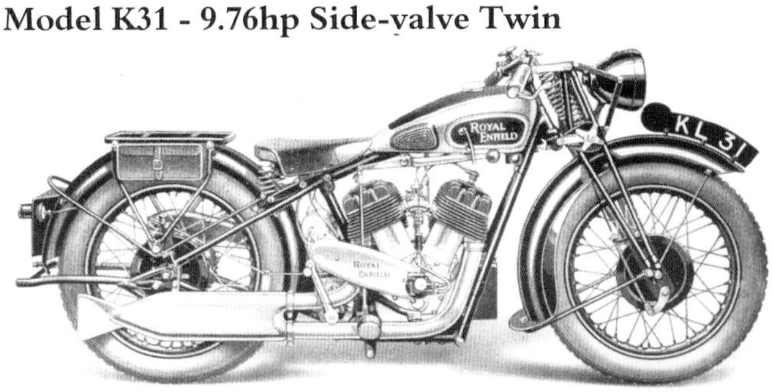

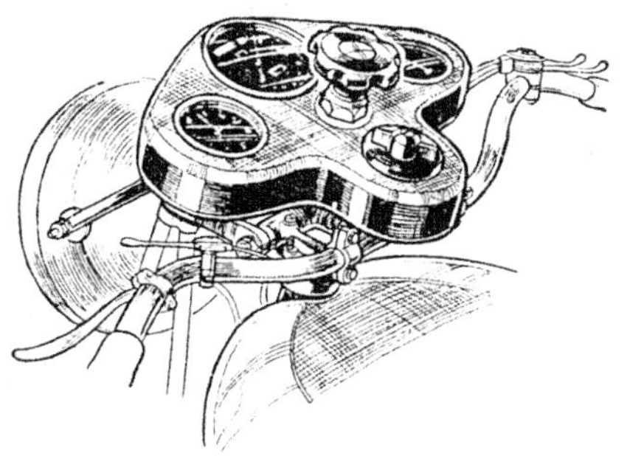

For an extra £3, the De Luxe models F, G and H and the the 9.76hp Model K Twin were available with the instruments set into a panel above the handlebar. The Smiths clock, if provided, cost an additional £1 10s

Model JF31 - 4.88hp Four-valve OHV

Type: Four-stroke Four-valve OHV
Capacity: 488cc Bore & Stroke: 85.5 x 85 mm
Gear ratios: Standard: 4.8, 6.4 & 11.5 to 1
Sidecar: 5.5, 8 & 14.6 to1
Brakes: 6½ in drums front and rear.
Length: 7ft 2in Width: 2ft 8in Saddle height: 27in
Petrol tank capacity: 2½ gallons
Finish: Black enamel and chrome. Tank in two shades of green cellulose
Price: £50 With Lucas Magdyno lighting set £55 10s

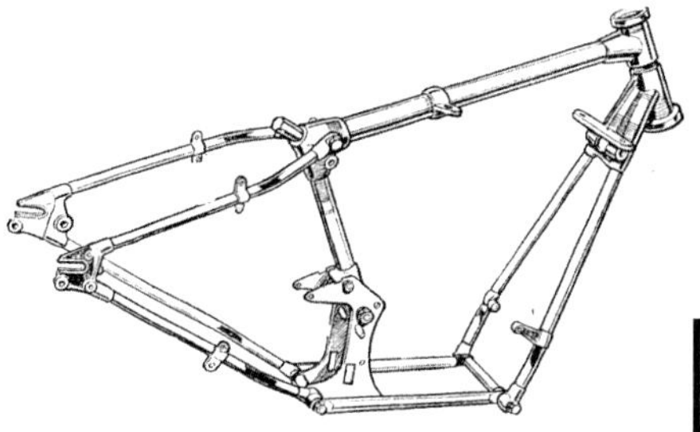

The new duplex cradle frame introduced for 1931 models.

PERFORMANCE FIGURES.

MODEL	Compression Ratio	Maximum Speed on Level		Petrol Consumption M.P.G.		Oil Consumption M.P.G.	
		Solo	Sidecar	Solo	Sidecar	Solo	Sidecar
A. 225c.c. 2-Stroke	6½	47-50	—	100	—	1500	—
C. 346c.c. S.V.	5	50-53	—	100	—	1500	—
C.S. 346c.c. O.H.V. Light	5½	60-63	—	100	—	1500	—
C.S.O.346c.c. O.H.V. Sports	5½	62-65	—	100	—	1500	—
G. 346c.c. O.H.V. De Luxe	5½	62-65	45-50	100	70	1500-2000	1200-1800
H. 570c.c. S.V. De Luxe	5	60-65	50-55	80	60	1500-2000	1200-1800
J. 499c.c. O.H.V. De Luxe	5¼	70-75	55-60	90	65	1500-2000	1200-1800
K. 976c.c. S.V. Twin	5	70-75	60-65	65	50	1000-1200	900-1000
L. 499c.c. S.V.	5	55-60	45-50	80	60	1500-2000	1200-1800
L.F. 488c.c. O.H.V. 4-Valve	6	75-80	60-65	90	65	1500-2000	1200-1800

NOTES: The above figures represent average performances of standard machines in good condition and after careful running-in. No guarantee is given that any particular machine will have the performance given above. Consumption figures assume moderate speeds on good roads.

In the case of machines supplied with a compression plate, the above compression ratios are with the plate removed. The ratio is one quarter to one half lower when the plate is in position.

High compression pistons giving a ratio of 6½—1, can be supplied for Models C.S.O., G. and J. These give an increase in maximum speed up to 5 m.p.h., and at the same time give improved acceleration. We do not, however, recommend their general use as they call for more use of the ignition and gear control, more frequent decarbonisation and possibly the use of anti-knock fuels.

Special Pistons giving higher compression ratios can be supplied for O.H.V. engines for racing purposes with special fuels, etc., but these are not suitable for use with ordinary petrol.

It was not normal practice for the Enfield Cycle Company to publish performance figures for the motor cycles produced by the company. However, for the 1932 season a pamphlet was made available quoting performance figures and power curves (see overleaf). The charts should also be applicable to the late Twenties models.

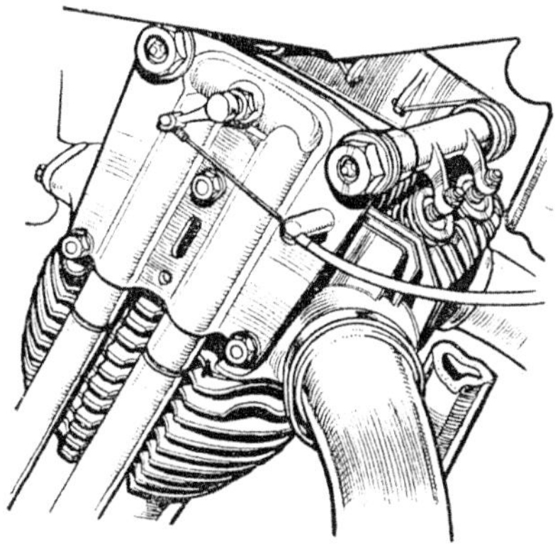

The cylinder head of the new 488 cc four-valve Model JF.

provided full weather protection and electric lighting for £23 17s 6d.

Detailed improvements were made to all models across the range. The de luxe 350cc Models F and G were redesigned with a modified engine crankcase to suit the new duplex cradle frame and employed dry-sump lubrication. The side-valve Model F also gained a

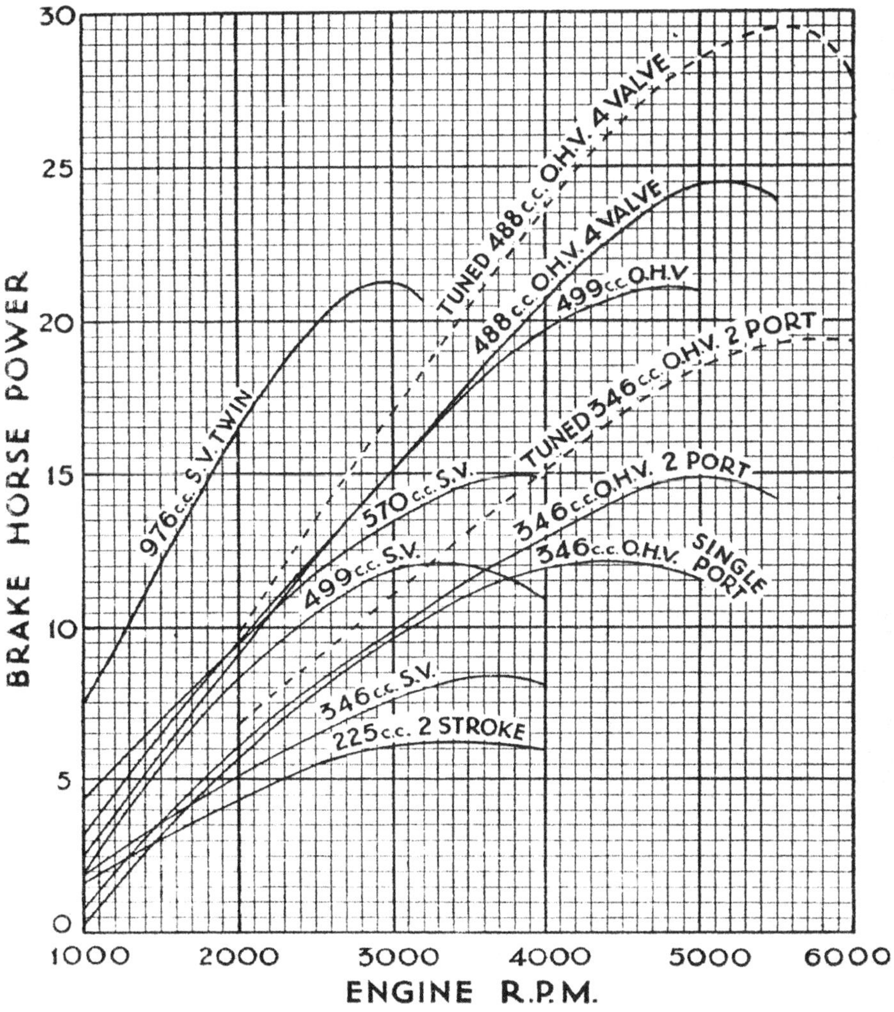

Power curves for Royal Enfield engines provided by the company in 1932.

1931 Sidecar Range	
Suitable Models F & G	
Model 14 Sports	£14 15s
Model 12 Touring	£14 15s
Suitable Models HA & JA	
Model 8 Sports	£19 10ss.
Model 10 Touring	£19 17s. 6d
Model 6 Delivery Box	£20 15s
Suitable for Model H, J & K	
Model 2 Touring	£22
Model 4 Sports	£21
Model 22 Two-seater	£27
Model K Combinations	
Model 185 Delivery Box	£76
Model 165 Dairyman's	£78 10s
Model 155 GPO Type	£84 10s

detachable cylinder head designed so that the valves lifted with the head to facilitate adjustment and maintenance. The Model J, which previously had 'square' dimensions of 85.5 x 85 mm (488cc), was now fitted with a new long-stroke engine with dimensions of 80 x 99.25 (499cc). Another entirely new addition to the range was the 5.70hp De Luxe Side-valve Model H31. The engine was of 570cc capacity (85.5 x 99.25 mm) and the frame was the same duplex design used for the 499cc ohv and the big twin models. The machine was designed as a dual-purpose mount with a big reserve of power for sidecar use. A last minute surprise at the Olympia Show in November 1930 was the introduction of the new Model JF, which boasted a four valve version of the 488cc overhead-valve engine fitted into the new duplex frame. It was priced at £50 with a three-speed gearbox or £52 with a four-speed box.

The prices of the 1931 models were reduced compared with the 1930 models. By comparison the range of sidecars was reduced in number and significant price increases imposed.

Birth of the Bullet

The most famous model name in Royal Enfield history came into existence in September 1932 with the announcement of the new single cylinder overhead-valve sports models, which were identified as Bullets. They were originally available in the 250, 350 and 500 cc classes, but it would not be long before a four-valve 500 model was added to the range. The machines were the forerunners of the modern machine which carries the Bullet name to this day and is built in India by Royal Enfield (India) at Chennai.

Death of Robert Walker Smith

The death of Robert Walker Smith, at the age of 77, occurred on February 6 1933. R.W.Smith had, jointly with Albert Eadie, founded the Enfield Cycle Co. It was Eadie who provided the commercial flair, but it was Smith who had provided the engineering guidance and innovation to ensure the commercial success of Royal Enfield bicycles and motorcycles. J.W.Davies had been appointed managing director in 1907, upon the departure of Eadie to BSA, and it was 1913 before

Robert Walker Smith

R.W.Smith assumed full responsibility for the company. He remained managing director until his death, whereupon he was succeeded by his eldest son, Frank Walker Smith, who took control as chairman and managing director. A second son G.H.Smith was also a director and secretary of the company and his two other sons were employed within the business.

Robert Walker and his sons were keen motorcyclists. This allowed them to take a hands-on approach in the assessment of problems, the introduction of innovations and the development of new models. A consequence was that Enfield Cycle Company became a closely knit community and, unlike the bigger firms in the British motorcycle industry, always maintained good management/worker relations.

The Move to India

In 1954 the Indian Army placed an order for 800 350 Bullet models to be finished in matt sand colour and to be ready for immediate action. Working seven days a week, the contract took several months to complete. This was due in part to the extensive full-throttle testing required to weed out the engines at risk of seizure. It was the Indian Government policy at this time that the country should manufacture products, rather than import from overseas. Accordingly, in 1956 a joint venture company, Enfield of India, was established, owned 51% by the Madras Motor Company and 49% by the Enfield Cycle Company, for the supply of additional machines. At first the Bullet was shipped from Britain in knocked-down kit form for assembly in Chennai, India. Subsequently the chassis components were produced in India with the engine shipped from Enfields in England. Later the engines were supplied in component form for assembly in India. Starting in 1957, Enfield of India acquired the machinery necessary to build components in India. Initially chassis components were produced and the engines shipped from Enfields in England. By 1962 the whole manufacture and assembly was carried out by Madras Motors in India.

Death of Frank Walker Smith

Frank Walker Smith, the eldest of R.W.Smith's four sons, was born on 7 August 1888. He was educated at Quinton College and Birmingham University before serving an apprenticeship with R.W.Ward and BSA. He joined the Enfield Cycle Company in 1909 and was appointed joint managing director with his father in 1914. He had served in the Royal Artillery Territorials prior to 1914 and in 1915 became a pilot in the Royal Flying Corps. Major Smith rejoined the Enfield Cycle Company in 1921 and was to remain joint managing director with R.W.Smith, until 1933 when he became sole managing director upon the death of his father. He became chairman in 1935, a post he held until his death on April 19 1962.

Demise of the UK Company

Subsequent to the death of Frank Walker Smith, on 1 August 1962 the Enfield Cycle Company was bought by the investment firm E & H P Smith, who at that time also owned Alpha bearings. In June 1966 the company sold the site in Redditch to the Redditch Development Corporation, Holmwood, but rented the site back to continue production. In March 1977, Enfield Industrial Engines Ltd became an independent business.

The whole of the share capital of Enfield Cycle Company Ltd was purchased by Norton Villiers Ltd on 28 March 1967 and the registered office moved to Marston Road, Wolverhampton. The Enfield Cycle Company Ltd vacated the works in Redditch during September-October 1967.

In 1967, the motorcycle side of the business, including 33 per cent of Enfield India, was sold for £82,500 to Norton Villiers; Enfield's diesel engine division and the pedal cycle and spares divisions were not included in the sale. Bicycle manufacture continued at the Royal Enfield Redditch factory until its closure in 1967. Production of motorcycles ended in 1970 and the original Redditch based company was dissolved in 1971.

Major Frank Walker Smith, C.B.E., J.P., T.D., M.I.Mech.E.

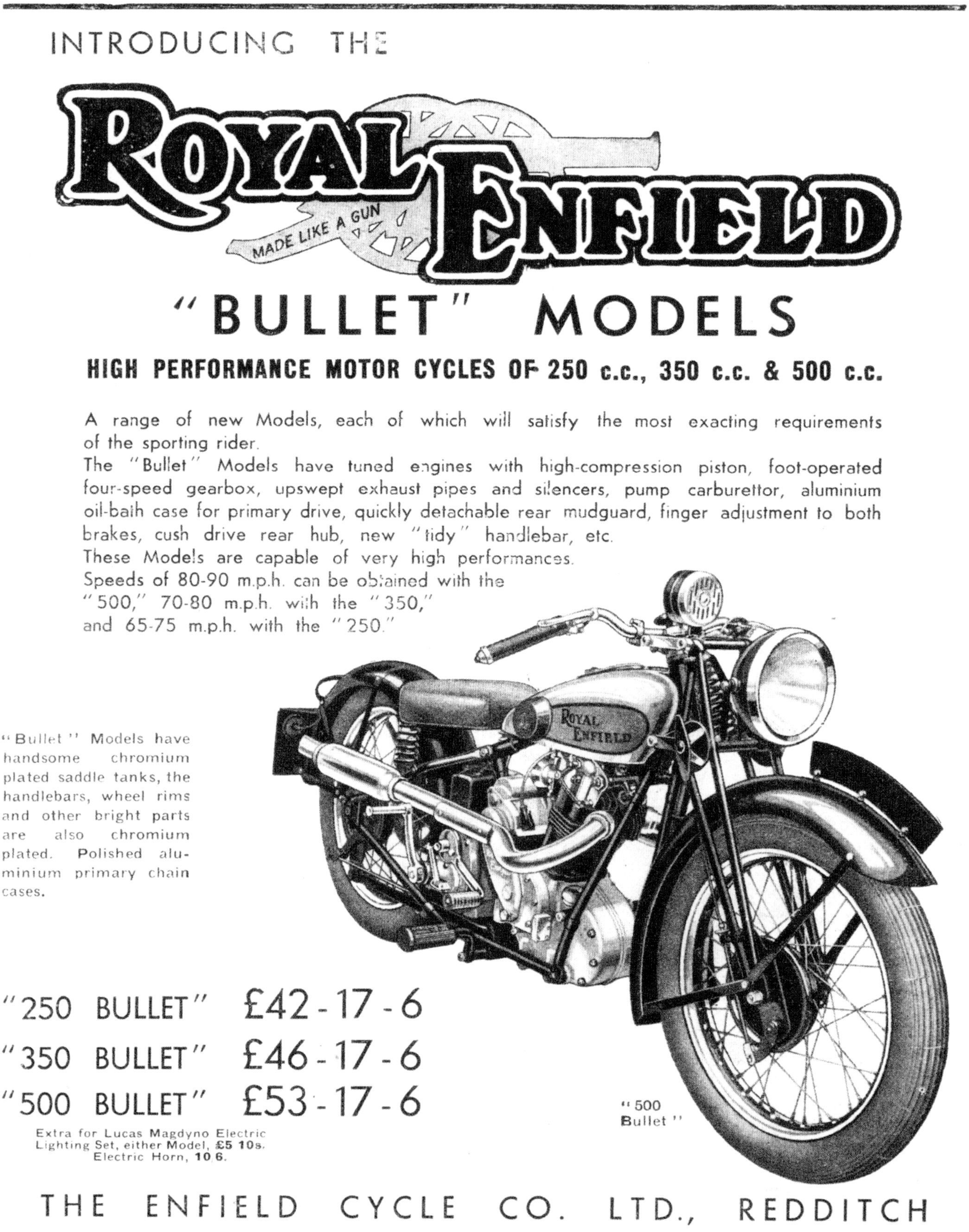

The now famous 'Bullet' name was first used for a new range of sporting models announced in September 1932.

Motorcycle Production Models 1902 - 1930

Year Model	Model Number	Engine Type	Bore mm	Stroke mm	Capacity cc	Speeds	Weight lbs	Price
1902								
1½ hp Minerva		Single ioe						£50
1903								
2 hp Belt drive		Single ioe	65	72	239	1		£47 5s
2¼ hp Chain drive		Single ioe wc	70	72	277	1		£78 15s
1½ hp Minerva		Single ioe						
1904								
2¾ hp Belt drive		Single	65	72	239			£42
3½ hp Belt drive		Single	65	72	239			£45 5s
2½ hp Chain drive		Single						£42
1905								
2¼ hp Belt Drive		Single sv	70	72	277	1	90	
2¼ hp Chain drive		Single sv	70	72	277	2	111	
1910								
2¼ hp Lightweight	140	Twin sv	52	70	297	1	100	£39 18s
2¼ hp Lightweight	145	Twin sv	52	70	297	1	100	£42
1911								
2¼ hp Lightweight	140	Twin sv	52	70	297	1	100	£39 18s
2¼ hp Lightweight	145	Twin sv	52	70	297	1	100	£42
2¾ hp Lightweight	150	Twin sv	54	75	344	1	120	£43 1s
2¾ hp Lightweight	155	Twin sv	54	75	344	1	120	£45 3s
2¾ hp Lightweight	160	Twin sv	54	75	344	2	125	£52 10s
1912								
2¾ hp	150	Twin sv	54	75	344	1	120	£43 1s
2¾ hp	160	Twin sv	54	75	344	2	130	£52 10s
2½ hp Open Frame	170	Single sv	64	75	241	2	120	£52 !0s
6 HP Sidecar	180	Twin sv	76	85	771	2		£84
1913								
3 hp	140	Twin ioe	60	75	424	2		£52 10s
2¾ hp	160	Twin sv	54	75	344	2	130	£50
6 hp Sidecar	180	Twin sv	76	85	771	2		£84
1914								
3 hp	140	Twin ioe	60	75	424	2		£52 10s
3 hp TT Roadster	150	Twin ioe	60	75	424	2		£52 10s
2¾ hp	160	Twin sv	54	75	344	2		£45
6 hp Sidecar	180	Twin sv	76	85	771	2		£84

Year Model	Model Number	Engine Type	Bore mm	Stroke mm	Capacity cc	Speeds	Weight lbs	Price
1915								
2¼ hp	200	Single ts	64	70	225	2		£39 18s
3 hp	140	Twin ioe	60	75	424	2		£52 10s
3 hp TT Roadster	150	Twin ioe	60	75	424	2		£52 10s
6 hp Sidecar	180	Twin sv	76	85	771	2		£84
8 hp Sidecar	180	Twin sv	85.7	85	980	2		£86 2s
6 hp Combination De Luxe	190	Twin sv	76	85	771	2		£100
8 hp Combination De Luxe	190	Twin sv	85.7	85	980	2		£102 2s
6 hp Trade Combination	180	Twin sv	76	85	771	2		£84
1916								
2¼ hp	200	Single ts	64	70	225	2		£42
3 hp	140	Twin ioe	60	75	424	2		£52 10s
3 hp TT Roadster	150	Twin ioe	60	75	424	2		£52 10s
6 hp Sidecar	180	Twin sv	76	85	771	2		£89 5s
8 hp Sidecar	180	Twin sv	85.7	85	980	2		£91 7s
6 hp Combination De Luxe	190	Twin sv	76	85	771	2		£105
8 hp Combination De Luxe	190	Twin sv	85.7	85	980	2		£107 7s
6 hp Trade Combination	180	Twin sv	76	85	771	2		£84
1917								
2¼ hp	200	Single ts	64	70	225	2		
3 hp	140	Twin ioe	60	75	424	2		
6 hp Sidecar	180	Twin sv	76	85	771	2		
8 hp Sidecar	180	Twin sv	85.7	85	980	2		
6 hp Trade Combination	180	Twin sv	76	85	771	2		
1918								
2¼ hp	200	Single ts	64	70	225	2		
3 hp	140	Twin ioe	60	75	434	2		
6 hp Combination	180	Twin sv	76	85	771	2		
8 hp Combination	180	Twin sv	85.7	85	980	2		
6 hp Trade Combination	180	Twin sv	76	85	771	2		
1919								
2¼ hp	200	Single ts	64	70	225	2		£52 10s
3 hp	140	Twin ioe	60	75	424	2		£69 6s
3 hp TT Roadster	150	Twin ioe	60	75	424	2		£69 6s
6 hp Sidecar	180	Twin sv	76	85	771	2		£115
8 hp Sidecar	180	Twin sv	85.7	85	980	2		£117 10s
6 hp Combination De Luxe	190	Twin sv	76	85	771	2		£147
8 hp Combination De Luxe	190	Twin sv	85.7	85	980	2		£149 2s
6 hp Trade Combination	180	Twin sv	76	85	771	2		£126
1920								
2¼ hp	200	Single ts	64	70	225	2		see text
6 hp Sidecar	180	Twin sv	76	85	771	2		see text
8 hp Sidecar	180	Twin sv			980	2		see text
6 hp Combination De Luxe	190	Twin sv	76	85	771	2		see text
8 hp Combination De Luxe	190	Twin sv			980	2		see text
6 hp Trade Combination	180	Twin sv	76	85	771	2		see text
1921								
2¼ hp	200	Single ts	64	70	225	2		£65
8 hp Standard Combination	180	Twin sv	85.5	85	976	2	462	£160
8 hp Lighting Set. Comb.	190	Twin sv	85.5	85	976	2	515	£182
8 hp Trade Combination	180	Twin sv	85.5	85	976	2	427	

Year Model	Model Number	Engine Type	Bore mm	Stroke mm	Capacity cc	Speeds	Weight lbs	Price
1922								
2¼ hp	200	Single ts	64	70	225	2		£55
8 hp Standard Combination	180	Twin sv	85.5	85	976	2	462	£140
8 hp Lighting Set. Comb.	190	Twin sv	85.5	85	976	2	515	£158
8 hp Trade Combination	180	Twin sv	85.5	85	976	2	427	
1923								
2¼ hp	200	Single ts	64	70	225	2		£50
8 hp Standard Combination	180	Twin sv	85.5	85	976	2	462	£120
8 hp Lighting Set. Comb.	190	Twin sv	85.5	85	976	2	515	£135
8 hp Trade Combination	180	Twin sv	85.5	85	976	2	427	
8hp Special Trade Delivery	155	Twin sv	85.5	85	976	2		£140
1924								
2¼ hp	201	Single ts	64	70	225	2	155	£42
2¼ hp Open Frame	201A	Single ts	64	70	225	2	155	£42
2¼ hp Sports	200	Single ts	64	70	225	2	155	£42
2¼ hp	202	Single ts	64	70	225	2	155	£35
2¾ hp OHV Sports	351	Single ohv	70	90	346	2/3	206	£57
2¾ hp OHV Standard	350	Single sv	70	90	346	2/3	189	£49
2¾ hp OHV Super Sports	352	Single ohv	74	80	344	2		
8 hp Sidecar	180	Twin sv	85.5	85	976	2	476	£90
8 hp Sidecar (Lighting)	190	Twin sv	85.5	85	976	2		£110
8 hp Trade Combination	185	Twin sv	85.5	85	976	2		£90
8 hp Trade GPO Type	155	Twin sv	85.5	85	976	2		£120
1925								
2¼ hp	201	Single ts	64	70	225	2	160	£35
2¼ hp Open Frame	201A	Single ts	64	70	225	2	160	£36
2¼ hp Sports	200	Single ts	64	70	225	2	155	£35
2¼ hp	202	Single ts	64	70	225	2	155	£32 10s
2¾ hp OHV Sports	351	Single ohv	70	90	346	3		£55
2¾ hp OHV Standard	350	Single sv	70	90	346	3		£47
2¾ hp OHV Super Sports	352	Single ohv	74	80	344	3		£75
8 hp Sports Solo	182	Twin sv	85.5	85	976	3		£75
8 hp Sports Combination	180	Twin sv	85.5	85	976	3	476	£95
8 hp Sidecar (Lighting)	190	Twin sv	85.5	85	976	2		£105
8 hp Trade Combination	185	Twin sv	85.5	85	976	2		£90
8 hp Trade GPO Type	155	Twin sv	85.5	85	976	2	490	
1926								
2¼ hp	201	Single ts	64	70	225	2	160	£33 10s
2¼ hp Open Frame	201A	Single ts	64	70	225	2	160	£35
2¼ hp Sports	200	Single ts	64	70	225	2	155	£33 10s
3.46 hp OHV Sports	351	Single ohv	70	90	346	3	224	£50
3.46 hp OHV Sports Comb.	351	Single ohv	70	90	346	3	327	£60
3.46 hp OHV Touring Comb.	351	Single ohv	70	90	346	3	422	£62
3.46 hp Standard	350	Single sv	70	90	346	3	210	£45
3.46 hp Sports Combination	350	Single sv	70	90	346	3	313	£55
3.46 hp Touring Comb.	350	Single sv	70	90	346	3	418	£57
3.44 hp OHV Super Sports	352	Single ohv	74	80	344	3	224	£75
9.76 hp Sports Solo	182	Twin sv	85.5	85	976	3	308	£70
9.76 hp Sports Combination	182	Twin sv	85.5	85	976	3	434	£84
9.76 hp Combination	180	Twin sv	85.5	85	976	3	462	£84
9.76 hp De Luxe Comb.	190	Twin sv	85.5	85	976	3	490	£95
9.76 hp De-Luxe 2-seater Comb.	190	Twin sv	85.5	85	976	3		£99
9.76 hp Trade Combination	185/175	Twin sv	85.5	85	976	3		£84
9.76 hp Dairyman's Comb.	165	Twin sv	85.5	85	976	3	490	£90
9.76 hp Trade Combination	155	Twin sv	85.5	85	976	3		£109

Year Model	Model Number	Engine Type	Bore mm	Stroke mm	Capacity cc	Speeds	Weight lbs	Price
1927								
2¼ hp Standard	201	Single ts	64	70	225	2	160	£32 10s
2¼ hp Open Frame	201A	Single ts	64	70	225	2	160	£34
2¼ hp Sports	200	Single ts	64	70	225	2	155	£32 10s
3.46 hp Standard	350	Single sv	70	90	346	3	210	£45
3.46 hp OHV Sports	351	Single ohv	70	90	346	3	224	£50
3.44 hp OHV Super Sports	352	Single ohv	74	80	344	3	242	£75
4.88 hp Standard	500	Single sv	85.5	85	488	4	245	£52 10s
9.76 hp Sports	182	Twin sv	85.5	85	976	3	308	£68 10s
9.76 hp Sports Combination	182	Twin sv	85.5	85	976	3	434	£84
9.76 hp Standard	180	Twin sv	85.5	85	976	3	308	£66 10s
9.76 hp Combination	180	Twin sv	85.5	85	976	3	462	£84
9.76 hp De Luxe Comb.	190	Twin sv	85.5	85	976	3	490	£95
9.76 hp Trade Combination	185	Twin sv	85.5	85	976	3		£84
9.76 hp Dairyman's Comb.	165	Twin sv	85.5	85	976	3	490	£85
9.76 hp Trade Del. Comb.	155	Twin sv	85.5	85	976	3		£95
1928								
2¼ hp Standard	201	Single ts	64	70	225	2	160	£32 10s
2¼ hp Open Frame	201A	Single ts	64	70	225	2	160	£32 10s
2¼ hp Sports	200	Single ts	64	70	225	2	160	£32 10s
2¼ hp Standard 3-speed	202	Single ts	64	70	225	3	160	£35
2¼ hp Sports 3-speed	203	Single ts	64	70	225	3	160	£35
3.46 hp Standard	350	Single sv	70	90	346	3	230	£44 10s
3.46 hp Standard OHV	351	Single ohv	70	90	346	3	242	£50
3.46 hp OHV Super Sports	352	Single ohv	74	80	344	3	252	£78 15s
3.46 hp Sports OHV	355	Single ohv	70	90	346	3	244	£52 10s
4.88 hp Standard	500	Single sv	85.5	85	488	4	245	£48 5s
9.76 hp Standard	180	Twin sv	85.5	85	976	3	308	£67 10s
9.76 hp Sports	182	Twin sv	85.5	85	976	4	336	£73 10s
9.76 hp De Luxe Comb.	190	Twin sv	85.5	85	976	3		£93 15s
9.76 hp Trade Combination	185	Twin sv	85.5	85	976	3		£86 5s
9.76 hp Dairyman's Comb.	165	Twin sv	85.5	85	976	3		£88 10s
9.76 hp Trade Del. Comb.	155	Twin sv	85.5	85	976	3		£94 10s
1929								
2¼ hp Standard	201	Single ts	64	70	225	2	160	£28 17s 6d
2¼ hp Open Frame	201A	Single ts	64	70	225	2	164	£30
2.25 hp Standard	202	Single ts	64	70	225	3	160	£32 10s
2.25 hp Sports	203	Single ts	64	70	225	3	160	£37 15s
2.25 hp Side-valve	250	Single sv	64	70	225	3	190	£36
3.46 hp Standard	350	Single sv	70	90	346	3	230	£43
3.46 hp Standard OHV	351	Single ohv	70	90	346	3	238	£45 10s
3.46 hp Sports 2-port	355	Single ohv	70	90	346	3	304	£50
3.46 hp Super-Sports 2-port	352	Single ohv	70	90	346	3		£75
4.88 hp Standard	501	Single sv	85.5	85	488	3	280	£45 10s
4.88 hp De-Luxe	502	Single sv	85.5	85	488	3	308	£53 15s
4.88 hp Two-port	505	Single ohv	85.5	85	488	3	308	£52 10s
9.76 hp	182	Twin sv	85.5	85	976	3	345	£62 10s
9.76 hp Trade Combination	185	Twin sv	85.5	85	976	3		£77 10s
9.76 hp Dairyman's Comb.	165	Twin sv	85.5	85	976	3		£79 10s
9.76 hp Dairyman's Comb.	175	Twin sv	85.5	85	976	3		£77 10s
9.76 hp Trade Del. Comb.	155	Twin sv	85.5	85	976	3		£85
1930								
2.25 hp Two-stroke	A	Single ts	64	70	225	3	180	£32 10s
2.25 hp Two-stroke	AL	Single ts	64	70	225	3	197	£37 15s
2.25 hp Four-stroke	B	Single sv	64	70	225	3	188	£36
2.25 hp Four Stroke	BL	Single sv	64	70	225	3	208	£41 5s

Year Model	Model Number	Engine Type	Bore mm	Stroke mm	Capacity cc	Speeds	Weight lbs	Price
1930 Continued								
3.46 hp Lightweight	C	Single sv	70	90	346	3	203	£39 15s
3.46 hp Lightweight	CL	Single sv	70	90	346	3	220	£45
3.46 hp Lightweight OHV	CO	Single ohv	70	90	346	3	230	£43 5s
3.46 hp Lightweight OHV	COL	Single ohv	70	90	346	3	249	£48 15s
4.88 hp Standard	D	Single sv	85.5	85	488	3	272	£45 10s
4.88 hp Standard	DL	Single sv	85.5	85	488	3	295	£51
4.88 hp Standard Two-port	E	Single ohv	85.5	85	488	3	309	£49
4.88 hp Standard Two-port	EL	Single ohv	85.5	85	488	3	332	£49
3.46 hp De Luxe	F	Single sv	70	90	346	3	309	£45
3.46 hp De Luxe	FL	Single sv	70	90	346	3	332	£50 10s
3.46 hp De Luxe Two-port	G	Single ohv	70	90	346	3	283	£51 10s
3.46 hp De Luxe Two-port	GL	Single ohv	70	90	346	3	306	£57
4.88 hp De Luxe	H	Single sv	85.5	85	488	3	276	£48 10s
4.88 hp De Luxe	HL	Single sv	85.5	85	488	3	299	£54
4.88 hp De Luxe	HA	Single sv	85.5	85	488	3		£45 10s
4.88 hp De Luxe	HAL	Single sv	85.5	85	488	3		£51
4.88 hp De Luxe Two-port	J	Single ohv	85.5	85	488	3	313	£55
4.88 hp De Luxe Two-port	JL	Single ohv	85.5	85	488	3	336	£66 10s
4.88 hp De Luxe Two-port	JA	Single ohv	85.5	85	488	3		£49 17s 6d
4.88 hp De Luxe Two-port	JAL	Single ohv	85.5	85	488	3		£66 10s
9.76 hp Side-valve Twin	K	Twin SV	85.5	85	976	3	326	£62 10
9.76 hp Side-valve Twin	KL	Twin SV	85.5	85	976	3	347	£68

Notes

1. With the exception of the De Luxe models and the 1930 season models the prices quoted are for the basic model and any extras, such as lighting sets, will add to the price. For the 1930 season the models with electric lighting sets are identified by the additional model letter 'L'.

2. The price may change during the year and the model specifications may change from year to year

3. Prior to the 1927 season it was the practice to provide the combinations as specific models and quote prices for the motorcycle combination. From the 1927 season onwards it was usual to list the solo versions of all machines with suitable passenger sidecars for each model listed and priced separately. Details and prices of the various sidecars are given in the text of the book. It remained the practice to list the commercial sidecars as separate machines.

Chronology

1815 George Townsend, the son of an Alcester needle manufacturer, born.
1851 George Townsend builds his own needle making mill, named Givry Works.
1857 Robert Walker Smith born.
1863 Albert Eadie born.
1871 Givry Works had become a major employer with 170 workers.
1879 Death of George Townsend, aged 64.
The Givry Works taken over by his son George Jr. and George's half-brother Foster Townsend.
1881 Foster acquires one of the first 'boneshakers'.
1885 William Cook working in his shed at nearby Headless Cross invents a saddle constructed using a single length of wire for the frame and two springs. Townsend patents the saddle and markets it as the Townsend Saddle and Spring, The company enters the bicycle parts trade.
1888 Townsend manufactures his own bicycles, which are marketed under the Ecossais name.
The Givry Works by now extended along half of one side of Enfield Road with the stamp shop on the other side of Wharrage Brook.
1889 First competition success for a Townsend safety bike ridden by du Cros of Irish Cycling Club.
Fred Shelton joined company from school and proved an excellent cyclist.
1890 George Townsend & Co. incorporated as a Limited Liability Company.
1890 Financial problems result in employing financiers from Birmingham.
1891 By this time Townsend had a range of over two dozen different designs of cycles, which were exhibited at Crystal Palace.
1891 (November) Failure to agree with the financiers requirements results in the departure of the Townsend brothers and the financiers bringing in Albert Eadie and Mr R.W.Smith to manage the company.
Albert Eadie was sales manager for Perry & Co. Ltd. and Robert Walker Smith was assistant works manager at D Rudge & Co Ltd in Coventry.
1892 (April) Eadie and Smith threw in their comfortable well paid jobs to become managing director and works manager, respectively, of George Townsend & Co. Ltd.
1892 (13 August) George Townsend & Co. Ltd. renamed Eadie Manufacturing Co. Ltd.
1892 (Autumn) Contract to supply rifle components to the Royal Small Arms factory at Enfield, Middlesex.
1892 (October) The Enfield name first used for bicycles manufactured by the Eadie Manufacturing Co.Ltd. The name capitalised on the firearms association.
Bicycles marketed by the Enfield Manufacturing Co. Ltd., a subsidiary sales organisation.
1892 (17 December) Advertisement showing R.W.Smith's front driving gear for Enfield cycles.
1893 (24 February) Enfield Manufacturing Co. Ltd. registered to handle bicycle sales.
Headquarters at 166 Edmund Street, Birmingham.
1893 (February) Company trademark of field gun within shield first used in advertising.
1893 (June) Company trademark of field gun within shield registered.
1893 Slogan 'made like a gun' first used with image of a field gun.
1893 (November) Enfield cycles of superior quality appearing at the Stanley Show were the first to be given the Royal Enfield name.
1893 (11 November) The registered office of Enfield Manufacturing Co. Ltd. changed to 94 Snow Hill, Birmingham.
1895 (10 January) Registered office moved to 2 William Street, Redditch.
1895 (September) The first building of a new factory complex in Lodge Road, Redditch completed. The building occupied by the newly formed Albert Eadie Chain Co. Ltd. The Eadie Manufacturing Co. Ltd. moved in the following year.
1896 (1 July) The New Enfield Cycle Co. Ltd. registered. The company combined the Enfield Manufacturing Co. Ltd. and the cycle manufacturing activities of the Eadie Manufacturing Co. Ltd
1896 (31 August) The New Enfield Cycle Co. Ltd. formally takes over the Enfield Manufacturing Co. Ltd. and the cycle manufacturing activities of the Eadie Manufacturing Co. Ltd.
1897 Eadie sets up his own factory in Lodge Road to supply cycle components to the overseas market, in order to reduce the load on the New Enfield Cycle Co. Ltd. The company was identified as the New Eadie Manufacturing Company in order to distinguish it

from the original company, but the 'New' prefix was seldom used.

1897 (November) The New Enfield Cycle Co. Ltd. name shortened to the Enfield Cycle Co. Ltd.
With the move to other premises of the Eadie Manufacturing Co. Ltd., the whole of the Redditch works became the premises of the Enfield Cycle Co. Ltd.

1897 (8 January) The Enfield Manufacturing Co. Ltd formally wound up.

1897 First quadricycle built by R.W.Smith for his personal use.

1898 First powered vehicles produced and fitted with DeDion engines. Used for evaluation.

1898 All machines were now marketed under the Royal Enfield name and the motto 'Made like a Gun' used to summarise the quality of manufacture.

1899 Commercial manufacture of quadricycles and tricycles commenced.

1899 (September) First marketing of quadricycles and tricycles. Both 2¼hp quadricycles and 1¾hp tricycles available.

1899 The 'Flexible' spring frame design of Rucker and Philpott displayed on the Enfield stand at the Stanley Show.

1899 The description Modèle Riche first used to identify top of the range bicycles. The description would be subsequently abbreviated to Riche.

1900 (January) Walter Grew appointed manager of the motor department.

1900 (April) Edward Iliffe, accompanied by Walter Grew, driving his Royal Enfield Quadricycle, wins a Silver Medal in the Thousand Mile Trail.

1900 (7 August) R.W.Smith suffers an accident driving his 3½hp De Dion Bouton.

1900 (November) A three model range of tricycles and quadricycles displayed at the Stanley Show, including the Auto-Quad. The Enfield two-speed gear was introduced at the show.

1900 The Belgian engineer Jules Gobiet approaches R.W.Smith with his ideas to prevent motor cycle side-slip and is employed by the company.

1901 (January) Automobile Club de France exhibition in the Grand Palais, Paris. Ader, or S.I.T., 6hp car announced.

1901 (May) The French built Ader car become available on the British market. Some two or three months later three cars were imported by the Enfield Cycle Co Ltd for evaluation.

1901 (September) An Ader car, re-badged as a Royal Enfield entered in the Glasgow Reliability Trial.

1901 The Enfield Girder bicycle frame introduced at the Stanley Show.

1901 (November) First Royal Enfield motorcycle, based on Jules Gobiet's ideas to prevent side-slip, available with 1½hp Minerva engine mounted above front wheel.

1901 The 'Yellow Car' purchased for assessment.

1902 (November) New loop-frame designs of motorcycle available. Two models announced; a 2hp belt drive model and a 2¼hp water-cooled chain drive model.

1902 (July) Departure of Ernest Godbold, to be replaced by John Davis.

1903 Departure of Jules Gobiet and Walter Grew.

1903 Trevor Rapson joins the company and designs and puts into production a 10hp light car.

1903 The Fagan epicyclic two-speed hub manufactured by Eadie and introduced on Royal Enfield bicycles.

1903 (November) Enfield cars were exhibited for first time at the Stanley Show. Three cars and two motorcycles displayed.

1903 Manufacture of quadricycles and tricycles discontinued.

1904 (August) L.Hadley achieved first motorcycle race win for Royal Enfield.

1904 (September) 8hp Royal Enfield car entered in Reliability Trial organised by the Automobile Club of Great Britain and Ireland.

1904 (October) Appointment of W.M.Jenkins as works manager following departure of Trevor Rapson..

1904 (November) The new 20hp car was the first Royal Enfield to have a pressed steel frame.

1905 (September - November) New works manager W.H.Guillon designs new 15-20hp and 24-30hp cars which are exhibited at the Olympia Show in November 1905.

1906 The Enfield Cycle Co. Ltd. cease motorcycle production.

1906 (5 March) Enfield Autocar Company Ltd. Registered, to take over the portion of the Enfield Cycle Co. Ltd. concerned with motorcar manufacture.

1906 (August) Departure of W.H.Guillon to join the newly formed BSA motor car division. He was replaced by C.F.Pougher.
Guillon's departure appears to have been brief as he was subsequently involved in the design of Royal Enfield motorcycles.

1906 (October) A new 15hp car and a modified 24-30hp car, now identified as the 25hp were exhibited at the Olympia Show.

1906 (10 December) Registered office changed to Hewell Road, Redditch.

1907 Charles Yeomans Hopkins employed by Enfield Autocar Co to increase production and to determine whether the company was financially viable. He adopts a one-model policy based upon the 15hp model, now identified as the 18-22hp.

1907 The Duplex Girder bicycle frame introduced at the Stanley Show.

1907 (October) Albert Eadie announces his departure from Royal Enfield Co. Ltd. J.W.Davies appointed new general manager.
1907 The Eadie Company amalgamated with BSA. Albert Eadie joins the BSA board.
1907 (December) The Enfield Autocar Company placed in voluntary liquidation. Mr Hopkins responsible to the official receiver for the management of the company.
1908 (6 April) Auction of the Enfield Autocar Company's assets held at Grand Hotel, Birmingham; reserve price of £18,000 not achieved. The company subsequently purchased by Alldays & Onions. Production moved to Alldays Matchless Works in Birmingham.
1909 (November) New 2¼hp Model 140 Lightweight motorcycle revealed at Stanley Show. Designed by Monsieur Guillon and R.W.Smith, it was fitted with a 60 degree side-valve Motosacoche engine.
1910 The Lightweight 2¼hp Models 140 and 145 marketed and a new chain drive 2¾hp Lightweight model announced.
1910 Bicycle chain wheel design first featured the field gun trademark.
1910 (August) Eldest of R.W.Smith's four sons born. He would succeed his father as managing director.
1910 (November) Launch of 2¾hp Model 150 chain drive motorcycle with a slipping clutch; the first cush drive unit fitted to a motorcycle. Also launched at this time was the Model 160 with the Royal Enfield Patent Two-speed Gear and Free-engine.
1911 (June) Bert Colver's first race entry with a Royal Enfield at Brooklands.
1911 (November) 6hp JAP side-valve v-twin combination and an open frame model announced at the Olympia Show. The Royal Enfield Patent Cush Drive was also announced.
1912 Enfield Autorette announced.
1912 The Hunt End works, which had been sold with the Enfield Autocar Company, repurchased.
1912 (October) The Guillon designed 3hp ioe engine announced.
1912 (November) Introduction of the 3hp Model 140 at the Olympia Show.
1913 (May) Royal Enfield Cyclecar revealed.
1913 Enfield Autolette Light Car announced.
1913 Launch of 3hp Model with 425cc ioe v-twin engine. The first British motorcycle with dry sump lubrication and automatic geared pump. Patented design of R.W.Smith and W.Guillon.
1913 Trade delivery sidecar first announced.
1914 Enfield Autolette replaced by Alldays Midget and Enfield Nimble Nine.
1914 The aluminium chain case, providing full enclosure of the primary drive was introduced on the 6hp models.
1914 A sports version, the T.T. Roadster, of the 3hp ioe machine made available.
1914 (August) The first public appearance of the Enfield two-stroke.
1914 (November) Model 200, two-stroke, announced at the Olympia Motor Cycle Show.
1914 (December) First details published of Enfield sidecar machine gun.
1915 8hp Twin Model 180 version of the 6hp Twin made available.
1915 De Luxe versions of the 6hp and 8hp combinations available with either acetylene lighting or Lucas dynamo electric lighting set.
1915 (17 August) The Royal Enfield factories requisitioned for war work.
1916 (16 November) All non-essential vehicle production ended as a result of wartime restrictions.
1918 Plans announced for the Enfield-Allday Bullet.
1920 Launch of Vickers manufactured 8hp 976cc v-twin Wolseley engine designed by Royal Enfield.
1922 Special trade delivery combination produced under GPO contract.
1923 (July) New drop frame Ladies Model of 225cc two-stroke revealed.
1924 New models for the current season included a 225cc two-stroke drop frame model and a 350cc single with either a side-valve or overhead-valve JAP engine.
1924 (August) The announcement made of a new Enfield 8hp engine to replace the Wolseley engine manufactured for the Enfield company by Vickers.
1925 The 8hp twin, only previously available with side-car attached, could now be purchased as a sporting solo.
1925 Lightweight sidecars made available for smaller capacity machines.
1925 (April) Announcement of a new Enfield designed 350cc side-valve engine to replace the JAP engine.
1925 Alldays placed in hands of receiver and sold. Re-registered New Alldays & Onions Ltd on 8 July.
1925 (August) Tony Wilson-Jones joins Enfields to manage the racing team.
1925 The Enfield Cycle Company purchase the return of the Enfield name and trade mark from Enfield Allday Motors (1923).
1926 The popular 225cc two-stroke range updated with new engine and frame.
1926 Charlie Young wins both the 350cc and 600cc South African Tourist Trophy Races with his 344cc Royal Enfield.
1927 New overhead valve 350 machines available with Enfield 346cc engine or JAP 344cc super sports engine.
1927 New 500cc model available with Enfield side-valve engine.
1927 The 8hp Sports model was the first Enfield machine fitted with a saddle tank.

1927 The Enfield works riders win the Manufacturers Team Prize in the 1927 Isle of Man TT.
1928 A new 225cc lightweight four-stroke machine added to the range.
1929 The saddle tank fitted across the range with the exception of the 2.25hp Standard two-stroke and drop-frame model.
1930 A new lightweight 3.46hp side-valve model introduced with an inclined, or sloper, engine layout which had become fashionable. With the exception of the overhead-valve and side-valve 4.88hp models sloper layout was used across the range of four-stoke singles. Dry sump lubrication was introduced on these models.
1932 The death of Robert Walker Smith.
1933 Major Frank Smith takes over the management of the Enfield Cycle Company.

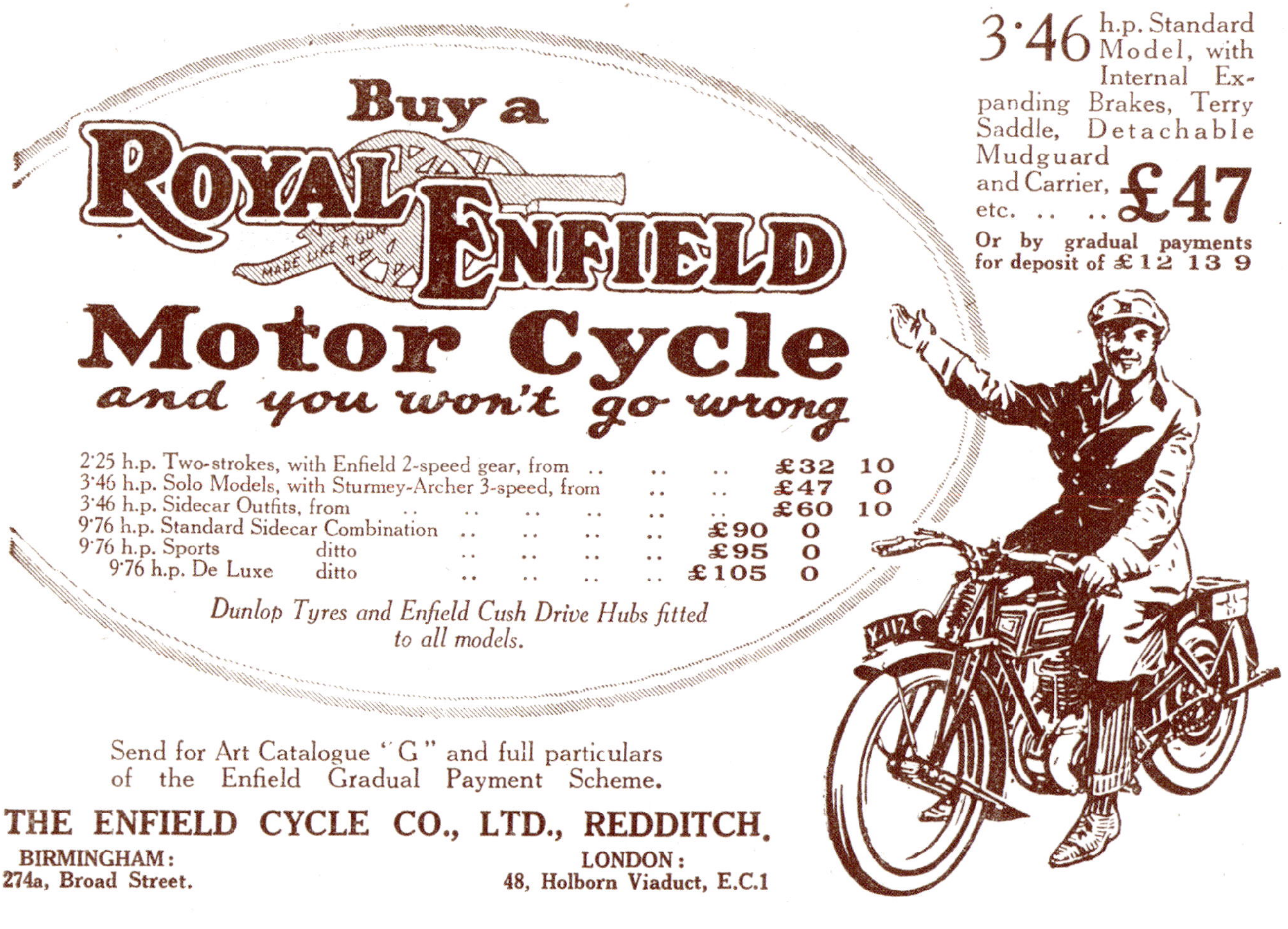

An advertisement from August 1925

Bibliography

AND

Acknowledgements

In compiling this book on the early history of Royal Enfield cycles, motorcars and motorcycles I have referred extensively for facts and figures to articles and reviews in the contemporary press. The magazines to which I have referred include Autocar, Motor Cycling, Motor Sport, The Motor Cycle and Motor Trader. I have made use of brochures and other material held by the National Motor Museum Library at Beaulieu, the Redditch Library, the Royal Enfield Owners' Club Archive, the Vintage Motor Cycle Club Library and the Vintage Sports Car Club Library. In addition to cycle and motorcycle sales material, the Redditch Library holds several of the Enfield Works photo-albums and scrapbooks, which were donated to the library by Ivor Mutton. Ivor had worked for a metal treatment company in Redditch, which had close links with the Enfield factory. He was the recognised expert on early Royal Enfield motorcycles and was presented with the photo-albums and scrapbooks by Barry Smith, the company secretary, when the factory closed. The Redditch library also holds unpublished works by William J Solloway on Early Redditch Motor-Cars and Enfield personalities. I also benefited from the early business records, including details of the Enfield tricycles and quadricycles, which were donated by the Enfield Cycle Company Ltd to the Royal Automobile Club and are now stored at the Churchill Archive Centre, Churchill College, Cambridge. I would like to record my thanks to the staff at the museums and libraries who have assisted me in my research.

I have made Internet searches and, in particular, have benefited from the information contained on the Graces Guide to British Industrial History-Website.

I have made reference to the following books:

Bradford, Anne. The Story of the company and the people who made it great: 1851-1969 (Amulree Publications, 1996)

Clew, Jeff. JAP: The Vintage Years (Foulis, 1985)

Culshaw, David & Horrobin, Peter. The Complete Catalogue of British Cars (Macmillan,1974)

Hartley, Peter. The Story of Royal Enfield Motor Cycles (Patrick Stephens, 1981)

Koerner, Steve. The Strange Death of the British Motor Cycle Industry (Crucible Books, 2102)

Orchard, C.J. & Madden, S.J. British Forces Motorcycles 1925-45 (Alan Sutton Publishing Ltd, 1925)

May, Gordon G. Royal Enfield, By Miles the Best (RG Publishing, 2014)

Painting, Norman. Alldays & Onions, A brief History (Landmark Publishing, 2002)

Walker, Mick. Royal Enfield: The Complete Story (Crowood, 2003)

Whiteley, William. The Quest for King Dick (William Whiteley, 2017)

I would like to record my thanks to Donald Curren, my son Richard and Nick Jeffery who willingly accepted the unenviable task of checking through all that I have written.

I would like to thank my wife, Janet, for all her support and understanding during the long period of research and writing the manuscript.

Index